Shifting Cultivation and Secondary Succession in the Tropics

To my late father Anthony Digun Aweto, who sacrificed so much
in order to ensure I had a good education.

Shifting Cultivation and Secondary Succession in the Tropics

Albert O. Aweto

Department of Geography
University of Ibadan, Ibadan, Nigeria

CABI is a trading name of CAB International

CABI	CABI
Nosworthy Way	38 Chauncey St.
Wallingford	Suite 1002
Oxfordshire, OX10 8DE	Boston, MA 02111
UK	USA
Tel: +44 (0)1491 832111	T: +1 800 552 3083 (toll free)
Fax: +44 (0)1491 833508	T: +1 (0)617 395 4051
E-mail: info@cabi.org	E-mail: cabi-nao@cabi.org
Website: www.cabi.org	

A catalogue record for this book is available from the British Library, London, UK.

Library of Congress Cataloging-in-Publication Data

Aweto, Albert O.
  Shifting cultivation and secondary succession in the Tropics / Albert O. Aweto.
     p. cm.
  ISBN 978-1-78064-043-3 (hardback)
  1. Shifting cultivation--Tropics. 2. Plant succession--Tropics. 3. Agriculture--Tropics. I. Title.

  S602.87.A94 2012
  631.5'818--dc23

                    2012015290

ISBN: 978 1 78064 043 3

Commissioning editor: Claire Parfitt
Editorial assistant: Chris Shire
Production editor: Tracy Head

Typeset by Columns Design XML Ltd, Reading, UK.
Printed and bound in the UK by MPG Books Group.

Contents

Preface

Shifting cultivation, involving rotational fallowing, is the dominant system of arable farming in the humid and sub-humid tropics, where the bulk of the total food output is produced by small-scale farmers, the majority of whom are shifting cultivators. The system of agriculture is, therefore, vital to food security and to meeting the Millennium Development Goal of reducing hunger and malnutrition in most tropical countries.

The system of shifting cultivation is an ecologically sound and effective strategy of arable land management, especially in areas where rural population density is low, as it depends on the natural processes of organic matter and nutrient cycling to restore the fertility of soils which declined during cropping. In the wake of rapid population increase in the tropics, particularly during the last two to five decades, fallow periods have been considerably shortened and the system of shifting cultivation is breaking down and becoming unsustainable. This brings to the fore the need to study the system of shifting cultivation in order to intensify it in an ecologically sound and sustainable manner, or replace it – where ecological, pedological and sociological conditions permit – with a more intensive system of farming. Several books and journal articles have been published on the subject of shifting cultivation, particularly since 1960 when Nye and Greenland published their seminal book, *The Soil Under Shifting Cultivation.*[1] It is pertinent to observe that such books or journal articles are area-specific, concentrating on particular countries or regions, or even on cultural practices by certain ethnic groups. This has tended to hinder a holistic and pan-tropical understanding of the system of shifting cultivation. It is also significant to note that published works on shifting cultivation, including recent works of the past 5 years or so, are scattered in a multiplicity of sources, including several scores of different journals. The present book attempts to document and systematize findings on shifting cultivation on a pan-tropical basis, drawing on major findings in the literature in the last five decades.

This book adopts a novel, perhaps even a unique approach. Most published research works on the subject have adopted a dichotomous methodology to the study of shifting cultivation, in which the soil and vegetation components of the bush fallow ecosystem are considered independently of one another. Foresters and ecologists have tended to focus primarily on the nature, composition, and dynamics of fallow vegetation; that is, secondary succession, without quantitatively characterizing the soils underneath fallow vegetation. In contrast, agronomists have tended to focus primarily on fallow soil dynamics, and hence on the process of soil fertility restoration during the fallow period,

---

1   Nye, P. and Greenland, D.J. (1960) *The Soil Under Shifting Cultivation.* Commonwealth Bureau of Soils, Harpenden, UK.

without matching soil changes with changes in fallow vegetation. The current book examines the processes of secondary succession and soil fertility restoration under bush fallow within an integrative framework. It goes beyond merely juxtaposing the discourse on secondary succession and the process of soil fertility restoration in bush fallow, and adopts a soil–vegetation system approach to the study of the bush fallow. This approach recognizes that the soil and vegetation components of the bush fallow – an important component of the shifting cultivation cycle – as open and interdependent systems which exert reciprocal effects on one another. The interrelationships between fallow soil and vegetation are explored using simple bivariate correlations and stepwise multiple regression, drawing mainly on the author's work in south-western Nigeria, although reference is also made to published works in South America. This approach made it possible to identify the salient characteristics of fallow vegetation which enhance the process of soil fertility restoration under bush fallow, and those of fallow soil which enhance the regeneration of fallow vegetation.

Another novel and distinctive feature of the book is that it uses the core–periphery analogy to analyse the process and stages of secondary vegetation development and soil fertility restoration under bush fallow. Trees and shrubs are foci of soil fertility restoration during the fallow period and were compared to cores of industrialized regions; the areas outside their canopies were compared to less-developed peripheral zones of industrialized regions. Based on the core–periphery analogy, four evolutionary stages of secondary succession were identified and discussed.

Other features of the book that may be of interest to the reader include: the discourse on the future of shifting cultivation and its intensification, including innovative methods such as palm fallows; infield composting; and *Quesungual* and various forms of agroforestry developed by shifting cultivators.

My interest in shifting cultivation dates back to when I was a child and used to accompany my parents to work on the farm. We would navigate a maze of meandering footpaths traversing a mosaic of bush fallow vegetation in different stages of succession, and ultimately reach our farm. This book is not borne out of my childhood experience as a shifting cultivator, but is the product of more than three decades of research in shifting cultivation and the associated process of secondary succession. It is my sincere hope that students, as well as researchers in geography, tropical agriculture, forestry, land use change and deforestation, and all who have a stake in tropical development, will find something useful in this book.

A.O. Aweto
Ibadan
2012

Acknowledgements

I would like to express my gratitude to many different people and organizations who contributed to the successful completion of this work. First, I would like to thank publishers, organizations and individuals who granted me permission to use copyright materials. Figure 1.4 is reprinted from *Soils of the Humid Tropics* (1972), courtesy of the National Academies Press (Washington, DC). I am grateful to Dr Murray Peel of the Department of Infrastructure Engineering (University of Melbourne, Australia) for the updated Koppen–Geiger climate map (Figure 1.2); the Food and Agriculture Organization for permission to use Figure 2.1; Taylor and Francis (Abingdon, UK) for permission to reprint Figure 3.1; Springer (Dordrecht, the Netherlands) for permission to reprint Figures 3.3 and 3.4; John Wiley (Chichester, UK) for Figures 5.2, 5.3, 5.7, and 8.3 and for the equations in Chapter 5; the Chairman, MAB National Committee of Nigeria for Figure 5.5; Bibiothèque Historique du Cirad (Nogent-sur-Marne, France) for Figure 8.1; and the World Agroforestry Centre (Nairobi, Kenya) for Figure 9.1.

I am indebted to the Technical Centre for Agricultural and Rural Co-operation (Wageningen, the Netherlands) for kindly sending me free copies of *Spore*, *LEISA* and *Farming Matters* for over 15 years, so that I could be well informed about agriculture and farming conditions in the developing world. I have also benefited immensely from the papers of the International Institute for Environment and Development, London, which were kindly provided at no cost.

The illustrations were drawn by Mr Joseph O. Olumoyegun and Mrs Martina Olufunsho. I am grateful to them for their valuable cartographic skills. Mrs Mercy Ubani and Mrs Funke Kobiowu kindly word-processed Chapters 1–4, and Mr Wale Olutayo handled Chapter 5. I am grateful to my son Tejiri for taking the photographs, and to Miss Caraline Harshman of the University of Wisconsin at Madison for processing the photographs. I would also like to thank my wife Ruth, for her patience and for constantly urging me to finish this book so that I can move on to do other things. I would also like to thank the staff of CABI, especially Claire Parfitt, Chris Shire and Tracy Head for their patience and for being very pleasant people to work with. Finally, I would like to thank God for His love and mercy and for giving me the strength and resources to complete this book.

1 The Tropics

1.1 Definition of the Tropics

This book focuses on shifting cultivation and the associated processes of secondary succession and soil fertility restoration in the tropics. It is important, therefore, to define the tropics and outline their main features, as the agricultural system of shifting cultivation is now largely restricted to that region.

The tropics constitute the part of the earth lying near the equator – the imaginary line passing through the centre of the earth midway between the two poles. Since the equator divides the earth into two hemispheres along an east–west axis, the tropics occupy a central part of the earth, extending north and south of the equator. A distinct feature of the tropics is that they are hot throughout the year, without a cold season such as winter in temperate regions. This is mainly because the sun's rays strike the earth more perpendicularly than they do elsewhere. Owing to the tilt of the earth's axis at an angle of 66.5°, the sun can only pass directly overhead and be at zenith between latitudes 23.5°N and 23.5°S of the equator; the 'overhead sun' is never experienced in temperate and polar regions. The part of the earth lying between the equator and latitude 23.5°N is the Tropic of Cancer, while its counterpart in the southern hemisphere, lying between the equator and latitude 23.5°S, is the Tropic of Capricorn (Fig. 1.1).

As pointed out earlier, the tropics are hot throughout the year, with no marked seasonal variation in temperature. Although the annual range of temperatures usually increases from the equator towards the region's poleward limits, within the tropics the diurnal range of temperatures is greater than the annual range (Nieuwolt, 1977). This again underscores the relative uniformity of temperatures throughout the year, causing the seasons to be more clearly defined using rainfall total, distribution, and variability (rather than temperature, as in extra-tropical latitudes).

Climatic conditions similar to those found within the traditional geographical limits of the tropics, defined by latitudes 23.5°N and 23.5°S, occur in the subtropics. Similarly, tropical vegetation types are found beyond the traditional limits of the tropics (Walter, 1971). On climatic and ecological considerations, the tropics can justifiably be regarded as extending from latitude 30°S to 30°N. Shifting cultivation is practised beyond the strict geographic limits of the tropics, for example in the Khasi hills in Meghalaya, north-eastern India (latitude 25.5°N). This book, while recognizing the

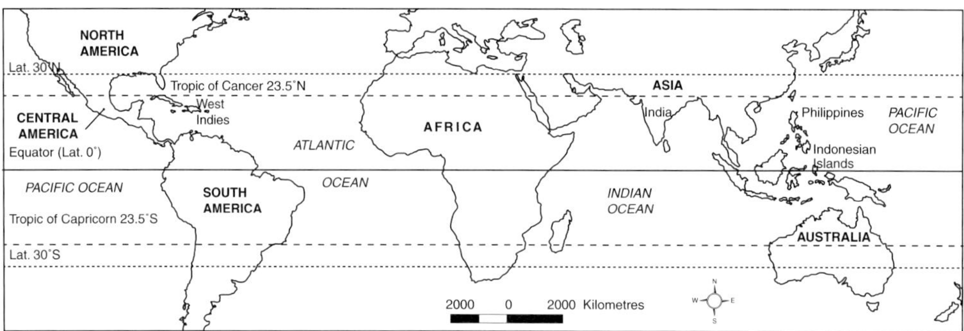

Fig. 1.1. The tropics: geographic setting.

traditional limits of the tropics to be latitudes 23.5°N and 23.5°S, draws on examples of shifting cultivation practices from the sub-tropics, where appropriate.

1.2 Climate

In spite of its relative uniformity of temperature conditions, a number of climate types can be recognized in the tropics on the basis of rainfall total, distribution, and variability. Koppen (1936), who related climate types to the distribution of major types of vegetation, recognized six types of climate in the tropics: (i) tropical rainforest (Af); (ii) tropical monsoon (Am); (iii) tropical savanna (Aw); (iv) hot desert (BWh); (v) tropical steppe or semi-desert (BSh); and (vi) warm temperate (mesothermal) (Cw, Cf, and Cs).

1.2.1 Tropical rainforest climate

Tropical rainforest climate occurs in areas lying near the equator, usually between latitude 8°N and 8°S of the equator. The annual rainfall usually exceeds 1500 mm and is distributed over about eight or more months of the year. The annual range of temperature is very small, being usually under 4°C. At high elevation, temperatures are considerably reduced due to altitudinal effect, as exemplified by Quito in South America at an elevation of 2879 m above mean sea level. Here there is a mean annual temperature of 15°C and an annual range of

0.6°C. Unlike the adjoining lowlands, plant growth is very slow in high elevations in the tropical rainforest environment, owing to the persistently low temperatures. It takes 11 months for maize to mature in Bogotá at an altitude of 2800 m in the Colombian Andes (Sanchez, 1976) compared to about 4 months in the adjoining plains of the Vichada and Meta rivers. In equatorial latitudes, rainfall is usually well distributed over 10 or more months, and plant growth is virtually continuous throughout the year, as there is no pronounced period of soil moisture deficit. The main areas that experience a tropical rainforest climate include the Congo basin in Central Africa, eastern Madagascar, the Amazon basin in South America, and the Indonesian archipelago in South-east Asia (Fig. 1.2).

1.2.2 Tropical wet–dry (savanna) climate

A tropical wet–dry (savanna) climate char-acteristically occurs in areas lying between the limits of rainforests and the semi-deserts. The rainfall total is usually much less than in areas experiencing a rainforest climate. Also, the length of the wet season is reduced to about 5–6 months, or shorter as the desert is approached. Owing to the marked seasonality of rainfall distribution, water deficits and droughts are serious constraints on crop and animal pro-duction in this climatic zone. Because of the marked effect of continentality associated with most areas experiencing a savanna climate, the annual range of temperatures is much greater

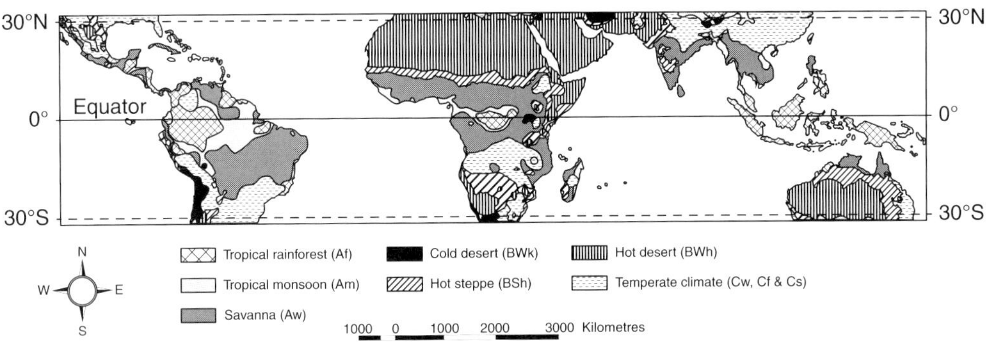

Fig. 1.2. Koppen–Geiger's climate map of the tropics. (After Peel *et al.*, 2007.)

than in a rainforest climate and can reach 15°C or more. Annual rainfall is usually considerably less than in a rainforest climate and usually varies between about 500 and 1200 mm. A tropical savanna climate occurs over much of the interior of West Africa south of the Sahara, East Africa, and Southern Africa; in the Orinoco basin in Venezuela; much of the interior plateau of Brazil south of the Amazon; and in parts of tropical Asia, especially India.

1.2.3 Desert and semi-desert climates

These occur at the poleward margin of the tropical savanna climate. Semi-deserts abut on savanna climates at their wetter equator-ward margin, but give way to a hot desert climate at their drier poleward limit. Areas with a semi-desert climate, classified as steppe by Peel *et al.* (2007), usually receive an average annual rainfall of about 250–500 mm. Droughts occur frequently in semi-deserts, and pastoralism (usually nomadic livestock rearing) replaces sedentary arable agriculture as water shortage, exacerbated by high rates of evaporation, imposes a major constraint on plant growth. Deserts receive less than 250 mm of rainfall annually. Although the fringes of hot deserts receive some annual rainfall, very little rain falls in the heart of the desert, where there may be no rain for several years. Hot deserts proper extend beyond the limits of the tropics and are characterized by large diurnal and annual ranges of temperature. Hot deserts, with their

fringing areas of semi-desert, occur in North Africa (Sahara Desert); in the south-western part of southern Africa; along the west coast of Peru; in north-western India; and over much of the interior of Australia.

1.2.4 Monsoon climate

The monsoon climate is best developed in South-east Asia, especially in India, Bangladesh, and Myanmar. Although the rainfall total in monsoon areas may be as high as or higher than in areas with a rainforest climate, the rainfall is concentrated into a few months, usually about 5 months during the year. On account of the marked seasonality in rainfall distribution, water deficit imposes a severe limitation on plant growth during winter, when the prevailing winds blow predominantly offshore.

1.2.5 Mild temperate (mesothermal) climate

A mild temperate climate occurs mainly in areas of high elevation where the altitude moderates the high temperatures associated with tropical lowlands. The main areas of mild temperate climate in the tropics include the high plateau of southern Africa and the highlands of south-eastern Brazil. The former is characterized by dry winters on account of its continental location, while the latter has no prolonged dry season.

1.3 Vegetation

The natural vegetation of the tropics is varied, in response to the gradient of decreasing rainfall from the equator to the deserts and considerable variations in soils, topographic, and substrate conditions. The very humid parts of the tropics lying near the equator are characterized by rainforest vegetation, and the drier poleward margins of the tropics by desert and semi-desert vegetation.

1.3.1 Tropical rainforest

Tropical rainforests are the most complex terrestrial ecosystems in the world in terms of structure and biological diversity. They usually consist of three layers of trees and a shrub layer that are not clearly differentiated from one another. The uppermost tree layer consists of trees that are over 30 m tall, some of which attain a height of 45 m or more. In the Amazon basin of South America, this upper tree layer is continuous due to relatively little human interference with the vegetation. In West Africa, where human interference with the vegetation is greater, the upper tree storey is discontinuous, having been reduced to a few scattered trees that tower above the general level of the forest. A remarkable feature of tropical rainforests is the diversity of the tree flora. The forests are not only densely stocked with trees, but also with several dozens of tree species per hectare, so it is misleading to talk about a dominant species. It is more realistic to talk about family dominance. In Indonesian and Malaysian rainforests, up to 50% of trees in the upper tree layer belong to the family *Dipterocarpaceae*. These forests are therefore characterized as dipterocarp forests, although several other tree families are represented in the lower tree strata. In some South American and some West African rainforests, the dominant families are *Leguminosae* and *Meliaceae*, respectively (Richards, 1996).

Another diagnostic feature of tropical rainforests is the profusion of epiphytes and climbers. The latter occur more commonly near the cut edge of the forest, often forming a dense tangle and giving an erroneous impression that the forest is difficult to penetrate. The interior of the forest is quite open and one can move freely without being hindered by climbers.

Tropical rainforest ecosystems have been shrinking fast due to lumbering, conversion of forests into tree plantations, agricultural, and other uses (Myer, 1990). Agricultural expansion has been blamed by Delacote (2007) as the major cause of deforestation in the tropics. Deforestation in the tropics has been a major concern to ecologists and politicians worldwide because of its adverse consequences, especially soil erosion and land degradation, reduced agricultural productivity, loss of biodiversity, increased flooding, and emission of carbon dioxide into the atmosphere. Takasaki (2007a) has suggested that promoting income-generating non-farm activities among poor farmers would reduce their dependence on land and help reduce deforestation. However, Takasaki (2007b) observed that improving the process of soil fertility restoration and promoting soil conservation measures such as terracing, construction of windbreaks, and the adoption of agroforestry among shifting cultivators is a more effective means of reducing deforestation than the promotion of non-farm activities. Small-scale farmers in the tropics, including shifting cultivators, are more interested in issues of 'bread-and-butter', and hence in meeting their immediate needs, than in the efficacy of land conservation measures. Mwanukuzi (2011) observed that people in the Uporoto Mountains of south-west Tanzania abandoned effective methods of controlling land degradation, because these did not contribute towards meeting their immediate needs. It would seem, therefore, that an improvement in soil fertility will reduce deforestation if crops respond adequately to the increased availability of soil nutrients, leading to substantially increased crop yields, and possibly an enhanced income for the farmers. Fuelwood exploitation is also a major cause of deforestation, accounting for more than 80% of the domestic energy consumed in many countries in West and Central Africa (Aweto, 1995; FAO, 2001). In Madagascar and the Comoros, fuelwood exploitation is the greatest cause of deforestation (UNEP, 1999). Forest fires, which occur during periods of intense drought, have also caused considerable damage to forests in Brazil, Mexico, Ethiopia, and Indonesia (UNEP, 2002). Today, the largest area of fairly undisturbed rainforest occurs in the Amazon basin of South America (Fig 1.3). Large expanses of rainforest also occur in the Congo basin in Africa, and in Malaysia and Indonesian islands in South-east Asia.

1.3.2 Savanna

Savanna ecosystems are tropical grasslands. Unlike temperate grasslands, trees are an important element of tropical savannas, and the structural or physiognomic classification of savanna ecosystems is usually based on the presence and abundance of trees in them. A distinct herb layer, in which grasses and sedges feature prominently, is a hallmark of savanna vegetation. Savanna ecosystems occur between the drier margins of tropical rainforests and semi-deserts. They are most extensive in tropical Africa, where large areas in West, East, and Southern Africa are under savanna vegetation. In tropical America savanna vegetation occurs in Guyana, the Llanos of Colombia and Venezuela, and in Brazil, especially in the interior plateau of Mato Grosso. Areas under savanna vegetation in Asia are largely restricted to eastern India, and parts of Vietnam, Myanmar (Burma), Sumatra, Papua New Guinea, and Kalimantan. Fairly extensive areas of savanna vegetation also occur in northern Australia.

Regionally, the structure and floristic composition of savanna vegetation varies considerably, especially between the American, African and Asiatic savanna formations, and between the more humid savannas towards the equator and the drier and semi-arid savannas at the margins of hot deserts. The humid and sub-humid savannas near the drier limits of rainforest ecosystems are densely wooded, with tall trees that may attain 10–15 m in height. Dense, tufted, perennial grasses 2–3 m tall also occur. In contrast, semi-arid savannas near desert margins are less wooded and trees are much smaller, commonly 3–5 m tall. In response to periods of pronounced drought, trees in semi-arid savanna are frequently xeromorphic, with small leaves and thorns, while annuals largely replace the dense, tall grasses of the more humid savanna formation. Acacia trees, *Commiphora africana*, and other drought-resistant trees commonly feature in semi-arid savanna in Africa. In spite of its superficial uniformity, the structure and floristic composition of savanna vegetation varies within limited areas due to differences in topography and biotic factors such as grazing, cultivation and regeneration status of savanna vegetation after cropping (Otieno *et al.*, 2011), and the frequency and intensity of burning.

In Australian savannas, the *Eucalyptus* features prominently in the tree flora. Other trees such as species of *Acacia* are also present in Australian savannas. Grasses in the savannas

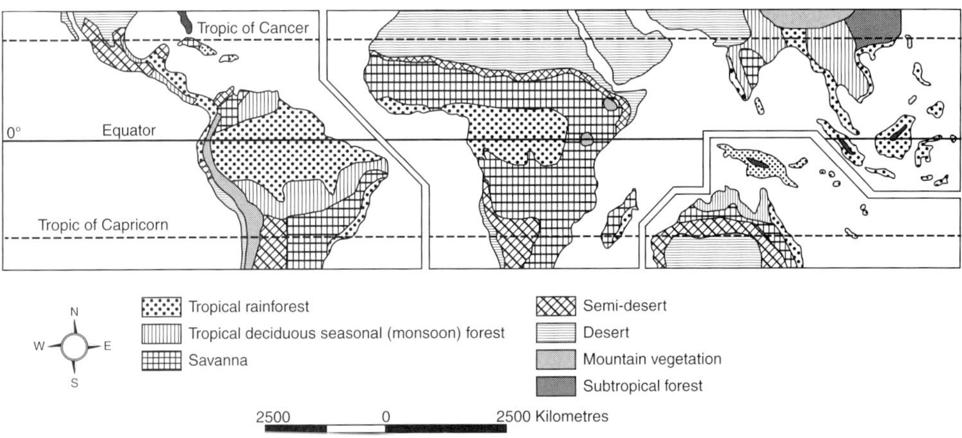

Fig. 1.3. Vegetation map of the tropics. (Adapted from maps in Eyre, 1968.)

include *Astrebla*, *Themeda*, and *Triodia*. The composition of savanna vegetation in South America varies considerably. The savannas of the Orinoco basin are associated with poorly drained soils found in deltas or river flood plains. Such savannas, known as Llanos, are treeless, except for the occurrence of a few solitary palms such as *Mauritia minor*. In contrast, the well-drained savannas of the Brazilian plateau are well-wooded and trees include *Curatella americana* and *Byrsonima* spp., while grasses include *Aristida pallens* and *Melinis minutiflora*. Fire is an important ecological factor in the savanna ecosystem. Most savannas are burnt deliberately or accidentally, usually annually. Savanna vegetation is burnt to clear land for cultivation, to destroy crop pests, or to facilitate sprouting of fresh grasses for livestock. As a result of frequent burning, most savanna trees have thick barks to withstand fires, and gnarled stems. Fire plays an important role in maintaining savannas in their current state and many authorities regard them as fire or anthropogenic climax. In fact, burning is endemic in African savannas. Dayamba *et al.* (2011) argued that complete fire exclusion in African savannas is impracticable and that emphasis should be placed on management strategies designed to reduce fire severity.

1.3.3 Monsoon forest

The monsoon lands of the tropics, particularly tropical Asia, are characterized by forest vegetation which is less luxuriant than the rainforest ecosystem. This is mainly due to the fact that rainfall in monsoon regions is markedly seasonal, being concentrated in 5–6 months of the year. Owing to the occurrence of a pronounced dry season, during which moisture deficit imposes a limitation on plant growth, biodiversity in monsoon forests is considerably less than in rainforest. In addition, trees in the forest, especially those in the upper layer, shed their leaves as a physiological response to drought during the dry season. Hence, monsoon forests are also called tropical deciduous seasonal forests. Another feature of deciduous seasonal (monsoon) forests is the occurrence of a herb or field layer in which grasses feature prominently. The grass layer makes deciduous seasonal forests prone to burning during the dry season, when the grass layer dries up and larger trees shed their leaves. Deciduous seasonal forests are very extensive in South-east Asia, particularly in north-eastern and western India, central Myanmar, and parts of Cambodia and Thailand. In South and Central America, they occur at the margins of rainforests in western Honduras, Nicaragua, Costa Rica, and Brazil. They also occur in East and Central Africa, especially in the Democratic Republic of Congo, Malawi, Zimbabwe, and Zambia where they are also called *miombo* woodlands. The monsoon forests of South-east Asia are sources of valuable trees such as *Tectona grandis* (teak) and *Shorea robusta* (the sal tree).

1.3.4 Desert and semi-desert

Semi-desert ecosystems are characterized by a sparse cover of grasses in which annuals predominate and the landscape is dotted by stunted drought-resistant trees and shrubs. Extensive areas of semi-desert vegetation occur in East Africa, especially in Somalia, Ethiopia, and northern Kenya. Much of the northern part of the Sahel is a semi-desert that forms a transition to the Sahara Desert. In Southern Africa, the Namib and Kalahari deserts are fringed on the eastern part by semi-desert in which *Acacia* trees are an important element of the flora. Similarly, the Atacama and Chilean deserts in South America are fringed on their eastern margins by semi-deserts in which succulent plants and dwarf shrubs feature prominently. Much of the Chihuahan, Sonoran and Mohave deserts in the south-western part of North America feature semi-desert vegetation in which succulent plants such as *Euphorbia*, *Cactus* and *Agave* account for a significant proportion of the ground cover. Except where substrate conditions such as the occurrence of stony or rocky surfaces make plant growth impossible, hot deserts are usually characterized by a sparse vegetal cover. Annual grasses and other ephemerals become established after a period of rainfall and complete their life cycle in a matter of weeks. Annuals dominate the flora of deserts because of the low rainfall and unreliability of its distribution. Drought-resistant shrubs and trees, including stem succulents, also feature in the flora of hot desert ecosystems, but they are largely restricted to the floors of the dry valleys.

1.4 Relief

In terms of relief, the bulk of the land area in the tropics consists of plateaux that are separated from the surrounding oceans by relatively narrow coastal plains. These plains consist mainly of more recent sediments from the Tertiary or Quaternary periods, and are most extensive: in the Amazon basin of South America, where they extend over 2000 km inland; in the Niger Delta in Nigeria; and along the valleys of major rivers such as the Indus, Ganga, and Mekong in tropical Asia. The plateaux, which constitute over 80% of the continental masses within the tropics, consist of igneous and metamorphic rocks of Precambrian age that have remained fairly stable since the Cambrian period. Unlike the coastal plain which consists mainly of unconsolidated sediments, especially of alluvium and sandstone, the plateaux are made of hard crystalline rocks such as granite, gneiss, quartzite, and schists. Large areas of the plateau surfaces were subsequently covered by more recent sediments as a result of weathering of the original igneous and metamorphic rocks, coupled with marine transgression at different geologic periods.

Much of West and North Africa consists of a low plateau with an elevation of about 200–1000 m above mean sea level. In East and Southern Africa, the elevation of the plateau is higher, often ranging between 1000 and 3000 m. A system of north–south trending rift valley is entrenched into the East African Plateau, while volcanicity (associated with faulting that produced the rift valleys) has resulted in the formation of volcanic mountains such as Mount Kenya and Kilimanjaro.

The most important physiographic features of South America include the low-lying Amazon, Orinoco and Uruguay–Paraguay basins, the interior plateau, and the north–south trending Andes mountain range. The basins are generally less than 150 m in elevation, while much of the interior plateau, especially the Mato Grosso of Brazil, has an average elevation of about 200–1000 m above mean sea level. The most imposing relief feature of South America is the Andes mountain range; this rises precipitously from the Pacific coast to an elevation of over 5000 m. The Andes range, a young fold mountain of Tertiary age, extends northwards into Central America and North America.

As with the other continental masses lying within the tropics, tropical Asia is dominated by plateaux, the most important of which are the Deccan plateau of India and the Arabian plateau. Areas of Tertiary folding occur as southward extensions of the Himalayan fold mountain range in Myanmar Republic, Thailand, Malaysia, and the Indonesian archipelago. Tropical Australia consists of a narrow coastal plain and a low interior plateau that ranges in elevation between an average of 200–800 m above mean sea level.

1.5 Soils

Given the wide range of relief, climate, geological formations and vegetation types, soil conditions vary considerably within the tropics. This notwithstanding, some diagnostic features of tropical soils, which clearly differentiate them from those of temperate regions, can be recognized. Unlike in temperate regions where the processes of soil formation have been repeatedly interrupted by glaciation, pedogenic processes have operated largely uninterrupted for long geologic periods in the tropics. This is particularly so in the humid tropics, especially on old plateau surfaces such as the Brazilian shield and the plateau of West and East Africa.

The soils formed on these relatively stable plateau surfaces are very old, and their profiles depict features associated with advanced stages of soil formation. In the humid tropics, chemical weathering is much more intense than in temperate regions, on account of the higher temperatures and the availability of moisture throughout the year in the former. Consequently, the soils in the humid tropics are more deeply weathered than their temperate counterparts, and one major consequence of this is that their clay fractions are dominated by low-activity clays with low nutrient-adsorbing capacity, especially kaolinite. In contrast, the clays in temperate soils are predominantly montmorillonite, whose nutrient-adsorbing capacity is considerably higher. In general, as moisture availability decreases from the humid tropics (especially humid rainforest environments) to the hot deserts, the intensity of chemical weathering decreases significantly. Consequently, soils in semi-arid and arid parts of the tropics contain a much higher proportion of more active clays, especially montmorillonite, than soils in the humid tropics. The comparatively high proportion of montmorillonite in desert soils largely explains why some desert soils are fertile and highly productive under arable uses, if water is provided through irrigation.

Several classification schemes have been used for categorizing and characterizing the soils of the tropics, as has been done for extratropical areas. At least five major soil classification schemes have been widely used in the tropics (Sanchez, 1976). The use of different systems of classification for the same soil types has hindered dissemination of information on the major soil types and the extrapolation of research findings on a specific soil to other areas where the same soil type occurs. The US Soil Taxonomy (Soil Survey Staff, 1990) is used in this subsection in describing the major soil types of the tropics and in characterizing their distribution. A major advantage of the US Soil Taxonomy is that it is based on properties of soils that can be measured in the field or laboratory, rather by than inference regarding the mode of formation of the soil. Second, the nomenclature adopted in US Taxonomy gives an indication of the major characteristics of the soil. Eight soil orders were recognized for the soils of the tropics, namely: oxisols, ultisols, alfisols, inceptisols, entisols, aridisols, vertisols, and mollisols (Sanchez, 1976).

1.5.1 Oxisols

Oxisols are the most intensely weathered soils of the tropics, with very low weatherable mineral reserves. The soils are deep, usually red or yellow in colour. They are characterized by an 'oxic' horizon, dominated by kaolinitic clay mineral, from which silica has been leached, but rich in iron and aluminium oxides. Oxisols are of low fertility status. They occur under rainforest vegetation and occupy vast expanses of land in the Amazon basin in South America and the Congo basin in Central Africa (Fig 1.4).

1.5.2 Ultisols

Ultisols were also formed in humid environments, especially under rainforest vegetation. They are, however, less intensively weathered than oxisols. Consequently, they contain a higher proportion of weatherable mineral reserves. They are red or yellowish in colour and are relatively infertile and usually have a distinct subsoil horizon in which clay has accumulated – the argillic horizon. Ultisols are less extensive than oxisols in the tropics. They occur in north-eastern India, peninsular Malaysia, central Sumatra and coastal Borneo in Asia. In South America, ultisols occur in the coastal part of eastern Brazil, and also in parts of the Mato Grosso and Parana plateaux in Brazil and north-eastern Bolivia. In West Africa, ultisols occur in a belt stretching from southern Senegal to south-western Côte d'Ivoire. Ultisols also occur in north-eastern Australia.

1.5.3 Alfisols

Alfisols are inherently more fertile than oxisols and ultisols. As with ultisols, alfisols have a clay-enriched subsurface horizon known as the argillic horizon. The base saturation of the argillic horizon of alfisols usually exceeds 35%, being higher than that of ultisols. Alfisols usually form under forest and savanna vegetation, in areas characterized by soil moisture deficit during part of the year. Extensive areas of alfisols occur under savanna vegetation in West Africa and Zimbabwe in southern Africa. In South America, a large expanse of land in north-eastern Brazil is under alfisols, while in tropical Asia they occur mainly in eastern India.

1.5.4 Inceptisols

Inceptisols are young, immature soils that have been little altered by the process of soil

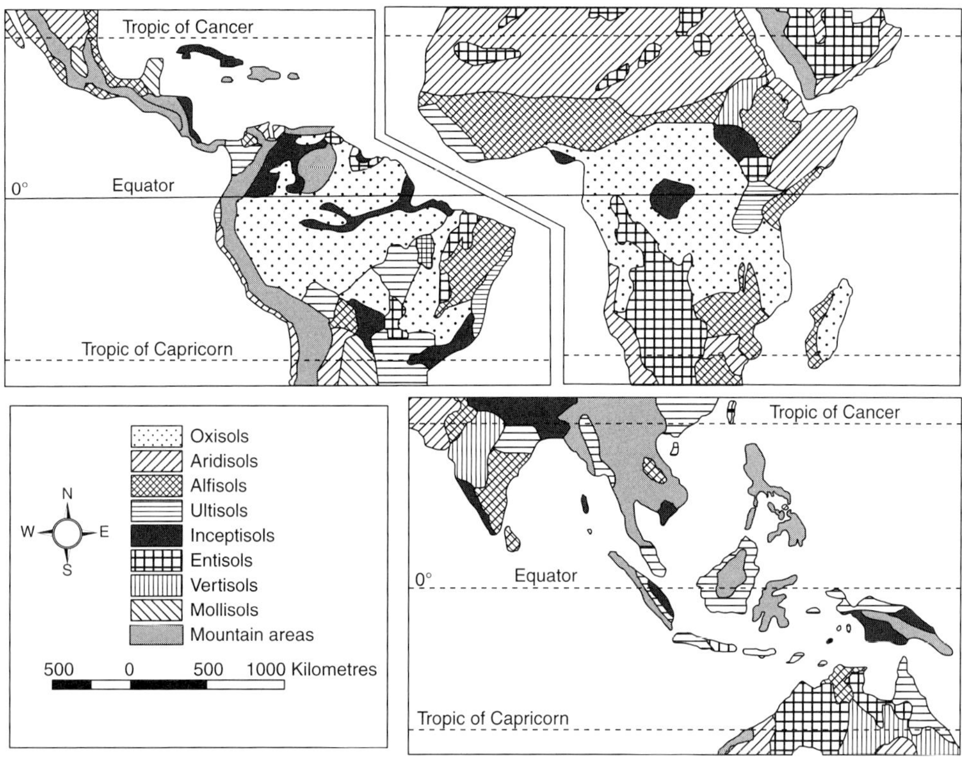

Fig. 1.4. Soil map of the tropics. (After Aubert and Tavernier, 1972, from Sanchez, 1976. Reprinted from *Soils of the Humid Tropics* (1972) with permission from the National Academy of Sciences, courtesy of the National Academies Press, Washington, DC.)

formation. They are characterized by one horizon – the cambic horizon – and usually no other horizon can be recognized. The cambic horizon is differentiated from the parent material from which the soil was formed in terms of structure, and may also be characterized by the removal or transformation of mineral materials (Foth, 2006). The profiles of inceptisols do not usually reveal features associated with advanced stages of soil formation such as a horizon of clay accumulation, as in the case of alfisols. Inceptisols are usually associated with volcanic parent materials and alluvial floodplains, both of which are usually characterized by rich fertile soils. Inceptisols developed on alluvium occur in the Niger Delta in Nigeria, southern Sudan, and the Amazon floodplain in South America. They developed on volcanic parent materials in Indonesia, where they support intensive arable farming and high population density on account of their high fertility.

1.5.5 Entisols

Entisols are mainly recent deposits of weathered materials that show very little evidence of horizon differentiation. Some entisols occur on barely weathered rocks and have very limited agricultural potentials in view of their very shallow depth. Others are associated with recent alluvial deposits along rivers, while yet another group consists of recent sand dunes, such as occur in deserts and sandy beaches, that do not reveal evidence of soil profile development and differentiation. Sandy entisols known as psamments are excessively drained. They contain very low levels of organic matter and their capacity to retain plant nutrients and moisture is very low. They occur over extensive areas characterized by sand dunes in the Sahara, Arabian, Kalahari, and Australian deserts.

1.5.6 Vertisols

These are tropical swelling clays. Their clay fractions are dominated by 2:1 clay minerals, which swell appreciably when the soils are wetted, but shrink considerably, forming prominent cracks when dry. Usually, their clay fractions contain a significant amount of smectite up to a depth of 1 m or more (Esu, 2010). They are usually dark to black in colour and generally form under savanna and semi-arid tropical and subtropical climates, especially from rocks rich in bases such as basalt. Vertisols are usually of moderate fertility and they occur in the Deccan plateau in India, Sudan and Ethiopia in Africa, and also in part of north-eastern Australia.

1.5.7 Aridisols

These are the soils formed in hot deserts of the tropics and subtropics. Many desert soils are characterized by a marked accumulation of soluble salts such as sodium chloride or calcium carbonate in their surface horizon. Owing to the very sparse vegetal cover, these soils contain very low levels of organic matter. However, they may contain significant amounts of active clays, especially 2:1 clay minerals, which enhance their nutrient-adsorbing capacity. Consequently, many desert soils are agriculturally productive under irrigation when soil salinity problems are properly managed. Extensive areas characterized by aridisols occur in the hot deserts of the tropics and subtropics including the Sahara, Arabian, Kalahari, Peruvian Atacama, and Australian deserts.

1.5.8 Mollisols

Mollisols are among the most productive soils in the world. They are formed under temperate grasslands and are usually characterized by a surface horizon that is rich in plant nutrients. Mollisols can be more appropriately referred to as temperate soils, although limited pockets can be found in the tropics; in fact only 1% of the land area of the tropics is covered by this soil (Sanchez, 1976). Pockets of mollisols occur in Mexico in Central America and Paraguay in South America.

1.6 Socio-economic Conditions

Just as climatic conditions in the tropics differ from those of temperate regions, so do socio-economic conditions. With the exception of Australia, the northern part of which extends into the tropics, the countries in the tropics are developing countries. They share a number of characteristics that clearly set them apart from the industrially developed countries of Europe and North America. The basic characteristics of tropical countries include low per-capita income and low level of industrialization, with 50–80% of the population engaged in agriculture as the primary means of livelihood, among others. The basic socio-economic features of tropical countries are highlighted in the subsections that follow.

1.6.1 Low per-capita income

Many tropical countries including India, Bangladesh, Burundi, Liberia, Sierra Leone, Haiti, Nicaragua, Vietnam and Guatemala have a low gross national product (GNP) per capita of less than US$500. These countries are among the poorest countries in the world. The very low GNP per capita of most tropical countries is mainly attributable to the low level of industrialization and to the fact that agriculture, the largest employer of labour, is mainly subsistence in nature. The farmer cultivates small plots of land, using simple implements such as hoes and cutlasses. In savanna regions where animal husbandry assumes considerable importance, oxen may be used to plough the land for cultivation. This notwithstanding, agricultural productivity is very low, as farmers rarely use inputs such as fertilizers to increase yields. In countries such as India and Mauritania, animal dung and crop residue that could have been returned to the soil as organic fertilizers are harvested and utilized as fuel due to the prevailing levels of grinding poverty. The levels of GNP per capita are not uniformly low in the tropics as there are considerable variations from country to country. Nations such as the Bahamas, Bahrain, Kuwait, Singapore and Saudi Arabia are characterized by comparatively high GNPs per capita that exceed US$5000.

1.6.2 Low level of industrialization

Apart from a few countries such as Brazil, Singapore, Taiwan and Malaysia that have made significant progress towards industrialization, the level of manufacturing in most tropical countries is abysmally low and this broadly explains why a large proportion of the population is engaged in agriculture. Owing to the low level of industrialization, the contribution of manufacturing to the gross domestic product (GDP) is low. For instance, in Ghana and Nigeria, which are judged to have made some progress in the bid to industrialize in the West African sub-region, manufacturing accounts for less than 15% of GDP. Most countries are dependent on the export of agricultural products or minerals, usually with minimal processing and value added prior to export. A number of countries are overly dependent on a single commodity from which they obtain the bulk of their foreign exchange to execute development projects. Nigeria depends mainly on crude oil exports for 90% of its export earnings. Other countries such as Zambia, Mauritania and Gambia depend on the exports of copper, iron ore and groundnuts, which account for about 60–95% of their foreign exchange earnings. Countries that depend mainly on a single commodity are faced with the vagaries of change in commodity price in the world market. Zambia, for instance, has been in major economic crisis particularly since the 1980s, due to the low price of copper.

1.6.3 Demographic characteristics

Tropical countries are characterized by high fertility rates and declining mortality rates. The rate of natural increase in population is generally much higher in countries in the tropics than in their temperate counterparts. In West African countries such as Benin, Burkina Faso, Mauritania, Nigeria and Niger, the population growth rate varies between 2.8% and 3.0% compared to growth rates of 0.1–0.4% in countries such as France, the Netherlands, Switzerland and Belgium (Population Reference Bureau, 2001). Population growth rates have declined in Asia generally, compared to the rates in African countries. In Asiatic countries such as Vietnam, Thailand, Singapore, Indonesia and Malaysia, the rate of natural increase in population varies between 0.8% and 2.0%. The implication of the high population growth rates in tropical countries is that a significant proportion of the population consists of children aged 14 years and under.

1.6.4 Political instability

Many tropical countries, especially in South America and Africa, have been characterized by political instability, with frequent changes in government and sometimes military incursion into politics and governance. Political instability is, however, not exclusive to the tropics, as the experience of the former Yugoslavia in the temperate region clearly shows. Tropical countries such as Benin Republic, Chad and Venezuela have been plagued by problems of political and economic stability. Political instability has stalled economic development in countries such as Somalia, Benin Republic, and Nicaragua. In contrast, countries that have been stable politically, especially Costa Rica, Botswana and Mauritius, have achieved remarkable economic development within a few decades.

1.6.5 Poverty

In view of the generally low GNP per capita in many tropical countries, especially in Southeast Asia and Africa, poverty is widespread and militates against meaningful economic development. Poverty is accentuated not only by a poor resource base but more importantly by corruption, bad government policies, and political instability – especially wars, which not only disrupt economies but also result in forced movement of people. Poverty exacerbates the problems of environmental degradation. The problem of dependence on fuelwood leading to enhanced rates of deforestation in parts of East and West Africa has already been referred to. Owing to the problems of poverty in the tropics, especially in tropical Africa and Asia, people depend mainly on traditional fuels, especially fuelwood and charcoal, in order to meet their domestic energy requirements. The domestic utilization of commercial fuels, especially in rural areas, is relatively unimportant. In some cases, commercial fuels are either too expensive for the urban and rural poor, or they cannot even afford to buy electrical appliances or gas or kerosene stoves to reduce dependence on traditional fuels. In many countries in tropical Africa such as Burkina Faso, Burundi, Chad, Ethiopia, Liberia, Malawi and Tanzania, traditional fuels (mainly fuelwood) account for up to 90% of the total energy consumed. In South and Central America, where GNP per capita is generally higher than in tropical Africa, the utilization of commercial fuels is more popular. In most countries in South and Central America, commercial fuels account for over 70% of the energy utilized. Except in a few countries such as Cambodia, Laos and Myanmar, commercial fuels generally account for up to 70% of the energy consumed in countries in tropical Asia. In countries such as Saudi Arabia and Singapore traditional fuels are not consumed, indicating a very high level of utilization of commercial fuels.

1.6.6 Diseases

Finally, it is important to briefly draw attention to the problems of diseases as they also affect social and economic development. In the humid tropics, diseases such as malaria are endemic, causing the death of an estimated 1–2 million people annually on a pan-tropical basis. Many people living in urban slums and rural areas do not have access to safe and clean potable water. About 10 million people die annually in developing countries as a result of water-borne diseases (Miller Jr, 2000). Diseases and morbidity not only reduce human lifespan but also retard economic development through loss of manpower. Furthermore, the debilitating effects of diseases cause loss of valuable human hours. This is particularly so in the tropics where

medical facilities, especially in the rural areas, are often woefully inadequate. It is instructive to observe that the HIV/AIDS pandemic in Africa is taking its toll on economic development. In 2001, 28.1 million sub-Saharan Africans had HIV/AIDS, representing 70% of the people infected worldwide (UNEP, 2002). The HIV/AIDS pandemic threatens to undermine economic development in Africa, and countries with very high rates of infection, such as Botswana, may experience reversals of economic gains achieved so far.

1.7 Agriculture

Agriculture is of pivotal importance as an employer of labour in tropical countries, given the relatively low level of industrialization. Farming systems and crops produced vary considerably within the tropics in response to variations in climate, soils, relief, vegetation, culture, and diet. In particular, there is a sharp contrast between the more humid rainforest regions and the drier semi-arid savannas towards the poleward limit of the tropics. In the former, agriculture is dominated by the culture of tree crops such as bananas, cocoa, oil palm and rubber, and root crops such as yams and cassava. This amounts to an almost virtual exclusion of livestock farming, as the natural vegetation, which is dominated by trees, does not provide suitable herbage or forage for livestock. In contrast, due to the much lower rainfall and the reduced length of the growing season in the drier parts of the tropics, the culture of grains such as maize, millet and sorghum (which have a short maturation time) is more important, and livestock farming assumes considerable significance in the agricultural economy due to the presence of savanna (grassland) vegetation. Also, the much drier climate in the less humid parts of the tropics frees them from tsetse fly, a vector that spreads the sleeping sickness disease of cattle. The following broad agricultural systems can be recognized in the tropics: (i) shifting cultivation; (ii) permanent cultivation of field crops, especially rice; (iii) plantation agriculture; and (iv) livestock production. Shifting cultivation, including all forms of rotational bush fallowing, is the dominant system of arable crop production and soil fertility management in the humid and sub-humid tropics. The system is widespread in tropical Africa and America, but less so in tropical Asia, where sedentary systems of rice culture have been developed in the densely settled floodplains of major rivers.

1.7.1 Shifting cultivation

Shifting cultivation is a system of agriculture that involves cultivating a piece of land for a few years, then subsequently leaving it uncultivated or rested for a much longer period, so that the natural vegetation that develops on it restores the soil fertility which declined during cropping. This system of agriculture is ecologically sound, especially where population density is low. It is, however, unsustainable in areas with high population densities that are experiencing a population explosion (Shampa *et al.*, 2010). The agricultural landscape in shifting cultivation areas consists of scattered cultivated fields and a mosaic of regrowth (fallow) vegetation in different stages of regeneration. In areas where shifting cultivators live in permanent settlements, as in most parts of the rainforest of West Africa, the farmers grow tree crops such as cocoa, coffee, rubber and, occasionally, oil palm on smallholdings in order to earn cash to meet the basic necessities of life. Shifting cultivators may also engage in the hunting and trapping of animals, and fishing, in order to supplement their protein intake.

1.7.2 Permanent cultivation of field crops

Permanent cultivation of field crops has been developed on river floodplains where the yearly deposition of alluvium helps to replenish soil fertility. In many parts of tropical Asia, swamp rice has been grown for several centuries in river floodplains, although some form of rotation of other crops with swamp rice is often practised. The prevailing anaerobic conditions in water-logged rice fields usually results in anaerobic decomposition of soil organic matter; this is usually slower than aerobic decomposition of organic matter in dry, well-drained soils (Webster and Wilson, 1980). This factor, together with

the annual accumulation of alluvium in the floodplains, would undoubtedly have enhanced the productivity of floodplain soils, making possible their sustained utilization for crop production for centuries. The culture of swamp rice has helped to sustain very high population densities in the floodplains of major rivers in South-east Asia, but it is not very popular in tropical America and Africa. In Africa, swamp rice cultivation is gaining ground in the inland delta region of the Niger in Mali, and also in the Gambia River and the floodplains of rivers in Madagascar. The point needs be stressed here that in both tropical Africa and America, there are cultural obstacles to the widespread acceptance of swamp rice culture. In both continents, root crops such as cassava and yams, and cereals such as maize, millet and sorghum feature prominently in the diet of the indigenous people, rather than rice.

Continuous cultivation of field crops also occurs on rich and fertile volcanic soils. In Java, Indonesia, fertile volcanic soils are used for the production of cassava, maize and sweet potatoes on a continuous basis, in addition to the production of tree crops such as coffee, tea, rubber, and palm oil. Semi-continuous cultivation of field crops, especially sweet potatoes, is a feature of parts of the southern highlands of Papua New Guinea, also characterized by base-rich soils, where this crop may be grown up to 10 years continuously without a marked decline in productivity (Sillitoe and Shiel, 1999).

1.7.3 Plantation agriculture

Plantation agriculture is a highly specialized and capital-intensive form of farming that involves the establishment of monocultures over large estates that often cover thousands of hectares of land. Plantations usually have their own settlements and infrastructural facilities such as roads to service them, and the initial processing of the crop or produce takes place in the plantation prior to export. They require a lot of labour that may not be met readily by the population in the adjoining area, thus necessitating the use of migrant labour; this leads to social problems and the erosion of the cultural values of the indigenous people. The advantages of plantation agriculture are numerous and include the development of remote areas through the provision of jobs for the indigenes, and stimulating economic development through the development of infra-structure such as roads. In addition, plantations ensure a steady and dependable supply of high-quality produce as raw materials for industries, or as exports to earn foreign exchange.

Plantations have been established in different parts of the tropics and include those both of tree crops such as rubber, oil palm, cocoa and coffee, and of non-tree crops such as tobacco, sisal and cotton. Large commercial plantations of rubber have been established in Liberia, Nigeria, Malaysia and Brazil, and of coffee in Colombia, Brazil and Indonesia. Banana plantations are an important feature of the agricultural economy of the Caribbean, and bananas are important exports of Colombia, Costa Rica, Honduras, and Panama. Oil palm plantations occur in Malaysia, Indonesia and Nigeria, although the bulk of palm produce of Nigeria comes from groves of wild palms and small-scale holdings of native farmers. Small-scale farmers also produce coffee in Brazil, the world's leading producer and exporter of the crop (Watson and Achinelli, 2008). Very large plantations of clove, a tree that yields spice, are located on the islands of Zanzibar and Pemba in Tanzania, and account for the bulk of the world output. Sisal is a non-tree cash crop and is grown in plantations in Kenya and Tanzania. Although plantations have several advantages including economy of scale, they are not without major drawbacks. First, plantation agriculture is bedevilled by the problem of commodity price fluctuation in the world market. A major drop in the price of a certain crop in the world market would spell doom for plantations producing the crop. Second, since plantations are monocultures, outbreak of diseases or the attack of pests may substantially reduce the output of a plantation crop.

1.7.4 Livestock production

Livestock keeping is an important occupation in the savanna land of the tropics where the vegetation provides suitable herbage for feeding livestock. In the semi-arid savannas toward the poleward limits of the tropics, such as the Sahel in West Africa, stoking density is usually considerably lower than in sub-humid savannas. Due to droughts and seasonal changes in weather and vegetation, and hence in the availability of herbage and water for livestock, pastoral nomadism is an important feature of the drier savanna regions and herdsmen often move considerable distances with their livestock in search of water and grazing (McCarthy and Di Gregorio, 2007; Adriansen, 2008). In Brazil, considerable areas of forest have been cleared and replaced with pastures for livestock production, especially beef cattle. Although poultry and a few livestock such as sheep and goats are kept in the rainforest region, livestock production systems are best developed in the drier savanna zones of the tropics, which are relatively free from tsetse fly infestation. Although livestock can provide farmyard manure for intensifying arable farming, livestock production is beyond the scope of this book and will not be considered further.

References

Adriansen, H.K. (2008) Understanding pastoral mobility: the case of Senegalese Fulani. *The Geographical Journal* 174, 207–222.

Aubert, G. and Tavernier, R. (1972) Soil survey. In: *Soils of the Humid Tropics.* National Research Council (US) Committee on Tropical Soils, National Academy of Sciences, Washington, DC, pp. 17–44.

Aweto, A.O. (1995) A spatio-temporal analysis of fuelwood production in West Africa. *OPEC Review* 19, 333–347.

Dayamba, S.D., Savadogo, P., Sawadogo, L., Zida, D. Tiveau, D. and Oden, P.C. (2011) Dominant species' resprout biomass dynamics after cutting in the Sudanian savanna-woodlands of West Africa: long term effects of annual early fire and grazing. *Annals of Forest Science* 68, 555–564.

Delacote, P. (2007) Agricultural expansion, forest products as safety nets, and deforestation. *Environment & Development Economics* 12, 235–249.

Esu, I.E. (2010) *Soil Characterization, Classification and Survey*. HEBN Publishers Plc, Ibadan, Nigeria.

Eyre, S.R. (1968) *Vegetation and Soils*. Edward Arnold, London.

FAO (2001) *State of the World's Forests 2001*. Food and Agriculture Organization, Rome.

Foth, H.D. (2006) *Fundamentals of Soil Science*. John Wiley, New York.

Koppen, W. (1936) Das geographische system der climate. In: Koppen, W. and Geiger, R. (eds) *Handbuch der Klimatologie*. Vol. 1, Gerbruder Borntraeger, Berlin, pp. 1–44.

McCarthy, N. and Di Gregorio, M. (2007) Climate variability and flexibility in resource access: the case of pastoral mobility in northern Kenya. *Environment & Development Economics* 12, 403–421.

Miller, G.T. Jr (2000) *Living in the Environment*. Brooks/Cole Publishing Company, Pacific Grove, California.

Mwanukuzi, P.K. (2011) Impact of non-livelihood-based land management on land resources: the case of upland watersheds in Uporoto Mountains, south-west Tanzania. *The Geographical Journal* 177, 27–34.

Myer, N. (1990) Tropical forests. In: Legget, J. (ed.) *Global Warming*. The Greenpeace Report, Oxford University Press, Oxford, pp. 372–399.

Nieuwolt, S. (1977) *Tropical Climatology*. John Wiley, London.

Otieno, D.O., K'Otuto, G.O., Jakli, B., Schrottle, P., Maina, J.N., Jung, E. and Onyango, J.C. (2011) Spatial heterogeneity in ecosystem structure and productivity in a moist Kenyan savanna. *Plant Ecology* 212, 769–783.

Peel, M.C., Finlayson, B.L. and McMahon, T.A. (2007) Updated world map of the Koppen–Geiger climate classification. *Hydrology and Earth System Sciences* 11, 1633–1644.

Population Reference Bureau (2001) *2001 World Population Data Sheet*. Population Reference Bureau, Washington, DC.

Richards, P.W. (1996) *The Tropical Rainforest*. Cambridge University Press, Cambridge.

Sanchez, P.A. (1976) *Properties and Management of Soils in the Tropics*. John Wiley & Sons, New York.

Shampa, B., Swanson, M.E., Shoaib, J.U.M. and Haque, S.M.S.S. (2010) Soil chemical properties under modern and traditional farming systems at Khagrachari, Chittagong hill tracts, Bangladesh. *Journal of Forestry Research* 21, 451–456.

Sillitoe, P and Shiel, R.S. (1999) Soil fertility under shifting and semi-continuous cultivation in the southern highlands of Papua New Guinea. *Soil Use and Management* 15, 49–55.

Soil Survey Staff (1990) *Keys to Soil Taxonomy*. SMSS Technical Monograph No. 19, Blacksburg, Virginia.

Takasaki, Y. (2007a) Dynamic household models of forest clearing under distinct land and labor market institutions: can agricultural policies reduce tropical deforestation? *Environment and Development Economics* 12, 423–443.

Takasaki, Y. (2007b) A model of shifting cultivation: can soil conservation reduce deforestation? *Agricultural Economics* 35, 193–201.

UNEP (1999) *Western Indian Ocean Environment Outlook*. United Nations Environment Programme, Nairobi.

UNEP (2002) *Global Environment Outlook 3*. Earthscan, London.

Walter, H. (1971) *The Ecology of Tropical and Subtropical Vegetation*. Oliver & Boyd, Edinburgh, UK.

Watson, K. and Achinelli, M.A. (2008) Context and contingency: the coffee crisis for conventional small-scale coffee farmers in Brazil. *The Geographical Journal* 174, 223–234.

Webster, C.C. and Wilson, P.N. (1980) *Agriculture in the Tropics*. Longman, London.

2 Shifting Cultivation: Definition, Basic Features and Types

2.1 Definition

Shifting cultivation is a low-input system of arable farming that is practised in large areas of the humid and sub-humid tropics. It is the dominant system of farming in the humid tropics (Tchienkoua and Zech, 2000; Aweto, 2001), with the exception of tropical Asia where continuous cultivation, especially of swamp rice, is a significant feature of the agricultural economy. Today, shifting cultivation is practised mainly in the forest and savanna areas of South America and tropical Africa and in a few geographically restricted, scattered areas in tropical Asia such as north-eastern India and parts of Sumatra and the Malayan peninsula (Fig. 2.1). Shifting cultivation, as its name implies, is characterized by temporary, periodic or intermittent cultivation. A plot of land is cultivated for a short time, usually 1–3 years (depending on rural population density and hence the amount of land available to the farmer). Thereafter, the land is left uncultivated and allowed to revert to fallow or 'bush' in order to regenerate soil fertility naturally through the process of vegetation colonization and development. The period during which the land is left uncultivated or unutilized for other productive purposes is called the fallow period. It is usually much longer than the cropping period and may vary between 5 and 10 years, occasionally reaching 20 years. Long fallow periods of 20 years were common in most parts of the tropics three to five decades ago, but due to rapid rises in population, fallow periods have generally been reduced to less than 10 years. In French Guiana, South America, a country characterized by a very low population density, Rossi *et al.* (2010) reported fallow periods of up to 25 years in the southern region. Sillitoe (1983) observed that farmers in the highlands of Papua New Guinea may cultivate the same site for well over a decade. During the fallow period, they cultivate other plots to meet their family food requirements. Ultimately, they usually return to cultivate the plots under bush fallow when their fertility status is judged to have been adequately restored.

Although shifting cultivation is the most widespread agricultural system in the humid and sub-humid tropics, the system is largely replaced by nomadic or sedentary pastoral systems in the semi-arid and arid parts of the tropics. Some forms of shifting cultivation persist in semi-arid areas, where pastoralists sometimes grow field crops as a side-line to animal rearing in order to meet part of their food requirements. Pearl millet and cowpea are grown in the semi-arid Sahelian zone of West Africa, and bush fallow vegetation (an important component of the shifting cultivation cycle) is used for natural restoration of soil fertility (Klaij, 1990). Shifting cultivation is also an important feature of the agricultural economy of semi-arid

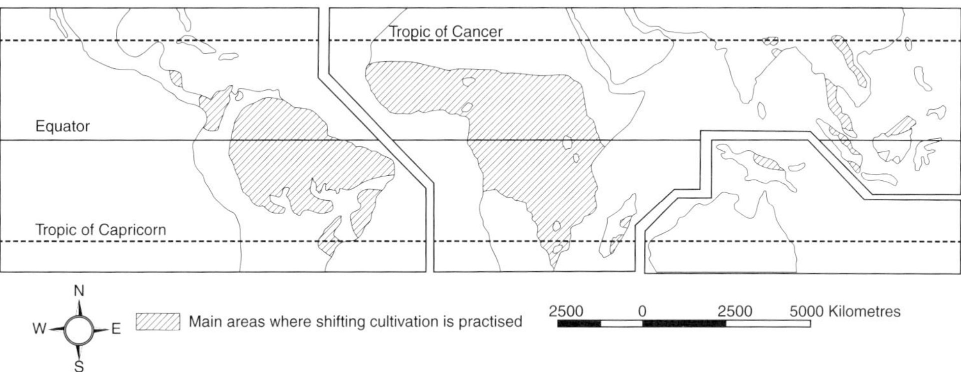

Fig. 2.1. The main areas under shifting cultivation in the tropics. (Adapted from FAO, 1984.)

north-eastern Brazil, where cassava, beans and maize are grown in areas receiving 400–800 mm of rainfall (Sampaio *et al.*, 1990).

As pointed out above, shifting cultivators do not abandon their farms permanently at the cessation of cropping when they leave them fallow. They usually return to recultivate the fallowed plot some time later. During the early stages of the evolution of the shifting cultivation system, farmers cultivated the fields in certain areas while shifting or rotating the cultivated fields and sequentially fallowing them after cultivation. When the fertility of the area within easy reach of their houses had been exhausted, they moved with their households to new areas where they built new houses, while utilizing the more fertile soil of the new location for farming. This relocation was to ensure a more effective exploitation of the soil and vegetation resources for farming and to ensure that time and energy expended in moving from their homes to cultivated fields were minimized. Some authorities (e.g. UNESCO, 1952; Manshard, 1974) restrict the use of the term 'shifting cultivation' to this primordial form of arable farming whereby farmers relocate cultivated fields as well as their houses. At the present time, farmers' houses are more permanent and settlements are fixed, and it is rare for farmers to relocate their dwellings when they shift or relocate their cultivated plots. In this book, 'shifting cultivation' is used in line with the usage of Nye and Greenland (1960) to imply 'shifting field agriculture' (Morgan, 1969), whether or not farmers' houses/settlements are

relocated alongside rotation or shifting of cultivated plots. In this context, shifting cultivation embraces all forms of temporary or periodic cultivation, spatial relocation or shifting of cultivated fields, and the use of bush fallow vegetation to restore soil fertility after a period of cropping. Spatial relocation of farmers' houses or settlements was not considered crucial in the definition of shifting cultivation adopted here. This is because the processes of soil fertility restoration in bush fallows and vegetation development in 'abandoned' farm plots (secondary succession), which are major themes of this book, are independent of and, at best, tangentially related to the spatial relocation of farmers' houses and settlements.

Ruthenberg (1980) and Upton (1996) distinguished between shifting cultivation and bush fallow systems on the basis of the intensity of land use. According to Ruthenberg (1980), bush fallow systems are a more intensive form of arable land use than shifting cultivation. He further observed that when more than 33% of the available arable and temporarily used land is cultivated, the agricultural system is bush fallow system; shifting cultivation is taking place when less than 33% of the land is cultivated annually. This book does not make a distinction between shifting cultivation and the rotational bush fallow system. This is primarily because the two are essentially the same, both being low-input systems of arable farming that involve the rotation or shifting of cultivated plots and the use of bush fallow vegetation for restoring soil fertility after cultivation.

2.2 Characteristics of Shifting Cultivation

The shifting or rotation of cultivated plots will be examined further here and in the section that follows. This is primarily because it is the shifting or spatial relocation of cultivated plots that, perhaps, more appropriately characterizes the system of agriculture and clearly sets it apart from permanent or continuous cultivation. The major characteristics of shifting cultivation are summarized in Table 2.1; these will now be examined briefly. The cropping phase of the shifting cultivation cycle will be discussed in Chapter 3, while soil and vegetation changes occurring during the fallow period are the themes of Chapters 4 and 5, and will not be considered further here.

The shifting cultivator may cultivate up to three or more fields at the same time, while leaving the rest fallow. Field sizes are usually very small, especially in the rainforest region. Farms are usually well under 1 ha in the rainforest region of West Africa, sometimes being 0.2–0.3 ha, and occasionally may be up to 1 or 2 ha. In the rainforest of Sarawak, Malaysia, the average size of shifting cultivation plots is about 1 ha but farm sizes usually range between 0.2 and 3 ha (Bruun *et al.*, 2006). In the savanna region, farms are usually larger because of the greater ease of land clearing, and fields may be up to 3 ha or larger. Usually, it is necessary for the farmer to cultivate several fields in order to meet the food requirement of the household throughout the year. As pointed out previously, the shifting cultivator usually abandons a farm temporarily after cultivating it for a few years, and goes on to cultivate other plots. Herein lies a major distinction between continuous farming in temperate regions and shifting cultivation in the tropics. While the former is characterized by permanent cropping (with occasional grass leys) and by rotation of crops on the same land with high input of fertilizers, the latter involves periodic cultivation and the shifting of fields from time to time. The spatial relocation, or rotation and shifting of cultivated plots is, perhaps, the most diagnostic and universal feature of the system of agriculture known as shifting cultivation.

Shifting cultivation depends on the process of natural cycling of nutrients for the restoration of soil fertility, which declines rapidly during cropping. The system does not usually involve the use of manure or inorganic fertilizers for the maintenance of soil fertility. This is partly because most shifting cultivators are subsistence farmers with meagre resources, which makes fertilizer application technically and economically unfeasible. In this regard, shifting cultivation also differs strikingly from continuous farming in temperate regions, as the

Table 2.1. Characteristics of shifting and continuous cultivation.

Agricultural system	Main characteristics
Shifting cultivation	*Temporary or periodic cultivation
	*Shifting or rotation of cultivated fields and plots
	*Slashing and burning of vegetation before cropping
	*Use of fallow vegetation to restore soil fertility at the end of cropping
	Minimal use of inorganic fertilizers, herbicides and pesticides
	Use of simple implements such as machetes, axes, hoes and digging sticks for land clearing and cultivation
	Intercropping
	Low crop yield
Continuous cultivation	Cultivated fields are fixed
	Cultivation is yearly with occasional grass ley
	Crops are rotated
	High input of inorganic fertilizers, pesticides and herbicides
	Use of tractors and ploughs; farm operations may be mechanized
	High crop yield

*Diagnostic features of shifting cultivation.

latter typically depends heavily on the application of fertilizers and on the use of herbicides and pesticides, which are barely utilized by shifting cultivators.

Another distinctive feature of shifting cultivation is the burning of the cut slash of vegetation prior to cropping. After clearing a site preparatory to cultivation, the felled trees, slashed climbers and shrubs are allowed to dry for a few months before the dried vegetation slash is burnt to 'clear' the site for cultivation and to release ash for fertilizing the soil (Fig. 2.2). Burning the dried slash of cleared vegetation is a diagnostic and important feature of shifting cultivation in the tropics. The practice of burning the dried slash of vegetation is so widespread in shifting cultivation areas that many authorities (e.g. Sirois *et al.*, 1998; Fijisaka *et al.*, 2000) refer to the system of farming as 'slash-and-burn agriculture'. Other authorities (e.g. Dove, 1985; Eden, 1993) refer to shifting cultivation as 'swidden cultivation' or 'swidden agriculture' to emphasize vegetation slash burning. The word 'swidden' means burnt field or plot. Burning vegetation slash is such a pervasive feature of shifting cultivation that it is only in a few areas where climatic conditions make burning difficult that cultivation takes place without prior burning of slashed vegetation. Such exceptional areas include the Atlantic lowland rainforest of Costa Rica, where frequent rains and the shortness of the dry season make burning of slashed vegetation difficult (Jordan, 1987).

The implements used by shifting cultivators for land clearing and cultivation are simple implements such as machetes, hoes, axes and digging sticks. The use of tractors or ploughs is rare among shifting cultivators. Farms are small because the farmers depend on human energy and simple implements for land clearing and cultivation. The use of the digging stick for planting is popular in South-east Asia, while hoes are widely used in tropical Africa. Hoes are frequently used for making mounds or ridges for crops such as yams and cassava, although the cassava cuttings are often planted in shallow, diagonal holes in the ground.

Finally, shifting cultivators usually practise intercropping, with several crops (sometimes up to 12 or more) planted in a small field (Fig. 2.3). Intercropping has several ecological advantages as will be pointed out in Section 2.4.1, but is a major hindrance to mechanization of farm operations, as the different crops mature and are harvested at different times.

2.3 Why Fields are Shifted

It is clear from the foregoing that the shifting or rotation of cultivated plots is a basic and diagnostic feature of the system of shifting cultivation. It is important, therefore, to examine the reasons for the shifting of cultivated plots. Soil fertility declines rapidly following the removal of 'natural' cover of forest and savanna vegetation and at the inception of cultivation in the tropics. Soil nutrients are rapidly depleted after a few years of cropping mainly due to leaching, erosion and nutrient removal in harvested crops (Nye and Greenland, 1960; Szott *et al.*, 1999). In response to the rapid decline in soil nutrients, crop yields decline appreciably and the farmer is forced to abandon the cultivated site after a few years, usually after 1–3 years, to cultivate areas under old bush fallow vegetation and occasionally under mature forest where higher yields can be obtained. In exceptional cases, the land is cultivated for a much longer period prior to fallowing. In the Nguru mountains of Tanzania, Mwampamba and Schwartz (2011) observed that the mean duration of cultivation was slightly over 10 years.

Another factor that may compel the farmer to abandon a cultivated site after a few years is the difficulty of coping with the problem of weeds. Usually, weed regrowth becomes more pronounced the longer a site is cultivated (Beets, 1990). The problem of weeds contributes to a reduction in crop yields. In addition, farmers have to expend more time and energy working on their cultivated plots in order to ensure that crops are not stifled by weeds. Consequently, farmers are forced to abandon cultivated plots when the labour requirement for weeding is becoming excessive. The problem of weeds may be further compounded by that of crop pests and diseases. Usually, pests become more numerous the longer a site is cultivated, once the incidence of infestation begins. Also, the continued cultivation of a site ensures the availability of host plants for pathogenic organisms. Discontinuation of cropping and shifting to a new

Fig. 2.2. A small burnt clearing, with tree stumps to facilitate regeneration of trees after cropping. Not all the cut slash of the fallow vegetation was consumed by the fire; branches and twigs of trees can be observed on the ground.

Fig. 2.3. Intercropping of cassava with yams and maize in a small plot. The wooden stakes in the foreground are used for training yam vines.

site that is free of crop pests and diseases is a cheap and effective way of controlling pests and pathogenic organisms during the shifting cultivation cycle.

Apart from nutrient uptake by cultivated crops, a major factor responsible for soil nutrient decline during cropping is erosion and leaching. Abandoned fields are colonized by fallow vegetation (Fig. 2.4), which not only assists in checking the process of soil erosion but also helps to recycle nutrients leached into the subsoil back to the topsoil (Nye and Greenland, 1960; Smith, 1993). Sheet erosion is usually significant on undulating surfaces, especially during the first few months following inception of cultivation. Soil nutrient loss through erosion is a major factor of soil nutrient diminution in hilly or mountainous areas with pronounced slopes, such as northern Thailand where shifting cultivators cultivate relatively steep slopes. The decision to abandon a cultivated field after a few years in such areas is often intimately related to the need to control or reduce soil nutrient loss due to erosion using forest fallow. Erosion con-

trol and watershed conservation are important functions of the fallow period and the re-generating secondary forest (Rerkasem *et al.*, 2009; d'Oliveira *et al.*, 2011). Fallow vegetation also helps to conserve biodiversity (Scales and Marsden, 2008) and provides a wide range of products, including fruits and medicine, to farmers and rural communities. The function of fallow vegetation is discussed in Section 5.1.

2.4 Forms of Shifting Cultivation

The practice of shifting cultivation varies considerably regionally and even locally due to a range of factors. These include the nature of the environment, especially the nature of the climate, soils, vegetation and topography; and also cultural factors including the dietary preferences and the indigenous knowledge and technology of the people. Nearness of the area in question to an urban centre and accessibility and the availability of markets will also largely determine whether shifting cultivators grow

Fig. 2.4. A young forest fallow vegetation of under 1 year with forbs; scattered trees left by the farmer during cultivation; the remains of a previous crop, a cluster of plantains, towards the right.

cash or export crops in order to earn cash to buy tools such as cutlasses and hoes for farming; and to buy clothes, pay taxes and pay school fees for their children. In very remote areas, shifting cultivators basically engage in subsistence farming supplemented with fishing and hunting in order to produce their food requirements. In accessible areas, shifting cultivators produce cash crops such as cocoa, oil palm and rubber as a sideline to food crop production as in the rainforest zone of West Africa. Even in relatively less accessible areas such as the mountains of northern Thailand, the Hmong people cultivate crops such as rice and maize, and also opium, as their main cash crop (Kunstadter and Chapman, 1978). Different forms of the system of shifting cultivation can be identified on the basis of variations in the shifting cultivation cycle, especially how land is prepared for cultivation, and the length of the period of cropping relative to that of fallow. In respect of land preparation for cultivation, one can recognize different variants of shifting cultivation on the basis of whether the cut slash of fallow vegetation is burnt, used as mulch, or ploughed into the soil as green manure prior to cropping. Also, one can distinguish between sub-types of the system of farming on the basis of types of crops grown and whether or not they involve the integration of trees or livestock. Similarly, one can differentiate between those forms that involve the relocation of settlements after the soil fertility in

the areas within walking distance of farmers' homes is exhausted (migratory shifting cultivation) and sedentary forms in which the farmers live in permanent settlements. In the subsections that follow, the main features of shifting cultivation cycle in forest and savanna lowlands of the tropics will be described before specific variants of the system are considered. Slash-and-burn agriculture is the predominant form of shifting cultivation in the lowland areas of the tropics that are characterized by forest and savanna vegetation, and is the theme of the next subsection.

2.4.1 Slash-and-burn agriculture in forest and savanna lowlands

The basic features of shifting cultivation in the forest and savanna lands of the lowland tropics are essentially similar. This is usually so on well-drained land in areas with a more or less distinct dry season which permits the thorough drying up of the cleared vegetation biomass, making it possible for farmers to burn the cut slash of vegetation before cultivation. The main features of the slash-and-burn type of shifting cultivation cycle are depicted in Fig. 2.5. The process of shifting cultivation is usually initiated when farmers select suitable areas of forest, secondary forest regrowths, or savanna woodland or fallows to clear for cultivation. Simple imple-

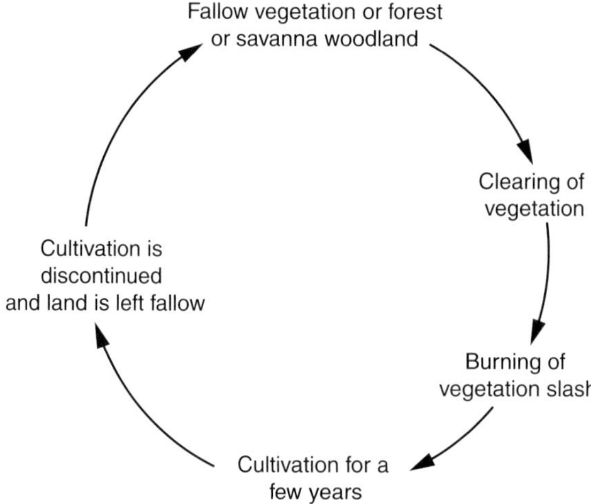

Fig. 2.5. The main features of the shifting slash-and-burn system in forest and savanna lowlands.

ments such as machetes and axes are used for cutting the trees at the height of about 1 m, leaving their stumps in the ground. Useful trees that yield timber, medicine, or fruits are usually left standing on the cleared plot during cropping because of their economic importance. Some of the larger trees of no obvious economic importance are felled using axes, while the very large trees are killed by piling the cut slash of the smaller trees at their bases and subsequently burning them. As ecological and environmental conditions vary locally and regionally, so do the main features of the shifting cultivation cycle.

Land clearing usually takes place during the dry season, a few months before the inception of the wet season, which is the growing season. This usually allows the cut slash of vegetation to dry thoroughly so that it can be burnt. The ash produced as a result of burning the dried slash of vegetation releases nutrients into the soil (Brand and Pfund, 1998; Giardina *et al.*, 2000), thereby helping to fertilize the soil before cultivation. Several crops including annuals, biennials and even perennials are planted mixed together, usually with no regular discernible pattern on the cultivated field. The intercropping of several crops on the same plot (sometimes up to 15 or more on the same plot) is an insurance against pest and diseases and an attempt to ensure food security. Richards (1985) has pointed that farmers in West Africa try to emulate the high species diversity of the rainforest ecosystem by planting mixture of several crop types on small plots or fields. This practice ensures a more efficient exploitation of soil nutrients, greater stability of the agricultural ecosystem and greater crop ground cover which reduces erosion. Usually hoes, cutlasses or digging sticks are used to cultivate crops such as cassava, maize, millet, rice, plantains, bananas and taro in tropical America, Africa and Asia. In tropical Africa, hoes are frequently used for making heaps for cultivating root crops such as cassava and yams. The use of digging sticks and machetes coupled with the preservation of stumps of trees on the cultivated field helps to preserve the soil's physical status during cropping. After a few years of cropping (usually 1–3 years), the farmer 'abandons' the cultivated plot and allows it to be colonized by fallow vegetation for 5–10 years in order to replenish its fertility. Where land is plentiful and population

density low, the land may be fallowed for much longer than 10 years, but in areas where population density is high and land shortage is becoming acute, fallows may be reduced to about 3 years. Eventually, the farmer returns to cultivate the fallowed plots when they have recovered their fertility sufficiently. The main features of the shifting cultivation cycle described above for tropical lowlands are summarized in Fig. 2.5. Slash-and-burn is the basic or primordial type of shifting cultivation and other forms of the system are modifications of this basic cycle or form.

2.4.2 The *chitemene* system

Chitemene is a variant of slash-and-burn agriculture (shifting cultivation) practised by the Bemba people of northern Zambia and other related ethnic groups of the Congo–Zambesi watershed of Central Africa (Araki, 1993). The area where *chitemene* is practised in Central Africa is characterized by 'miombo' woodland, a type of savanna vegetation in which the trees – *Brachystegia*, *Isoberlinia* and *Julbernardia* – feature prominently. The mean annual rainfall is about 1100 mm and the soils are mainly deeply and intensively weathered oxisols of low fertility status. *Chitemene* is similar to the basic or conventional shifting cultivation described above in that it involves the rotation or shifting of cultivated plots and burning the cut slash of vegetation so that the ash produced fertilizes the cleared plot, prior to cultivation. However, only a part of the cleared land is cultivated in the *chitemene* system and herein lies the fundamental difference between *chitemene* and conventional shifting cultivation. Branches of trees and the smaller boles of felled trees are collected and stacked together in the middle of cleared land. Sometimes, young men climb trees and cut or prune their branches from a wide area and these are stacked together in the middle of the cleared area and burnt. Only the burnt inner portion fertilized by the ash (infield) is cultivated, while the outer area left unburnt (outfield) is not cultivated in the *chitemene* system. This type of shifting cultivation depends on using the nutrients stored in the vegetation of a large area to fertilize a smaller area and is obviously less conservative of forest resources than conventional shifting cultivation. Allan (1965) has observed that *chitemene* is an adaptation to the poor and relatively infertile soils of the Central Plateau of Africa where the system evolved. Obviously, a lower number of people can be sustained by *chitemene* per unit area of land than by conventional shifting cultivation, other things being equal. In response to increasing population pressure, other, more intensive forms of arable farming such as green manuring are now being embraced by the Bemba people who traditionally practise *chitemene* (Stromgaard, 1991).

2.4.3 The *Hmong* system – a migratory shifting cultivation

The Hmong people live in southern China and several parts of South-east Asia, including Thailand. The account of shifting cultivation given below refers to that of the Hmong people in northern Thailand where they practise slash-and-burn agriculture with other ethnic groups such as the Karen and Lua'. The Hmong people live in mountainous areas in northern Thailand, where they cultivate relatively steep slopes. Shifting cultivation, as practised by these people, is notorious – not only because it results in environmental degradation, but more importantly because it involves the cultivation of opium as the major cash crop (Kunstadter and Chapman, 1978). Other crops produced by the Hmong include maize and rice. The people employ a system of deep hoe cultivation and clean weeding for crop production, especially for opium which requires considerable amount of labour. Clean weeding encourages erosion, this being exacerbated by the relatively steep slopes cultivated. The Hmong usually cultivate the same field for up to 5 years before it is abandoned due to a considerable decline in fertility, or when the invasion of noxious weeds and grasses necessitates shifting to more favourable sites where the labour requirements for weeding are considerably less. Generally, the people tend to prefer soils derived from limestone of higher base status that sustain cultivation for a longer period than soils of inherently lower fertility status (Kunstadter and Chapman, 1978). When a cultivated site is abandoned at the cessation of cropping, it may be fallowed for up to 40 years

for a natural dipterocarp forest to be re-established. In many instances, a forest may not be established as the cultivated site is invaded by grasses that hinder forest regeneration, even after several years. When this occurs, the original climax forest vegetation is replaced by a relatively stable savanna vegetation; this is an anthropogenic climax.

A major distinctive feature of the *Hmong* system of shifting cultivation is that villages and dwellings are not permanent. When the fertility of areas near the village is exhausted as a result of prolonged cultivation, the village community breaks up and the people move out in search of more fertile areas to cultivate. In this regard, the Hmong of northern Thailand conform to the 'traditional' view of shifting cultivation as a system characterized by the shifting of cultivated plots and the migration of the farming community from time to time.

2.4.4 Shifting cultivation in the Orinoco floodplain

The shifting cultivation cycle in the Orinoco floodplain in Venezuela is strongly influenced by the flood regime of the river, as cleared plots are flooded and cultivation cannot take place until the floodwater subsides (Fig. 2.6). The floodplain and the adjoining levees are submerged 2–3 m beneath floodwater during the rainy season, which lasts from May to October. The flooding of the Orinoco River has two major effects. First, it imposes a seasonality on cultivation, as the floodplain can only be cultivated after the floods have subsided. The period of flooding is synchronized with the period of peak rainfall. Second, the alluvium deposited by the river during flooding enhances the fertility of plots to be cultivated, an effect that is supplementary to that of ash fertilization. Traditional agriculture in the Orinoco has been described by Barrios (1996). The vegetation is mainly savanna with tall grasses, and the crops grown after the floodwater has subsided about November are cotton, cowpea, maize and cassava.

Site preparation for cultivation begins with the selection of appropriate fallow plots which are considered to have sufficiently rejuvenated their fertility, using suitable indicator plant species. Selected sites are cleared and the cut

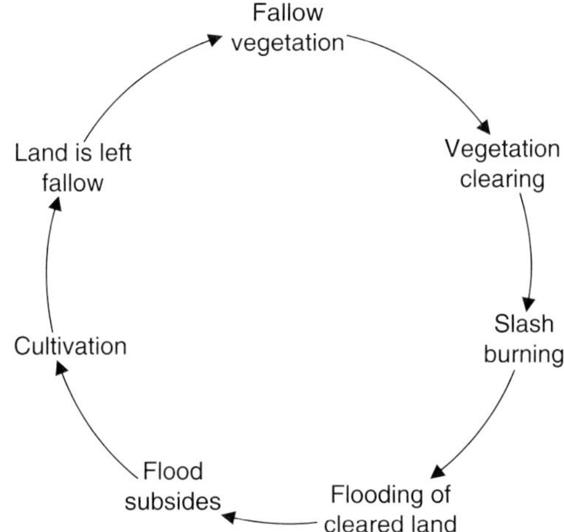

Fig. 2.6. Shifting cultivation cycle in the Orinoco floodplain.

slash of vegetation is allowed to dry before burning, prior to the onset of the rainy season in May. Unlike in most upland areas not prone to flooding, planting of crops does not take place soon after burning in the Orinoco floodplain. Rather, the burnt plot is covered by a dense mat of a creeping perennial grass, *Paspalum fasciculatum*, which sprouts quickly and grows vigorously in the burnt plot after the first few rains (Barrios, 1996). This grass, which invades the burnt cleared plot, is slashed by the farmers prior to flooding. The floodwater will carry part of the slashed grass biomass away, the remaining being buried under alluvium deposited by the floodwater. At the end of flooding, crops are planted on the floodplain soil. Usually, nutrients in the buried plant (grass) biomass and in the alluvium augment the nutrient status of the relatively infertile soils (entisols) that characterize the Orinoco River floodplain.

2.4.5 The slash–mulch system

This is a distinctive type of shifting cultivation that evolved in the Pacific lowlands of Central and South America, especially in a zone stretching from Ecuador to Panama, where there is no prolonged dry period to permit the burning of fallow vegetation (Kass and Somarriba, 1999). Consequently, native farmers do not burn the cut slash of vegetation prior to cropping. Rather, they leave it as mulch on the ground surface where it decomposes rapidly due to the wet and hot climatic conditions. Seeds are broadcast in the bush prior to vegetation clearing, or sown through the layer of cut vegetation mulch in the field after land clearing. The planted seeds germinate and emerge through the layer of vegetation slash in a matter of about 10 days. The main features of the cycle of this type of shifting cultivation, which is also known as slash–mulch, are depicted in Fig. 2.7. Although burning the cut slash of vegetation is regarded as an important feature of shifting cultivation, the slash–mulch cultivation system is an important form of shifting cultivation insofar as cultivation is periodic rather than permanent, and the cultivated plots are temporally abandoned after a few years of cultivation. Furthermore, bush fallow vegetation is used as the means of restoring the fertility status of soils that declined during cultivation, as with other forms of shifting cultivation.

Although the system was developed in very humid areas, it has spread to the drier parts of Central America, and even the dry areas of Mexico (Thurston, 1997) where climatic conditions facilitate rather than retard the drying and burning of slash of cleared vegetation. The readiness with which native farmers in South and Central America embrace the slash–mulch system, including those in seasonally dry areas such as Honduras and Mexico, is due to the system having several advantages:

1. The cut slash of cleared vegetation is left on the soil surface as mulch, which decomposes slowly to replenish soil organic matter and nutrients, thereby helping to sustain soil productivity for a much longer period than in slash-and-burn cultivation systems. Vegetation slash burning results in rapid decomposition of plant debris on the field and plant nutrients are released into the soil in the ash. Unfortunately, most of the ash is washed or blown away and does not result in long-term soil fertilization. In contrast, vegetation slash used as mulch decomposes slowly over a period of several years, thereby guaranteeing a steady supply of nutrients to the soil. It also prevents a decline in soil organic matter, which is a major factor accounting for a decline in soil productivity during cropping. In fact, the decomposition of vegetation slash over time may lead to a gradual build-up of organic matter and this improves the fertility status of marginal soils, thereby permitting their utilization for longer periods of cropping.

2. It ensures that the soil is adequately covered, even before cultivated crops are in full foliage, and this reduces soil erosion to the minimum, even on steep slopes. In parts of Honduras, where rainfall distribution is markedly seasonal, farmers plant maize and sorghum on hill slopes using this system, which reduces soil erosion and also ensures that planted seeds are not washed away by torrential rain (Thurston, 1997).

3. It contributes to soil moisture conservation, and this helps to lengthen the period of cultivation in areas that are prone to drought.

4. It has considerable potential with respect to the intensification of shifting cultivation, particularly as it is capable of conserving or even improving soil organic matter and nutrients over time, helping to prolong the period of cropping, relative to the length of the fallow period. In fact, the system of cultivation developed by native farmers in western Honduras as a viable alternative to shifting cultivation, known as *Quesungual* (after the village of that name), is a variant of the slash–mulch system. The *Quesugual* slash-and-mulch system, which also incorporates an element of an agroforestry system, is discussed in Chapter 10, Section 10.4.

2.4.6 The plough-in-slash system

Shifting cultivation is not restricted to the lowlands of the tropics, or even to the humid and sub-humid tropics, where vast areas of land are under this system of agriculture. At high elevations on the slopes of mountains and intermontane plateaux, shifting agriculture is practised in montane forests and in grassland areas above the treeline. In the latter, it is a

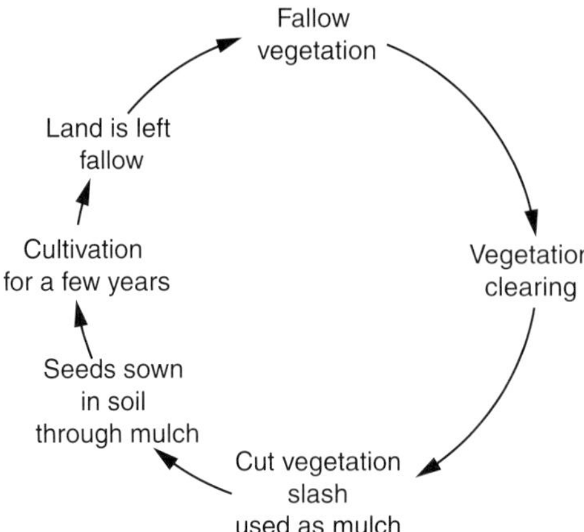

Fig. 2.7. The cycle of the slash–mulch form of shifting cultivation.

sideline to livestock farming. Abadin *et al.* (2002) have described the main features of shifting cultivation practised in the *paramo* (high plateau covered with grasses and shrubs at an elevation of about 3000–4800 m above sea level). In these ecosystems in the Venezuelan Andes, the cultivation of potatoes and cereals (including wheat and barley) for 1–3 years is alternated with fallow periods of 5–12 years.

As in other parts of the tropics, the shifting cultivation cycle begins with clearing mature fallow vegetation which is considered by the farmers to have adequately restored soil fertility. However, unlike most parts of the tropics, the cut slash of fallow vegetation is not burnt prior to cropping to release nutrients into the soil through ash fertilization. Rather, it is ploughed into the soil at a depth of about 20 cm and allowed to decompose (Fig. 2.8) for about 4–5 months, before the land is ploughed again to cultivate crops (Abadin *et al.*, 2002). This type of shifting cultivation can be characterized as 'plough-in-slash' and it depends on the use of the cut slash of fallow vegetation as green manure, ploughed into the soil, which helps to replenish soil organic matter and provide nutrients to crops planted after the fallow period. It has practical relevance to the intensification of shifting cultivation, as using the cut slash of fallow vegetation as mulch or green manure, instead of burning it, is a viable option for making slash-and-burn agriculture more sustainable.

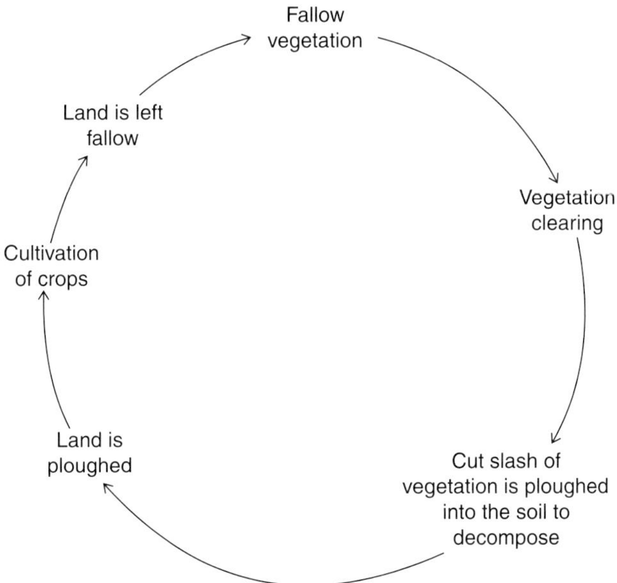

Fig. 2.8. The cycle of the plough-in-mulch form of shifting cultivation.

References

Abadin, J., Gonzalez-Prieto, S.J., Sarmiento, L., Villar, M.C. and Carballas, V.T. (2002) Successional dynamics of soil characteristics in a long fallow agricultural system of the high tropical Andes. *Soil Biology and Biochemistry* 34, 1739–1740.

Allan, W. (1965) *The African Husbandman.* Oliver and Boyd, Edinburgh, UK.

Araki, S. (1993) Effect on soil organic matter and fertility of the chitemene slash-and-burn practice used in northern Zambia. In: Mulongo, K. and Merckx, R. (eds) *Soil Organic Matter Dynamics and Sustainability of Tropical Agriculture.* Wiley-Sayce, Chichester, UK, pp. 367–375.

Aweto, A.O. (2001) Trees in shifting and continuous cultivation farms in Ibadan area, southwestern Nigeria. *Landscape and Urban Planning* 53, 163–171.

Barrios, E. (1996) Managing nutrients in the Orinoco floodplain. *Nature and Resources* 32, 15–19.

Beets, W.C. (1990) *Raising and Sustaining Productivity of Smallholder Farming Systems in the Tropics.* Agbe Publishing, Alkmaar, The Netherlands.

Brand, J. and Pfund, J.L. (1998) Site and watershed assessment of nutrient dynamics under shifting cultivation in eastern Madagascar. *Agriculture, Ecosystems and Environment* 71, 169–184.

Bruun, T.B., Mertz, O. and Elberling, B. (2006) Linking yields of rice in shifting cultivation to fallow length and soil properties. *Agriculture, Ecosystems and Environment* 113, 139–149.

d'Oliveira, M.V.N., Alvarado, E.C., Santos, J.C. and Carvalho Jr, J.A. (2011) Forest regeneration and biomass production after slash and burn in a seasonally dry forest in the southern Brazilian Amazon. *Forest Ecology and Management* 261, 1490–1498.

Dove, M.R. (1985) *Swidden Agriculture in Indonesia.* Mouton Publishers, Berlin.

Eden, M.J. (1993) Swidden cultivation in forest and savanna in lowland southwest Papua New Guinea. *Human Ecology* 21, 145–166.

FAO (1984) *Improved production systems as an alternative to shifting cultivation.* FAO, Rome.

Fijisaka, S. Escobar, G. and Veneklass, E.J. (2000) Weedy fields and forests: interactions between land use and the composition of plant communities in the Peruvian Amazon. *Agriculture, Ecosystems and Environment* 78, 175–186.

Giardina, C.P., Sanford, R.L. and Dockersmith, I.C. (2000) Changes in soil phosphorus and nitrogen during slash-and-burn clearing of a dry tropical forest. *Soil Science Society of America Journal* 64, 399–405.

Jordan, C.F. (1987) Shifting cultivation: slash and burn agriculture near San Carlos de Rio Negro, Venezuela. In: Jordan, C.F. (ed.) *Amazonian Rain Forest: Ecosystem Disturbance and Recovery.* Springer-Verlag, New York, pp. 9–23.

Kass, D.C.L. and Somarriba, E. (1999) Traditional fallows in Latin America. *Agroforestry Systems* 47, 13–36.

Klaij, M.C. (1990) Impact of management practices on pearl millet yield and the sandy soils of the Sahel. In: Stewart, J.W.B. (ed.) *Soil Quality in Semiarid Agriculture, Vol. 11: Local and Regional Concerns on Soil Quality.* Saskatchewan Institute of Pedology & University of Saskatchewan, Saskatoon, Canada, pp. 204–210.

Kunstadter, P. and Chapman, E.C. (1978) Problems of shifting cultivation and economic development in northern Thailand. In: Kunstadter, P., Chapman E.C. and Sabhasri, S. (eds) *Farmers in the Forest.* University Press of Hawaii, Honolulu, Hawaii, pp. 3–23.

Manshard, W. (1974) *Tropical Agriculture.* Longman, London.

Morgan, W.B. (1969) Peasant agriculture in tropical Africa. In: Thomas, M.F. and Whittington, G.W. (eds) *Environment and Land Use in Africa.* Methuen, London, pp. 241–272.

Mwampamba, T.H. and Schwartz, M.W. (2011) The effects of cultivation history on forest recovery in fallows in the Eastern Arc Mountain, Tanzania. *Forest Ecology and Management* 261, 1042–1052.

Nye, P. and Greenland, D.J. (1960) *The Soil Under Shifting Cultivation.* Commonwealth Bureau of Soils, Harpenden, UK.

Rerkasem, K., Lawrence, D., Padoch, C., Schmidt-Vogt, D., Ziegler, A.D. and Bruun, T.B. (2009) Consequences of swidden transitions for crop and fallow diversity in Southeast Asia. *Human Ecology* 37, 347–360.

Richards, P. (1985) *Indigenous Agricultural Revolution.* Hutchinson & Co., London.

Rossi, J.P., Celini, L. Mora, P., Mathieu, J., Lapied, E., Nahmani, J., Ponge, J.-F. and Lavelle, P. (2010) Decreasing fallow duration in tropical slash-and-burn agriculture alters soil macroinvertebrate diversity: a case study in southern French Guiana. *Agriculture, Ecosystems and Environment* 135, 148–154.

Ruthenberg, H. (1980) *Farming System in the Tropics.* Clarendon Press, Oxford, UK.

Sampaio, E.V.B., Salcedo, I.H. and Tiessen, H. (1990) Agriculture in Brazil's semiarid northeast: a nutrient budget under shifting cultivation. In: Stewart, J.W.B. (ed.) *Soil Quality in Semiarid Agriculture. Vol. 11: Local and Regional Concerns on Soil Quality.* Saskatchewan Institute of Pedology & University of Saskatchewan, Saskatoon, Canada, pp. 323–331.

Scales, B.R. and Marsden, S.J. (2008) Biodiversity in small-scale tropical agroforests: a review of species richness and abundance shifts and the factors influencing them. *Environmental Conservation* 35, 160–172.

Sillitoe, P. (1983) *Roots of the Earth: Crops in the Highlands of Papua New Guinea.* Manchester University Press, Manchester, UK.

Sirois, M.C., Margolis, H.A. and Camire, C. (1998) Influence of remnant trees on nutrients and fallow biomass in slash and burn agroecosystems in Guinea. *Agroforestry Systems* 40, 227–246.

Smith, P. (1993) Soil and water conservation. In: Rowland, J.R.J. (ed.) *Dryland Farming in Africa.* CTA & Macmillan, London, pp. 142–171.

Stromgaard, P. (1991) Soil nutrient accumulation under traditional African agriculture in the miombo woodland of Zambia. *Tropical Agriculture (Trin.)* 68, 74–80.

Szott, L.T., Palm, C.A. and Buresh, R.J. (1999) Ecosystem fertility and fallow function in the humid and subhumid tropics. *Agroforestry Systems* 47, 163–196.

Tchienkoua and Zech, M.C. (2000) The effect of cultivation and fallowing on phosphorus pools on ferralitic soils in central Cameroon. In: Floret, C. and Pontanier, R. (eds) *La Jachere en Afrique Tropicale.* John Libbey Eurotext, Paris, pp. 204–211.

Thurston, H.D. (1997) *Slash/Mulch Systems.* Westview, Boulder, Colorado.

UNESCO (1952) *Report of the Commission on World Land-use Survey for the Period 1949–1952.* UNESCO, Worcester, Massachusetts.

Upton, M. (1996) *The Economics of Tropical Farming Systems.* Cambridge University Press, Cambridge.

3 Soil Dynamics during Cultivation

The changes that take place in the soil during cropping are of pivotal importance in at least two major respects. First, the rate of decline in soil physical, biological and chemical status will largely determine the length of cropping prior to fallowing, and sometimes the sequence of crops planted. When soil fertility does not decline dramatically within the first 1–2 years of cropping, cultivation may be continued into the third year or longer. Usually, the more demanding crops such as plantains, yams, sugarcane and maize are planted when soil fertility is still high, soon after establishing fields from old secondary fallow regrowth vegetation or densely wooded savanna or secondary forests. Also, a wider range of crops – sometimes up to 16 or more – is intercropped when soil fertility is still high; the number of cultivated crops is reduced considerably, and less demanding crops such as potatoes and cassava are cultivated when soil fertility has declined. Second, soil fertility status at the cessation of cropping will largely determine the vigour of fallow vegetation regeneration, and hence the rate of soil fertility restoration during the fallow period.

The changes that occur in soil properties during cropping result from the interplay of several distinct processes including: (i) site clearance and disruption of the nutrient cycle; (ii) slash burning and ash addition to the soil; (iii) decline in soil organic matter and nutrients; and (iv) deterioration in soil physical status. In the discussion that follows, examples of soil changes during the cultivation phase of shifting cultivation are drawn from major agroecological zones – rainforest, savanna and monsoon forest – taken from the major continents of the tropics. Shifting cultivation is also practised in floodplain soils in parts of the Amazon basin, although the system is more typical of well-drained upland soils. This chapter also briefly examines soil dynamics under shifting cultivation in river floodplains that are seasonally flooded.

3.1 Effects of Vegetation Clearing

Clearing of vegetation by shifting cultivators, prior to cropping, has been described in Section 2.4.1. As pointed out earlier, land clearing does not involve the use of tractors but relatively simple implements such as machetes. Trees are not clear-felled; rather, the smaller trees are cut at the height of about 1–1.3 m, leaving the stumps in the ground, while some of the very large trees may be left standing or eventually killed by firing if they are not of much economic importance to the farmer. This method of land clearance disrupts the topsoil minimally, and therefore has the benefit of conserving soil physical and nutrient status. Land-clearing techniques of native farmers who practise shifting cultivation are much more conducive to ensuring long-term agricultural sustainability than the 'modern' or 'technologically advanced' land-clearing methods involving the use of tractors and bulldozers. Alegre *et al.* (1986) demonstrated that in the Amazon basin of Peru, land clearing using bulldozers equipped with shear or straight blades had a more adverse effect on soil physical properties (such as bulk density, infiltration capacity, aggregate stability) and soil organic matter status than traditional clearing methods of slash-and-burn adopted by shifting cultivators. Similar results were obtained by Opara-Nadi *et al.* (1986), who compared slash-and-burn methods with various mechanized land-clearing methods in southern Nigeria. They also emphasized the superiority of traditional slash-and-burn land-clearing techniques over mechanized methods in terms of the ability to conserve soil physical status and productivity.

3.1.1 Effects on microclimate

In spite of their comparative advantage in conserving soil physical and nutrient status, slash-and-burn land-clearing techniques invariably modify site characteristics and transform the original ecosystem. Land clearing results in increased receipt of solar radiation at ground level as a result of removal of the vegetation cover. Consequently, soil temperatures are generally higher in clearings than in adjoining mature forest ecosystems, and this accelerates the rate of soil organic matter decomposition and mineralization. Soil temperatures were on average 2–4°C higher in the 0–20 cm soil layer in pastures, established by clearing the forest ecosystem in the Amazon basin, than in the adjacent forest (de Souza *et al.*, 1996). Ghuman and Lal (1987) also observed that cleared sites in the forest zone of southern Nigeria had elevated soil and air temperatures compared to adjacent rainforest ecosystems, where air and surface soil temperatures were 3–6°C lower. The prevalent higher soil and air temperatures following vegetation clearance led to a more pronounced drying-out of the soil compared to uncleared sites. This encourages a more rapid mineralization of soil organic matter when the soil is subsequently wetted, resulting in increased availability of plant nutrients, especially nitrogen (Birch, 1958).

3.1.2 Effects on the soil

Site exposure resulting from the clearing of vegetation causes deterioration of soil physical and chemical status. This is mainly due to the disruptive effects of vegetation clearing on organic matter and nutrient cycles and the high temperatures in cleared sites, which enhance the rate of organic matter decomposition and mineralization in the soil. The study by Cunningham (1963) indicated that following 3 years of forest clearance and attendant site exposure (without cultivation) in Ghana, soil physical status deteriorated substantially, resulting in soil compaction, and a considerable decline in soil porosity and in water-stable aggregates. In addition, in the 0–5 cm layer, soil organic carbon, organic phosphorus, total nitrogen, exchangeable potassium and cation exchange capacity (CEC) declined to 30–65% of their initial level under forest vegetation. Following the clearing of vegetation, surface runoff and erosion assume increased significance and partly account for the deterioration in soil physical and chemical status in shifting cultivation fields.

3.1.3 Nutrient and organic matter cycles

Site clearance also disrupts litter supply to the soil (Ross, 1998), and this, coupled with the effects of erosion and leaching prior to when cultivated crops are in full foliage, adversely affects nutrient cycling in the ecosystem. In order to appreciate the disruptive effects of vegetation clearance on nutrient cycling, the processes of nutrient cycling in forest, savanna and an arable agroecosystem are examined in the next section.

3.1.4 Forest nutrient cycle

Nutrient cycling in tropical forests varies regionally in response to variations in climate, soils and underlying parent material, and in the floristic composition and successional status of the forests. Owing to considerable differences between mangrove forest, rainforest and monsoon forest ecosystems, marked differences exist between them in respect of the quantities of nutrients cycled and the pattern and the relative significance of pathways of nutrient transfer between major compartments of the ecosystem. Figure 3.1 shows the main features of nutrient cycling in forest and savanna ecosystems. The biomass and quantities of nutrients stored in the above-ground biomass of different tropical forest ecosystems vary considerably, especially in relation to climate, successional status and type of forest (Table 3.1).

Decomposition and mineralization of litter are slower in montane forests than in lowland rainforest, and consequently the rate of nutrient recycling to the soil is slower in the former. Generally, forests on moderately fertile and base-rich soils (e.g. alfisols) cycle relatively large amounts of nutrients, while forests on poorer soils such as oxisols and ultisols cycle smaller amounts of nutrients efficiently (Vitousek and Sanford, 1986). The cycling of nutrients in tropical forests varies not only regionally but also locally due to variations in soils and floristic composition of the vegetation. Koutika *et al.* (1999) have shown that the decomposition of soil organic matter in topsoil of the eastern part of the Amazon basin of Brazil is influenced by soil texture, while in Cameroon, Songwe *et al.* (1997) observed that

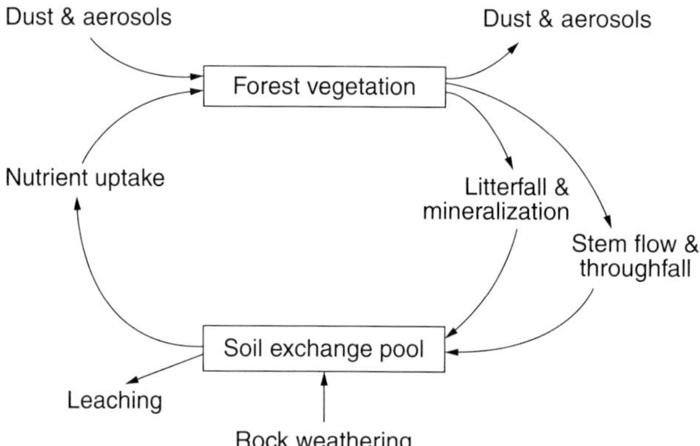

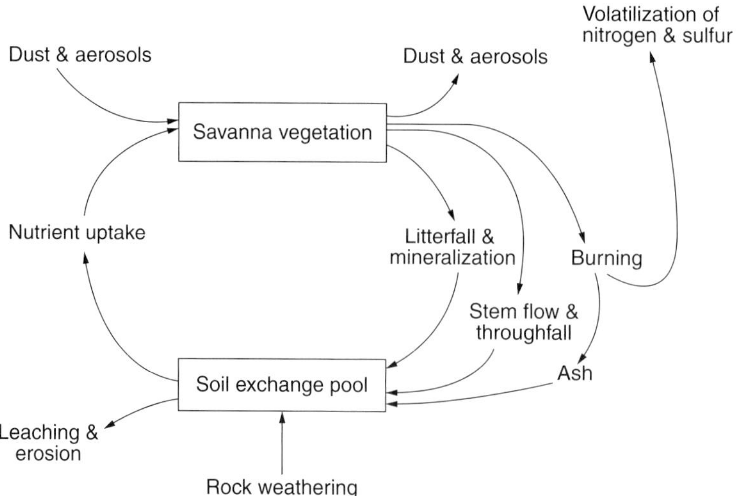

Fig. 3.1. Nutrient cycling in rainforest ecosystems (above) and in savanna ecosystems (below). (After Aweto, 2001.)

variations in tree species within the same rainforest community resulted in differential rates of litter decomposition, and hence in the rate of nutrient cycling.

In spite of the variations in the pattern of nutrient cycling in tropical forests, some generalizations can be made about the process. First, as pointed out by Proctor (1987), most of the nutrients are in the soil rather than in the plant biomass. The implication of this is that cutting and burning a forest does not usually result in an immediate and overwhelming loss of nutrients from the cycling pool. Second, the cycling of nutrients under mature forest has been characterized by Nye and Greenland (1960: 40) as 'a nearly closed cycle'. This does not imply that the forest ecosystem is a nearly closed system, as it receives considerable input of solar energy and moisture from the atmosphere and nutrients from the soil and weathered rocks. Instead, the nearly closed nutrient cycle implies that the transfer of nutrients between the soil and vegetation components of the tropical forest ecosystem is such that the loss of nutrients from the soil–vegetation system is minimal. Vitousek (1984) observed that nutrient cycling in tropical forests is 'tight' or 'efficient'. This is because nutrients transferred from the vegetation to the soil through litter fall and mineralization, and through stemflow, canopy drip and throughfall are rapidly absorbed by plant roots, mycorrhizal fungi and decomposers. Hence, the loss of nutrients through leaching is negligible. Forests on nutrient-deficient soils, especially spodosols and psamments, have efficient nutrient-conserving mechanisms, including high root/shoot ratios and roots heavily infested with mycorrhizal fungi that facilitate direct transfer of nutrients from decomposing litter to the roots of higher plants (Jordan, 1985). Nutrient cycling in monsoon forests is usually less efficient than in rainforest ecosystems. Owing to the prolonged dry season in monsoon forests, they are prone to fires – a tendency reinforced by the occurrence of a distinct grass/herb layer in these forests which dries up and becomes flammable.

Burning in monsoon forests leads to volatilization of nitrogen and sulfur stored in the plant biomass, especially the herb and litter layer, and release of plant nutrients into ash, the bulk of which will be lost from the ecosystem through leaching and erosion when the torrential monsoon rains set in. In this regard, monsoon forests are somewhat similar to savanna ecosystems that are usually burnt annually.

Leaf and twig litter are the main means of recycling nutrients from forest vegetation biomass to the soil, rainwash and throughfall usually being of secondary importance. In tropical forests, especially rainforests, litterfall usually accounts for about 60–80% of the turnover of major nutrients such as nitrogen, calcium, phosphorus and magnesium recycled from the aerial parts of the vegetation to the soil. As with forest biomass, the quantities of litter generated by tropical forests and the quantities of nutrients recycled to the soil via litterfall vary considerably (Table 3.2) in response to changes in climate, relief geology and soils. The table clearly shows that the quantities of litterfall and nutrients returned to the soil via litterfall are considerably smaller in montane rainforests and tropical deciduous seasonal (monsoon) forests compared to the more luxuriant lowland rainforests which are characterized by larger biomass.

3.1.5 Savanna nutrient cycle

The savanna nutrient cycle (Fig. 3.1) is less tight and efficient than that of the rainforest ecosystem, with considerable annual loss of nutrients from the soil–vegetation system of the savanna ecosystem. The inefficient cycling of organic matter and nutrients in savanna ecosystems is due to a number of factors.

Savanna ecosystems in most parts of the tropics are burnt annually, deliberately or accidentally. Savanna vegetation is usually burnt by herdsmen to facilitate the sprouting of fresh grasses for livestock towards the end of the dry season, when fresh grazing is difficult to find, or may be burnt as part of land preparation for cropping. In many cases, burning may be accidentally triggered by humans or through electrical discharge by lightning. Even the savannas in parts of Southern Africa that are protected against burning by government policy, such as in Botswana, are not immune from the ravaging effects of accidental fires. Burning results in the destruction of plant biomass, while

Table 3.1. Above-ground biomass and nutrient storage in tropical forest ecosystems.

Location and type of forest	Above-ground biomass (t ha^{-1})	Nitrogen (kg ha^{-1})	Phosphorus (kg ha^{-1})	Potassium (kg ha^{-1})	Calcium (kg ha^{-1})	Magnesium (kg ha^{-1})	Source
Lowland rainforest, Panama	316	–	158	3020	3900	403	Golley *et al.* (1975)
Lowland rainforest, Venezuela	398	1980	291	1820	3380	313	Hase and Folster (1982)
Montane rainforest, New Guinea	310	683	37	668	1270	187	Edwards and Grubb (1982)
40-year, secondary forest, Kade, Ghana	233	1690	112	753	2370	320	Greenland and Kowal (1960)
Seasonal deciduous (monsoon) forest, Varanasi, India	205	673	379	–	2004	–	Misra (1972)

Table 3.2. Litterfall and quantities of nutrients returned to the soil via litter in tropical forests and savanna ecosystems.

Type of eco-system	Location	Litterfall (t ha^{-1} yr^{-1})	Nutrients returned to soil in litter (kg ha^{-1} yr^{-1})					Source
			Nitrogen	Phosphorus	Potassium	Calcium	Magnesium	
Lower montane rainforest	Puu Maala, Hawaii	5.2	36	1.3	–	84	–	Vitousek *et al.* (1995)
Lower montane rainforest	Sabah, Malaysia	4.8	42	1.1	–	27	–	Proctor *et al.* (1989)
Lowland rainforest	Bakundu, Cameroon	13.6	176.8	22	164.6	242.1	44.8	Songwe *et al.* (1988)
Lowland rainforest	Atherton, Australia	9.0	136	13.1	77.8	229	28.4	Brasell *et al.* (1980)
Monsoon forest	Varanasi, India	7.7	126.7	10	–	184.6	–	Mistra (1972)
Semi-arid savanna	Keur Maktar, Senegal	1.9	27	1.6	13	33	5	Bernhard-Reversat (1987)

nitrogen and sulfur stored in vegetation are volatilized and lost from the ecosystem.

The ground cover of savanna vegetation is much less than that of rainforest ecosystems. Consequently, considerable amounts of nutrients are lost from savanna soils annually, as a result of leaching and erosion. This is particularly so in semi-arid savannas such as the Sahel of West Africa and the thorn savanna of north-east Brazil where the vegetal cover is very sparse, allowing large quantities of soil to be blown away by the wind. In general, vegetation cover decreases from sub-humid savannas adjoining forest regions to the semi-arid savannas at the desert fringe. Consequently, the efficiency of nutrient cycling generally decreases from the sub-humid savannas which are densely stocked with trees to the arid savannas, which are characterized by sparse vegetal cover. Trees enhance the process of nutrient recycling from the subsoil to the topsoil (Young, 1997). They also help to accumulate organic matter and nutrients underneath their canopies (Belsky *et al.*, 1993; Aweto and Dikinya, 2003) and are therefore vital to the process of nutrient cycling in savanna ecosystems.

As the biomass of savanna vegetation decreases towards the poleward limit of the tropics, in response to increasing aridity, so do the quantities of nutrients it cycles. In humid savannas adjoining forest ecosystems, litterfall from trees accounts for a major part of the nutrients recycled from the vegetation to the soil, although the herb layer also plays a significant role. Owing to the considerably reduced density and biomass of trees in semi-arid savannas, litterfall assumes less significance, while the herb layer and plant roots become increasingly important as means of nutrient recycling from the vegetation to the soil.

On account of its smaller biomass, the quantities of nutrients stored in savanna vegetation are generally considerably much smaller than those immobilized in rainforest ecosystems (Table 3.3). Consequently, smaller quantities of nutrients are returned to the soil via litterfall in savanna ecosystems (Table 3.2). This is particularly so in semi-arid savannas. Annual litterfall in a semi-arid savanna at Keur Maktar, Senegal, was 1.9 t ha^{-1}, which returned 27 kg of nitrogen, 2 kg of phosphorus, 13 kg of potassium, 33 kg of calcium and 5 kg of magnesium ha^{-1} to the soil (Bernhard-Reversat, 1987). The quantities of litter generated by the Bakundu rainforest in Cameroon and the quantities of nitrogen, phosphorus, potassium, calcium and magnesium returned to the soil via litterfall are 7–13 times higher than those reported for the semi-arid savanna at Keur Maktar, Senegal. Macro-nutrient input from the atmosphere is considerable in savanna eco-systems. Kellman (1989) observed that the

Table 3.3. Biomass and nutrient storage in the above-ground biomass of savanna ecosystems.

Location and type of savanna	Biomass (t ha⁻¹)	Nitrogen (kg ha⁻¹)	Phosphorus (kg ha⁻¹)	Potassium (kg ha⁻¹)	Calcium (kg ha⁻¹)	Magnesium (kg ha⁻¹)	Source
Semi-arid savanna, Dikeletsane, Botswana	44	487	12.9	135	1084	89	Tolsma *et al.*, (1987)
Sub-humid savanna, Ejura, Ghana	61.9	122.1	21.3	187	267.7	85.1	Nye and Greenland (1960)
'Derived' savanna, Varanasi, India	5.3	96.9	9.2	96.3	–	–	Misra (1983)
Humid pine (treeless) savanna, Mountain Pine Ridge, Belize	–	–	4.2	9.0	3.6	3.2	Kellman (1989)

atmospheric input of potassium may be adequate to meet the requirements of trees in Mountain Ridge Savanna in Belize.

3.1.6 Nutrient cycling in shifting cultivation agroecosystems

The main features of nutrient cycling during shifting cultivation are depicted in Fig. 3.2. The process of nutrient cycling during the cropping phase of the shifting cultivation cycle differs strikingly from that of a tropical rainforest ecosystem. While the rainforest is characterized by efficient nutrient cycling with minimal loss of nutrients, the cropping phase of shifting cultivation is marked by a steady and sustained loss of nutrients from the soil, making the process of nutrient cycling in the agroecosystem very inefficient. Prior to the inception of cultivation, the site is cleared, leaving a few trees standing, and the cut slash of vegetation is subsequently burnt. Trees are vital to the process of nutrient cycling, and land clearing results in their almost total destruction. The cycling of nutrients between the soil and vegetation components of the agroecosystem virtually comes to a standstill at the inception of cropping, except in cases where a sizeable number of trees is preserved and retained on the farm during land clearing. Because removal of the plant cover before farming causes the virtual collapse of the nutrient cycle, most of the mineral nutrients released by ash into the soil when the cleared vegetation is burnt cannot be taken up and stored in the plant biomass at the inception of cropping. Consequently, most of the nutrients released by the ash on the top of the soil are either blown away by the wind or washed away by surface runoff. In the *miombo* woodland of northern Zambia, where shifting cultivation is practised, Araki (1993) estimated that 70% of the ash derived from burning dried vegetation slash was blown away by the wind, not taking account of leaching and erosion losses during the rainy season. Also, in a dry forest on the Pacific coast of central Mexico, Giardina *et al.* (2000a) reported that immediately after burning, wind action resulted in 55% and 74% loss of the phosphorus and nitrogen in the ash, respectively.

Leaching and erosion losses of nutrients continue largely unabated until the cultivated

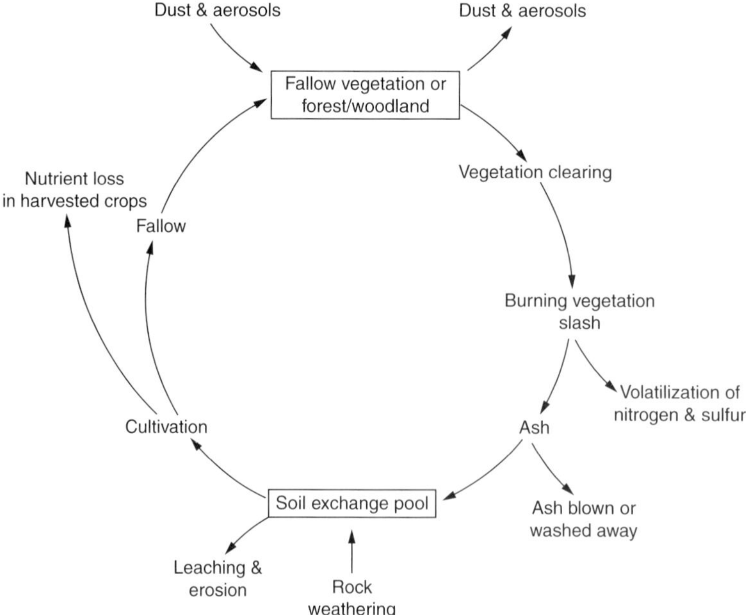

Fig. 3.2. Nutrient cycling during shifting cultivation.

crops are in full foliage and can afford the soil some measure of protection. Even so, leaching of nutrients continues, although at a substantially reduced rate compared with the inception of cropping when the field is largely bare. As should be expected, cultivated field crops such as maize, yams, cassava, sweet potatoes, plantains and millet (all commonly planted by shifting cultivators) take up nutrients from the soil and store them in their standing biomass. This process of nutrient immobilization by cultivated crops results in further depletion of soil nutrient capital. A substantial portion of the nutrients in cultivated crops is lost from the agroecosystem when crops are harvested. The fallow phase of the shifting cultivation cycle is restorative of soil nutrients, organic matter and physical status, and is the subject of Chapters 4 and 5. The fallow vegetation that develops on the cultivated plot at the end of cropping helps to accumulate nutrients and organic matter for a subsequent period of cultivation. It also helps to rejuvenate soil physical status, which deteriorates during cropping. The productivity and sustainability of the system of shifting cultivation largely depends on the fallow period and its efficacy in restoring soil physical, chemical and biological status. The shifting cultivation nutrient cycle, therefore, consists of two phases that alternate with and complement one another – the cropping phase during which soil nutrients decline, and the fallow, which restores nutrients to the topsoil.

3.2 Vegetation Slash Burning

As pointed out in Section 2.2, burning the dried cut slash of vegetation is a diagnostic feature of shifting cultivation. In the subsection that follows, the reasons for burning vegetation slash will be discussed briefly. Thereafter, the effects of slash burning on the soil will be examined.

3.2.1 Why vegetation slash is burned

Vegetation burning is a relatively cheap means of land clearing, especially in savanna environments where the destruction of the grass layer by fire virtually 'opens up' the land for cultivation. When dense forests and woodlands are cleared, much of the site is covered by logs of large trees and cut slash of the smaller trees, and this effectively reduces the space available to the farmer to cultivate. If the felled large trees and the cut slash of the smaller trees are allowed to decompose naturally, the process takes several months or a few years. The farmer therefore artificially accelerates the process of cleared vegetation decomposition, by burning it in order to clear the field of felled trees and plant debris to facilitate cultivation. Hence, as in savanna regions, the immediate aim of vegetation burning is to clear and open up the land for cultivation. Burning vegetation or the cut slash of cleared vegetation by shifting cultivators serves other purposes. Principal among these is the release of nutrients stored in the vegetation, mainly in mineral form in the ash, deposited on the soil when the dried cut slash of vegetation is burnt. Burning, therefore, enhances nutrient availability in the soil just prior to the inception of cropping, and this is beneficial to the first crops cultivated after site burning. In the humid tropics, especially areas with acidic low base soils such as oxisols and ultisols, burning usually has the effect of increasing soil pH, and hence soil base saturation and nutrient availability at the beginning of cultivation. Finally, burning may also have the beneficial effect of controlling crop pests and diseases, and possibly enhancing the mineralization of soil nitrogen. Burning helps to dry up the topsoil more completely and this encourages and enhances mineralization of soil organic matter to release mineral nutrients, especially nitrogen (a nutrient chronically limiting in most soils in the shifting cultivation areas of the tropics), when the soil is subsequently wetted at the inception of the growing season. One of the most beneficial effects of burning vegetation slash, from the point of view of the farmer, is the release of ash to fertilize the soil. This is discussed in the next subsection.

3.2.2 Effects on the soil

The quantity of ash released when vegetation slash is burnt varies with the intensity of the burn, the ecological zone, the nature of the vegetation, and hence the biomass of the slashed vegetation. Generally, more ash is produced when a mature rainforest is burned than when savanna vegetation with a much smaller biomass is burnt. Table 3.4 shows changes in soil properties following slash burning. As the table clearly shows, the effect of burning the dried cut slash of vegetation is ash fertilization resulting in a rise in soil pH and an increase in the levels of mineral nutrients in the topsoil layer, especially calcium, magnesium, potassium and phosphorus.

The study of Okonkwo (2010) in south-eastern Nigeria also showed that burning resulted in a substantial rise in soil pH, mainly due to effects of base cations, especially exchangeable calcium and magnesium, added to the soil via the ash. The quantity of ash released varies spatially over the burnt field or swidden, being considerably more in sites where trunks of trees and branches are piled prior to burning. The effects of slash burning on soil properties vary with the intensity of the burn and hence the temperatures attained by the soil during the process of burning, and this, in turn, is a function of the ecological zone in question. Andriesse and Schelhaas (1987) observed that soil surface temperatures did not exceed 250°C in the wet equatorial area of Sarawak in Malaysia, while in the drier monsoon forest of Thailand soil temperatures reached 300–700°C during slash burning. Towards the poleward limit of the tropics, climatic conditions are drier, allowing slashed plant biomass to dry out more completely; at the same time, the drier air encourages a more vigorous burning of the cut and dried vegetation slash. Giardina *et al.* (2000a) observed that in the dry forest on the Pacific coast of central Mexico at Jalisco (latitude 19°31'N) soil temperatures reached 300–800°C in the 0–1 cm layer. In contrast, near the equator in Sarawak (approximately latitude 3°0'N), Malaysia, the data of Andriesse and Schelhaas (1987) indicated that soil surface temperatures did not exceed 250°C during burning of rainforest slash. Nye and Greenland (1960) observed that slash burning increases soil nutrient capital through ash fertilization but does not substantially affect soil organic matter. It seems that it is only in humid tropics, especially in rainforest environments, that slash burning does not substantially affect soil organic matter level. At the burnt field level, it is only in sites where stems and branches of trees were not

accumulated before burning that the process of slash burning does not affect soil organic matter. Andriesse and Schelhaas (1987) reported that in the drier monsoon forest of Thailand and semi-arid (sub-humid) secondary forest in Sri Lanka, slash burning resulted in a 20% decline in soil organic carbon in the 0–75 cm layer of the soil, as well as slight decreases in total nitrogen, organic phosphorus and CEC. Ketterings *et al.* (2002) have shown that oven-heating forest soil in rubber agroforests in Sumatra, Indonesia, to a temperature of 600°C resulted in a 72% decline in organic carbon in the surface 0–5 cm layer, but the decline was up to 95% in the 5–15 cm layer with the same intensity of heating. Slash burning with the attendant elevated soil temperature affects soil organic matter, although this effect may not be manifested in a substantial decline in levels of soil organic matter. The chemical composition of organic matter can be changed through heating the soil (Fernandez *et al.*, 1997). Such a change in the chemical status of organic matter may result from its desiccation, thermal oxidation, mineralization, and even from soil sterilization. Giovannini *et al.* (1990) and Giardina *et al.* (2000a) have also pointed out that soil heating can enhance thermal mineralization of organic matter, and that in dry tropical forest with marked seasonality in rainfall distribution (such as occurs at Jalisco on the Pacific coast of Mexico), nitrogen and phosphorus input into the soil as a result of soil heating is considerably more than ash input of the two nutrients. Giardina *et al.* (2000b) have drawn attention to the fact that soil heating during slash burning is an important process in the transformation of soil nutrients from non-plant-available to plant-available forms, especially in dry and monsoonal tropical climates.

The practice of piling plant biomass prior to burning usually impacts negatively on soil chemical and biological status, mainly due to the very high soil surface temperatures. Apart from a decline in the level of topsoil organic matter and total nitrogen, Andriesse and Schelhaas (1987) reported that in Sri Lanka, piling biomass (mainly logs and large branches) before burning resulted in a marked increase in soil alkalinity, with the result that crops subsequently planted on the site failed to grow.

Table 3.4. Effects of vegetation slash burning on soil.

Type of vegetation and location	Soil layer (cm)	Status	Organic carbon (%)	Total nitrogen (%)	Calcium (cmol kg^{-1})	Magnesium (cmol kg^{-1})	Potassium (cmol kg^{-1})	pH	Phosphorus (mg kg^{-1})	Source
30-year monsoon	0–7	Pre-burn	2.49	0.27	3.36	2.67	0.64	6.6	9	Ramakrishnan and
forest, Burnihat, India	0–7	Post-burn	2.16	0.22	14.97	8.22	7.95	8.7	1	Toky (1981)
6-year bush fallow,	0–10	Pre-burn	26,500[a]	2,060[a]	757[a]	252[a]	153[a]	5.2	1	Gafur et al. (2004)
Chittagong Hill Tracts, Bangladesh	0–10	Post-burn	25,200[a]	1,890[a]	991[a]	285[a]	189[a]	5.3	1	
Miombo (savanna)	0–10	Pre-burn	2.07	0.15	1.38	1.67	0.25	5.2	12	Stromgaard (1991)
woodland, Kasama, Zambia	0–10	Post-burn	1.54	0.25	4.03	2.14	0.55	5.2	56	
40-year rainforest,	0–15	Pre-burn	1.67	0.16	4.65	1.08	0.37	5.0	7	Nye and
Kade, Ghana	0–15	Post-burn	1.76	0.16	11.25	2.00	1.41	7.2	19	Greenland (1960)
25–40-year secondary	0–5	Pre-burn	0.6	0.05	1.57	0.49	0.19	6.8	5	Andriesse and
forest, Vanathavillu, Sri Lanka	0–5	Post-burn	0.4	0.05	4.57	0.68	0.10	8.1	27	Schelhaas (1987)
17-year secondary	0–10	Pre-burn	–	–	0.26	0.15	0.10	4.0	5	Sanchez et al.
forest, Yurimaguas, Peru	0–10	Post-burn	–	–	0.59	0.29	0.32	4.5	17	(1983)

[a] Data are in kg ha^{-1}

Owing to the very high soil surface temperatures during slash burning in the dry forest in Jalisco, Mexico (referred to above), the soil microbial population was severely impacted upon in the topsoil. Dockersmith *et al.* (1999) reported low rates of nitrogen mineralization in the 0–10 cm layer of the Mexican dry forest soil, suggesting that soil microbes had been adversely affected by the elevated topsoil temperature. Soil fauna such as earthworms, termites and other surface and subsoil-dwelling animals may be severely impacted by the burn, as a result of exposure to high temperatures (Lal, 1987). Their diversity and populations may be drastically reduced, with an attendant reduction in the rate of mineralization of soil organic matter.

Finally, slash burning destroys seeds in the soil, thereby reducing the rate of growth of weeds at the inception of cropping. This is beneficial to the shifting cultivator who does not use herbicides to control weeds in the face of rapid weed growth and farm colonization. The results of Ewel *et al.* (1981) indicated that following slash burning of 8–9 year-old forest regrowth in the evergreen forest of Costa Rica, soil seed storage dropped to 3000 seeds m^{-2} compared to 8000 seeds m^{-2} in unburnt forest.

3.3 Organic Matter Decline

Soil organic matter declines rapidly during cropping, particularly during the first few years. Following conversion of natural ecosystems – especially tropical forests and woodlands – to agricultural ecosystems, soil organic carbon and hence levels of organic matter are reduced to about 25–50% of their initial levels (Lal, 2005). This is due to the disruption of litter supply to the soil following vegetation clearing and the effects of site exposure, at least before cultivated crops are in full foliage. The study of Brams (1971) showed that organic matter in both ferrallitic and alluvial soil in Sierra Leone declined rapidly to 62% of the initial level under rainforest during the first 2 years of cultivation. With further cultivation, the rate of organic matter diminution slowed down and tended to stabilize at a new equilibrium level well below the initial forest level after 3–5 years of continuous cultivation (Fig. 3.3). The decline in the level of organic matter is usually much faster in well-drained soil compared to poorly drained soil in river valley bottoms. In the latter, the prevalent anaerobic conditions caused by a high groundwater table tend to slow down microbial decomposition of soil organic matter. Consequently, organic matter may stabilize more quickly and at a higher level (relative to the initial level under forest or savanna vegetation) than on well-drained soil with similar texture, and in the same ecological and climatic zone. Usually, under shifting cultivation, fields are cultivated for a few (1–3) years before being left fallow. As a result, soil organic matter may not stabilize at the lower equilibrium level under cropping before the cultivated plot is fallowed. Due to repeated cultivation, the level of organic matter is usually gradually reduced to the 'cultivation equilibrium' level, much lower than the initial level under climax vegetation, if the periods of fallow are too short to enable adequate level of organic matter to build up in the soil.

Organic matter decline also occurs during cultivation of soils at high elevations in the tropics, in spite of the prevalent low temperatures. The data of Sillitoe and Shiel (1999) indicated that in the Was Valley of Papua New Guinea (with an elevation of 1980 m above mean sea level and an average annual temperature of 18°C with no marked seasonal variations) there was a considerable decline (32%) in the level of topsoil organic matter (0–10 cm layer) during the first year of cultivation. By the tenth year of cropping, organic matter levels decreased 55% relative to the level under montane forest. A parallel but much slower decrease in organic matter was observed in the subsoil layer (15–20 cm) of the montane volcanic soil of the Was Valley. The slower rate of organic matter decline in the subsoil during shifting or continuous cultivation is not unexpected, because – unlike the topsoil – the subsoil is shielded from erosion and solar radiation incidence. Furthermore, some organic matter may be transported or leached downwards from the topsoil to the subsoil. It is important here to point out that Sillitoe and Shiel (1999) did not monitor soil organic matter levels in the same site prior to forest cutting or during the subsequent period of cultivation. Rather, they compared the organic matter levels in different sites under varying periods of cultivation with a forest control site. Natural

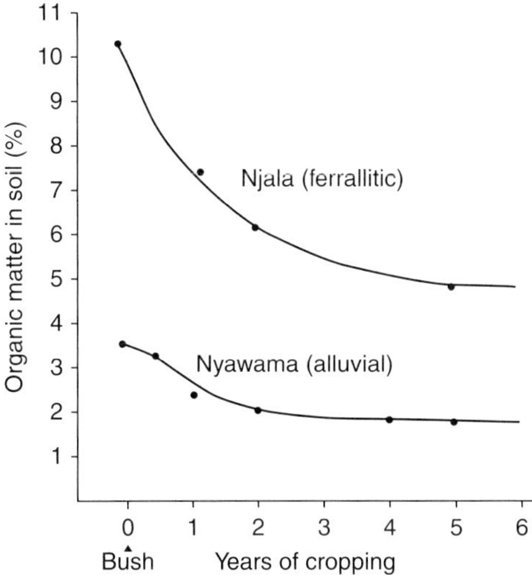

Fig. 3.3. Organic matter decline in the 2–18 cm layer of two Sierra Leonian soils during 5 years of continuous cropping. (After Brams, 1971.)

differences between the different sites compared could be erroneously attributed to the effects of varying lengths of cropping. This could have exacerbated or reduced the calculated rates of soil organic matter decline during cropping.

Organic matter decline is a distinctive feature of shifting cultivation and is due to a number of factors. These include a substantial reduction of litter supply to the soil during cropping, site exposure (Fig. 3.4), the attendant losses of organic matter through leaching and erosion, site burning prior to cropping and enhanced soil organic matter decomposition due to higher soil temperatures in cleared and cultivated sites before crops are in full foliage. Following 1–5 years of cultivation in different parts of the tropics, organic matter levels in the topsoil decreased 15–37% relative to their initial levels or levels under undisturbed forest control plots (Table 3.5). Rates of soil organic matter decline during cropping vary considerably by region, depending on climatic and ecological conditions, and locally on cultural practices, soil texture and drainage. As a result of 5 years of continuous cropping in the Blue Mountains of Jamaica, soil organic matter decreased by 31% in the 0–10 cm soil layer, relative to the forest level (McDonald *et al.*, 2002). The data of Sanchez

et al. (1983) indicated that in Yurimaguas, Peru, the same period of cultivation resulted in a 16% decline in the level of organic matter, relative to the pre-burn forest level. A study in south-western Nigeria, however, showed that soils cropped for an average of 2 years exhibited topsoil organic matter level reduced to below 50% of the forest equilibrium level (Aweto 1988). Most of the soils in Ijebu-Ode area of south-western Nigeria studied by Aweto (1988) have been under shifting cultivation for a very long period, having been cleared of the rainforest vegetation for over 100 years. Several decades of cultivation alternating with fallow and, in particular, the progressive shortening of the fallow period to 3–5 years within the last two to three decades, have resulted in drastic reduction in organic matter levels in slash-and-burn sites, relative to levels under natural rainforest. The decline in soil organic matter during cropping results in a decline in the population of soil microorganisms. Okonkwo (2010) reported that continuous cultivation in south-eastern Nigeria resulted in a decline in the populations of bacteria, fungi and actinomycetes and that this decline was associated with a general decline in soil fertility, and hence in soil organic matter levels. In volcanic soils in south Chile, Huygens

Fig. 3.4. A recent clearing prepared for cultivation in the foreground and a woody fallow in the background. The clearing is bare, except for some scattered stumps, making site exposure a major factor of organic matter and nutrient diminution during the first few months of cultivation.

Table 3.5. Soil organic matter decline during cropping.

Soil type and location	Soil layer (cm)	Length of cultivation (years)	Forest or initial organic matter level (%)	Organic matter level after cropping (%)	Soil organic matter decline (%)	Source
Alfisols cleared from woody fallow, Ibadan, Nigeria	0–11	4	3.50	2.50	29	Juo and Kang (1989)
Oxisols cleared from bush fallow, Kasama, Zambia	0–10	3.5	4.14	2.80	32	Stromgaard (1991)
Oxisols cleared from 30-year monsoon forest, Burnihat, India	0–7	1	4.98	4.14	15	Ramakrishnan and Toky (1981)
Mollisols cleared from secondary forest, Orange Walk District, Belize	0–12	3	3.60	2.90	20	Lambert and Arnason (1986)
Ultisols cleared from 17-year secondary forest, Yurimaguas, Peru	0–10	5	2.48	2.08	16	Sanchez *et al.* (1983)
Inceptisols cleared from secondary forest, Cinchona, Jamaica	0–10	5	21.25	13.33	37	McDonald *et al.* (2002)
Andisols cleared from bush fallow, Was Valley, Papua New Guinea	0–10	4	51.80	33.10	35	Sillitoe and Shiel (1999)

et al. (2011) observed that microbial biomass was three to five times higher in forest soil than in cultivated soil. This should be expected, as organic matter – the main source of food for microorganisms – declines during cultivation, with an attendant decline in soil microbial activities and soil biological status.

3.4 Nutrient Decline during Cropping

A major effect of the decline in the level of soil organic matter referred to in the preceding section is the attendant diminution of soil nutrients, because organic matter is the source and store of major plant nutrients such as nitrogen, phosphorus and sulfur (Brady and Weil, 2002). In addition, plant nutrients such as exchangeable calcium, potassium and magnesium are released into the soil when organic matter decomposes and mineralizes. Furthermore, soil organic matter is the major contributor and determinant of the CEC of most tropical soils, especially of the deeply weathered soils of the humid tropics that are dominated by low-activity clays, especially kaolinite. Hence, a reduction in soil organic matter levels leads to a lowering of the capacity of the soil to adsorb and retain nutrients, and this in turn leads to increased loss of nutrients from the soil through leaching and erosion. Other factors accounting for nutrient decline during the cropping phase of shifting cultivation include: (i) nutrient removal in harvested crops; (ii) leaching and erosion, especially during the first few months following slash burning before cultivated crops are in full foliage; and (iii) low quantity of litter generated and returned to the soil by cultivated crops.

Nutrient decline sets in immediately after site clearance and assumes increased significance after slash burning, when large quantities of nutrients, previously immobilized in the vegetation biomass, are released into the ground surface as ash. Nutrient diminution continues throughout the cropping period into the first few years of the fallow period (Jordan 1987; Roder *et al.*, 1997). There is a sustained and progressive loss of soil nutrients as the number of years in cultivation increases (Fig. 3.5). In areas where mature forests with large biomass are cleared, this loss of nutrients may be partly compensated for by nutrients released from decaying roots and logs on the field. In many instances, at the time the cultivated farm is abandoned and left fallow, soil nutrients in the cultivated plot are still higher than, or almost equal to, the levels in the pre-burn forest soil, as the studies of Zinke *et al.*(1978) and Sabhastri (1978) in Thailand, and Jordan (1987) in Venezuela (Fig. 3.5) have shown. However, in densely populated areas where mature forests have disappeared and farmers clear fallow vegetation that has been regenerating for 3–7 years, there is no large decaying plant biomass on cultivated fields to offset nutrient loss through erosion, leaching and plant uptake. In such areas, such as the forest region of southern Nigeria, soil nutrients are reduced to levels well below the levels in pre-burn forest soil. Aweto (1988) observed that in shifting cultivation plots in the Ijebu-Ode area of south-western Nigeria, nutrients (including exchangeable calcium, magnesium, potassium and total nitrogen) in the 0–10 cm layer of soil were reduced to 36–59% of their levels in pre-burn forest soil.

Obale-Ebanga *et al.* (2003) presented some intriguing data on the effects of eight decades of continuous annual slash-and-burn agriculture on the properties of zero-tilled vertisols in Diamare plain, northern Cameroon. Their results indicated that soil physical status – especially the proportion of water-stable aggregates and chemical properties including organic carbon and total nitrogen – improved, and that the impact of slash-and-burn agriculture on the soils appeared to be beneficial. They attributed the favourable impact of continuous slash-and-burn agriculture to the replacement of the natural savanna species by *Setaria pumila*, a grass species, which presumably has beneficial effects in improving soil physical and chemical status. The results of Obale-Ebanga *et al.* (2003) on the effects of slash-and-burn agriculture on the soil are not typical of shifting cultivation. Cropping in tropical and temperate regions usually results in the diminution of soil nutrients and organic matter. Hence, there is need to apply amendments – especially manure and inorganic fertilizers – to the soil during cropping, to replenish nutrients immobilized in cultivated crops. This is essential in order to ensure sustainability of production during prolonged or continuous cultivation.

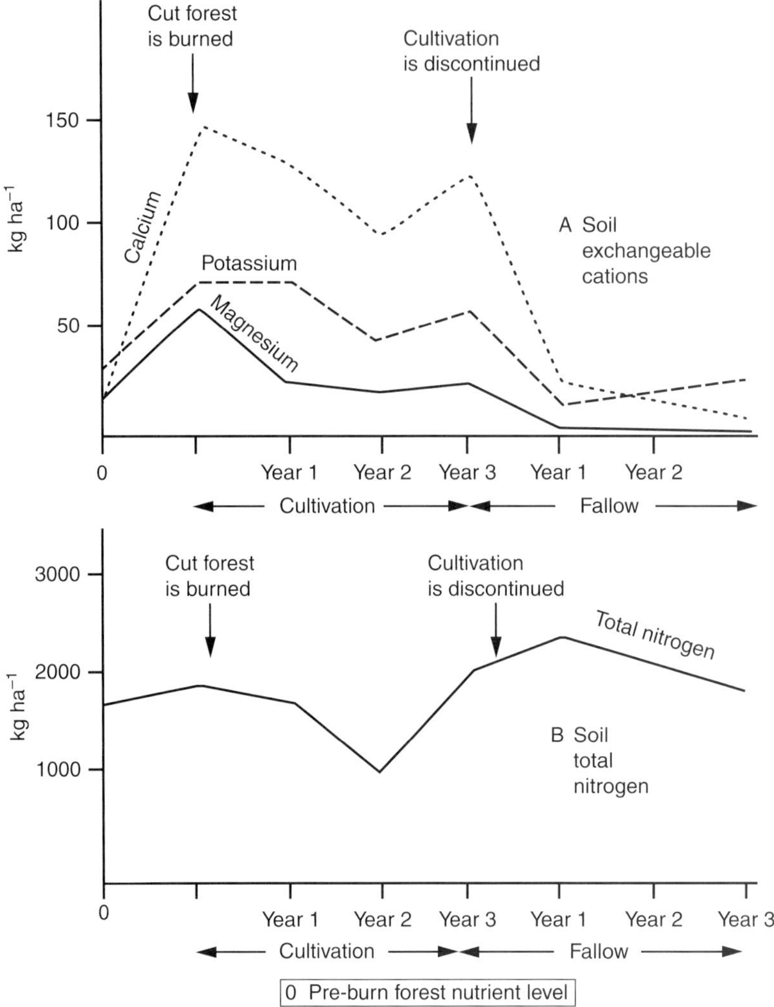

Fig. 3.5. Soil nutrient dynamics during the cropping and fallow phases of the shifting cultivation cycle after forest clearing, at San Carlos, Venezuela. (After Jordan, 1987.)

3.5 Decline in Soil Physical Status

Contemporaneous with the decline in soil organic matter during cropping referred to in Section 3.3 is the deterioration in the soil physical status. This is primarily because soil organic matter is a major determinant of the physical status of soils. Soil physical status is used here in an embracing sense to include the following:

1. Soil bulk density and the degree of soil compaction, and hence the ease with which plant roots can penetrate the soil.
2. The existence of continuous pores from the soil surface to the subsoil, to facilitate exchange of gases between plant roots and soil biota and the overlying air.
3. The capacity of the soil to transmit fluid to lower layers; that is, soil permeability and infiltration capacity.

4. The existence of abundant crumbs in the soil, and the stability and resistance of crumbs to raindrop impact and erosion.
5. The ability of the soil to retain water and provide water for plants.
6. Soil drainage, which also affects water availability in the soil.
7. The degree of soil aeration which largely depends on the number of pores in the soil, that is, soil porosity.

The various parameters of soil physical status listed above largely depend on soil organic matter, although they are also influenced by soil texture. In general, when soil organic matter level is high, the soil will have a good physical status suitable for plant growth. Soil organic matter even has the effect of mitigating soil physical conditions that make the soil less suitable for cropping, such as those induced by stickiness, plasticity and cohesion of clayey soil (Brady and Weil, 2002). Soil organic matter also helps to enhance the aggregation and water-holding capacity of coarse-textured soils, thereby improving their physical status for crop production.

Unfortunately, most studies on soil dynamics during shifting cultivation have tended to focus mainly on soil chemical properties, and the majority have ignored changes in soil physical properties over time. This is most likely due to the pervasive view that crop productivity under shifting cultivation is limited by low levels of soil nutrients. This observation is partly true, as shifting cultivation is practised mainly on deeply weathered soils of the humid tropics with low-activity clays. Soil fertility decline during the cropping phase of shifting cultivation is due not only to soil nutrient impoverishment, but also to the degradation of soil physical status. It is important to observe here that while problems of soil nutrient depletion can easily be addressed by applying appropriate organic manure or chemical fertilizers, it is not easy to restore a degraded soil's physical status to the original condition that promoted high productivity, such as occurs under forest or heavily wooded savanna. This underscores the need to include soil physical parameters in pedological studies on shifting or slash-and-burn agriculture.

Following the clearing of vegetation and the subsequent cultivation of field crops, soil physical condition deteriorates. Soil exposure and the attendant decline in soil organic matter and an increase in erosion are major factors responsible for degradation of soil physical status. The data of Cunningham (1963) indicated that after 3 years of forest clearance and site exposure (without cultivation) in Ghana, the physical status of an alfisol deteriorated considerably, resulting in a decline in total porosity from 52% to 42%, an increase in soil bulk density and a substantial decline in the proportion of water-stable soil aggregates. Raindrop impact and diminution of soil organic matter during cropping cause soil aggregates to break down, and the loose particles are washed downwards by percolating rainwater to seal up soil pores. This results in a decrease in the number of pores in the soil and an attendant decrease in infiltration capacity; in turn, this leads to increased runoff and erosion. In the 0–10 cm layer of forest soil at Yangambi, Democratic Republic of Congo, Mambani (1986) observed that land clearing by the slash-and-burn method followed by cultivation of corn and cassava resulted in a significant increase in soil bulk density from 1.19 g cm^{-3} to 1.32 g cm^{-3}. In cleared but uncultivated plots, soil bulk density was as high as 1.49 g cm^{-3}. In Peru, clearing 20-year secondary forest at Yurimaguas by the slash-and-burn method increased soil bulk density from 1.16 g cm^{-3} to 1.27 g cm^{-3}, while the infiltration capacity was reduced from the initial level of 64 cm h^{-1} to 51 cm h^{-1} in the 0–15 cm layer (Alegre *et al.*, 1986). As pointed out earlier, cultivation results in a decline in water-stable soil aggregates, and hence increases the proneness of soil to erosion. The proportion of water-stable aggregates that exceeds 1.0 mm in vertisols under forest vegetation in Ethiopia was 28.4%, compared to a mere 0.3%–0.5% in cropped soils (Spaccini *et al.*, 2002).

3.6 Erosion

Erosion is a major cause of degradation of soil physical and chemical status following the slash-and-burn clearance of vegetation and subsequent cultivation. Although tree stumps left on the farm help to reduce erosion during cropping, the erosion hazard can assume considerable proportions, especially on land with steep slopes. In the Blue Mountains of Jamaica, secondary forest clearance and the subsequent cultivation of field crops resulted in about a 20-fold increase in erosion (McDonald *et al.*, 2002). The data of Vaje *et al.* (2005) have shown that the average annual runoff from plots established on volcanic soils cultivated to maize, at the Lyamungu Research Station in the Kilimanjaro area of Tanzania, was 93.3 mm, while the average annual sediment loss was 22 t ha^{-1}. The average annual rates of loss of nitrogen, phosphorus and carbon from the cultivated runoff plots were 54.5, 6.5 and 484.5 kg ha^{-1}, respectively. The above data clearly show that there is considerable loss of sediments, organic matter and nutrients from cultivated land, especially in areas with steep slopes, resulting in soil impoverishment in a matter of a few months or years. When the topsoil is removed and the subsoil exposed as a result of accelerated erosion, soil fertility is substantially impoverished on a long-term basis, and the subsequent regeneration of fallow vegetation to rejuvenate soil physical and chemical status is also impaired.

3.7 Shifting Cultivation in River Floodplains

Shifting cultivation or slash-and-burn agriculture is traditionally associated with well-drained soils that usually occur in sites free from flooding. Soil organic matter and nutrients diminish rapidly as a result of a few years of cropping in such soils, hence the necessity to leave the land fallow and cultivate new sites. In contrast, the floodplains of rivers are characterized by alluvial soils whose fertility is rejuvenated yearly as a result of the deposition of sediments, organic materials and nutrients from areas upslope and upstream. Consequently, floodplains of major rivers, especially in tropical Asia, have been intensively used for continuous cultivation of arable crops (particularly rice) for several centuries. In tropical South America, where population pressure on land is con-

siderably less than in tropical Asia, river floodplains have been settled and their soil resources exploited for several centuries for arable farming.

However, in contrast to tropical Asia, alluvial soils in river floodplains are still utilized for shifting cultivation in South America. Zarin *et al.* (1998) studied soil dynamics during the cropping phase of the shifting cultivation cycle in alluvial soils subject to inundation of floodwater, on the island of Abaetetuba in Para, in the Brazilian Amazon basin. Their data indicated that following two harvests of rice and sugar cane, soil organic carbon and exchangeable potassium increased significantly by 69 and 31%, respectively, in the 0–10 cm layer of the cultivated plot. Soil organic matter accretion in the cultivated alluvial soil subjected to shifting cultivation is most likely due to deposition of organic detritus by floodwater on the cultivated plot. The increase in exchangeable potassium and the slight accretion in the exchangeable calcium and magnesium status of the cultivated soil were mainly due to the effects of slash burning coupled with receipts of nutrients via floodwater. The favourable changes in soil organic matter and the exchangeable potassium status of the cultivated slash-and-burn site were also observed in the forest control plot that was not burnt. This suggests that annual flooding is a major causative factor of organic matter and nutrient build-up in alluvial soils used for shifting cultivation. It is also important to note that in floodplain soils that are periodically flooded, the resulting poor aeration slows down microbial decomposition of soil organic matter considerably. This helps to conserve the organic matter in soils that are periodically inundated with floodwater.

The study of Adeboye *et al.* (2011) that compared irrigated and non-irrigated arable soils in the southern Guinea savanna zone in Minna, Nigeria, indicated that organic matter was significantly higher in the 0–10 cm layer of the irrigated soil. This should be expected, as irrigation water periodically reduces aeration and microbial activities in the irrigated soil. The results of Zarin *et al.* (1998) discussed above indicated that organic matter and soil nutrients can increase during cropping if there is an adequate input of sediments or other sources of nutrients, such as plant biomass or crop residue and household refuse. Shifting cultivators often cultivate areas around their houses (home gardens) continuously and fertilize them using household refuse such as ash from the kitchen, husks of grains and other discarded plant materials. The study of soil in home gardens of varying ages in Roraima, northern Brazil, showed that there was an increase in soil phosphorus, calcium, magnesium and potassium, and in the levels of zinc and iron, with increasing age of the gardens (Pinho *et al.*, 2011). This clearly shows that adequate manuring is vital to the intensification of shifting cultivation. Various strategies for intensifying shifting cultivation are discussed in Chapter 9.

References

Adeboye, M.K.A., Bala, A., Osunde, A.O., Uzoma, A.O., Odofin, A.Y. and Lawal, B.A. (2011) Assessment of soil quality using soil organic carbon and total nitrogen and microbial properties in tropical soils. *Agricultural Sciences* 2, 34–40.

Alegre, J.C., Cassel, D.K., Bandy, D. and Sanchez, P.A. (1986) Effect of land clearing on soil properties of an ultisol and subsequent crop production in Yurimaguas, Peru. In: Lal, R., Sanchez, P.A. and Cummings Jr, R.W. (eds) *Land Clearing and Development in the Tropics.* A.A. Balkema, Rotterdam, The Netherlands, pp. 167–177.

Andriesse, J.P. and Schelhaas, R.M. (1987) A monitoring study on nutrient cycles in soils used for shifting cultivation under various climatic conditions in tropical Asia. III. The effects of land clearing through burning on fertility level. *Agriculture, Ecosystems and Environment* 19, 311–332.

Araki, S. (1993) Effect on soil organic matter and soil fertility of the "chitemene" slash-and-burn practice used in northern Zambia. In: Mulongoy, K. and Merckx, R. (eds) *Soil Organic Matter Dynamics and Sustainability of Tropical Agriculture.* Wiley-Sayce, Chichester, UK, pp. 367–374.

Aweto, A.O. (1988) Effects of shifting cultivation on a tropical rain forest soil in southwestern Nigeria. *Turrialba* 38, 19–22.

Aweto, A.O. (2001) Impact of single species tree plantations on nutrient cycling in West Africa. *International Journal of Sustainable Development and World Ecology* 8, 356–368.

Aweto, A.O. and Dikinya, O. (2003) The beneficial effects of two tree species on soil properties in a semi-arid savanna rangeland in Botswana. *Land Contamination and Reclamation* 11, 339–344.

Belsky, A.J., Mwonga, S.M., Amudson, R.G., Duxbury, J.M. and Ali, A.R. (1993) Comparative effects of isolated trees on their undercanopy environments in high – and low – rainfall savannas. *Journal of Applied Ecology* 30, 143–155.

Bernhard-Reversat, F. (1987) Les cycles des elements mineraux dans un peuplement a *Acacia seyal* et leur modification en plantation d' *Eucalyptus* au Senegal. *Acta Oecologica* 8, 3–16.

Birch, H.F. (1958) The effect of soil drying on humus decomposition and nitrogen availability. *Plant and Soil* 10, 9–31.

Brady, N.C. and Weil, R.R. (2002) *The Nature and Properties of Soils.* Prentice Hall, Upper Saddle River, New Jersey.

Brams, E.A. (1971) Continuous cultivation of West African soils: organic matter diminution and the effects of applied lime and phosphorus. *Plant and Soil* 35, 401–414.

Brasell, H.M., Unwin, G.L. and Stocker, G.C. (1980) The quantity, temporal distribution and mineral element content of litterfall in two forest sites in tropical Australia. *Journal of Ecology* 68, 123–139.

Cunningham, R.K. (1963) The effect of clearing a tropical forest soil. *Journal of Soil Science* 14, 334–345.

de Souza, J.R.C., Pinheiro, F.M.A., de Araujo, R.L.C., Pinheiro Jr, H.S. and Hodnett, M.G. (1996) Temperature and moisture profiles in soil beneath forest and pasture areas in eastern Amazonia. In: Gash, J.H.C., Nobre, C.A., Roberts, J.M. and Victoria, R.L. (eds) *Amazonian Deforestation and Climate.* Wiley, Chichester, UK, pp. 125–137.

Dockersmith, I, Giardina, C. and Sanford Jr, R. (1999) Persistence of tree related patterns in soil nutrients following slash-and-burn disturbance in the tropics. *Plant and Soil* 209, 137–156.

Edwards, P.J. and Grubb, P.J. (1982) Studies of mineral cycling in a mountain rain forest in Papua New Guinea. V. Rates of cycling in throughfall and litterfall. *Journal of Ecology* 70, 649–666.

Ewel, J.J., Berish, C., Brown, B., Price, N. and Raich, J. (1981) Slash and burn impacts on a Costa Rican wet forest site. *Ecology* 62, 816–829.

Fernandez, I., Cabaneiro, A. and Carballas, T. (1997) Organic matter changes immediately after a wildfire in an Atlantic forest soil and comparison with laboratory soil heating. *Soil Biology and Biochemistry* 29, 1–11.

Gafur, A., Koch, C.B. and Borggaard, O.K. (2004) Weathering intensity controlling sustainability of ultisols under shifting cultivation in the Chittagong Hill Tracts of Bangladesh. *Soil Science* 169, 663–674.

Ghuman, B.S. and Lal, R. (1987) Effects of deforestation on soil properties and microclimate of a high rain forest in southern Nigeria. In: Dickinson, R.E. (ed.) *The Geophysiology of Amazonia.* John Wiley, Chichester, UK, pp. 225–244.

Giardina, C.P., Sanford, R.L. and Dockersmith, I.C. (2000a) Changes in soil phosphorus and nitrogen during slash-and-burn clearing of a dry tropical forest. *Soil Science Society of America Journal* 64, 399–405.

Giardina, C.P., Sanford, R.L., Dockersmith, I.C. and Jaramillo, V.J. (2000b) The effects of slash burning on ecosystem nutrients during the land preparation phase of shifting cultivation. *Plant and Soil* 220, 247–260.

Giovannini, G., Lucchesi, S. and Giachetti, M. (1990) Effects of heating on some chemical parameters related to soil fertility and plant growth. *Soil Science* 149, 344–350.

Golley, F.B., McGinnis, J.T., Clements, R.G., Child, G.I. and Duever, M.J. (1975) *Mineral Cycling in a Tropical Moist Forest Ecosystem.* University of Georgia Press, Athens, Georgia.

Greenland, D.J. and Kowal, J.M.L. (1960) Nutrient content of moist tropical forest of Ghana. *Plant and Soil* 12, 154–174.

Hase, H. and Folster, H. (1982) Bio-element inventory of a tropical evergreen seasonal forest on eutrophic alluvial soils, Western Llanos, Venezuela. *Acta Oecologia Plantarum* 3, 331–346.

Huygens, D., Roobroek, D., Cosyn, L., Salazar, F., Godoy, R. and Boeckx, P. (2011) Microbial nitrogen dynamics in south central Chilean agricultural and forest ecosystems located on an Andisol. *Nutrient Cycling in Agroecosystems* 89, 175–187.

Jordan, C.F. (1985) *Nutrient Cycling in Tropical Forest Ecosystems.* John Wiley, Chichester, UK.

Jordan, C.F. (1987) Shifting cultivation. Case study No 1: Slash and burn agriculture near San Carlos de Rio Negro, Venezuela. In: Jordan, C.F. (ed.) *Amazonian Rain Forests: Ecosystem Disturbance and Recovery.* Springer Verlag, New York, pp. 9–23.

Juo, A.S.R. and Kang, B.T. (1989) Nutrient effects of modification of shifting cultivation in West Africa. In: Proctor, J. (ed.) *Mineral Nutrients in Tropical Forest and Savanna Ecosystems.* Blackwell, Oxford, UK, pp. 289–300.

Kellman, M. (1989) Mineral nutrient dynamics during savanna-forest transformation in Central America. In: Proctor, J. (ed.) *Mineral Nutrients in Tropical Forest and Savanna Ecosystems.* Blackwell, Oxford, UK, pp. 137–151.

Ketterings, Q.M., van Noordwijk, N. and Bigham, J.M. (2002) Soil phosphorus availability after slash-and-burn fires of different intensities in rubber, agroforests in Sumatra, Indonesia. *Agriculture, Ecosystems and Environment* 92, 37–48.

Koutika, L.-S., Chone, T., Andreux, F., Burtin, G. and Cerri, C.C. (1999) Factors influencing carbon decomposition of topsoils from the Brazilian Amazon Basin. *Biology and Fertility of Soils* 28, 436–438.

Lal, R. (1987) *Tropical Ecology and Physical Edaphology.* Wiley, Chichester, UK.

Lal, R. (2005) Carbon sequestration for sustaining agricultural production and improving environment with particular reference to Brazil. *Journal of Sustainable Agriculture* 26, 23– 42.

Lambert, J.D.H. and Arnason, J.T. (1986) Nutrient dynamics in Milpa agriculture and the role of weeds in initial stages of secondary succession in Belize, C.A. *Plant and Soil* 93, 303–322.

Mambani, B. (1986) Effects of land clearing by slash burning on soil properties of an Oxisol in the Zairean basin. In: Lal, R., Sanchez, P.A. and Cummings Jr, R.W. (eds) *Land Clearing and Development in the Tropics.* A.A. Balkema, Rotterdam, The Netherlands, pp. 227–239.

McDonald, M.A., Healey, J.R. and Stevens, P.A. (2002) The effects of secondary forest clearance and subsequent land-use on erosion losses and soil properties in the Blue Mountains of Jamaica. *Agriculture, Ecosystems and Environment* 92, 1–19.

Misra, R. (1972) A comparative study of net primary productivity of dry deciduous forest and grassland of Varanasi, India. In: Golley, P.M. and Golley, F.B. (eds) *Tropical Ecology with Emphasis on Organic Production.* The University of Georgia, Athens, Georgia, pp. 239–279.

Misra, R. (1983) Indian savannas. In: Bourliere, F. (ed.) *Tropical Savannas.* Elsevier, Amsterdam, pp. 151–166.

Nye, P.H. and Greenland, D.J. (1960) *The Soil under Shifting Cultivation.* Commonwealth Bureau of Soils, Harpenden, UK.

Obale-Ebanga, F., Sevink, J., de Groot, W. and Nolte, C. (2003) Myths of slash and burn on physical degradation of savanna soils: impacts on vertisols in north Cameroon. *Soil Use and Management* 19, 83–86.

Okonkwo, C.I. (2010) Effect of burning and cultivation on soil properties and microbial population of four different land use systems in Abakaliki. *Research Journal of Agriculture and Biological Sciences* 6, 1007–1014.

Opara-Nadi, O.A., Lal, R. and Ghuman, B.S. (1986) Effects of land clearing methods on soil physical and hydrological properties in southwestern Nigeria. In: Lal, R., Sanchez, P.A. and Cummings Jr, R.W. (eds) *Land Clearing and Development in the Tropics.* A.A. Balkema, Rotterdam, The Netherlands, pp. 215–225.

Pinho, R.C., Alfaia, S.S., Miller, R.P., Uguen, K., Magalhaes, L.D., Ayres, M., Freitas, V. and Trancoso, R. (2011) Islands of fertility: soil improvement under indigenous homegardens in the savannas of Roraim, Brazil. *Agroforestry Systems* 81, 235–247.

Proctor, J. (1987) Nutrient cycling in primary and old secondary rainforests. *Applied Geography* 7, 135–152.

Proctor, J., Phillips, C., Duff, G.K., Heaney, A. and Robertson, F.M. (1989) Ecological studies on Gunung Silam, a small ultrabasic mountain in Sabah, Malaysia. II. Some forest processes. *Journal of Ecology* 77, 317–331.

Ramakrishnan, P.S. and Toky, O.P. (1981) Soil nutrient status of hill agro-ecosystems and recovery pattern after slash and burn agriculture (Jhum) in north-eastern India. *Plant and Soil* 60, 41–63.

Roder, W., Phengchanh, S. and Maniphone, S. (1997) Dynamics of soil and vegetation during crop and fallow period in slash-and-burn fields of northern Laos. *Geoderma* 76, 131–144.

Ross, S.M. (1998) Soil and vegetation effects of tropical deforestation. In: Goldsmith, F.B. (ed.) *Tropical Rain Forests: A Wider Perspective.* Chapman and Hall, London, pp. 119–174.

Sabhasri, S. (1978) Effects of forest fallow cultivation on forest production and soil. In: Kunstadter, P., Chapman, E.C. and Sabhasri, S. (eds) *Farmers in the Forest.* University Press of Hawaii, Honolulu, Hawaii, pp. 160–174.

Sanchez, P.A., Villachica, J.H. and Bandy, D.E. (1983) Soil fertility dynamics after clearing a tropical rainforest in Peru. *Soil Science Society of America Journal* 47, 1171–1178.

Sillitoe, P. and Shiel, R.S. (1999) Soil fertility under shifting and semi-continuous cultivation in the Southern Highlands of Papua New Guinea. *Soil Use and Management* 15, 49–55.

Songwe, N.C., Fasehun, F.E. and Okali, D.U.U. (1988) Litterfall and productivity in a tropical rainforest, Southern Bakundu Forest Reserve, Cameroon. *Journal of Tropical Ecology* 4, 24–37.

Songwe, N.C., Fasehun, F.E. and Okali, D.U.U. (1997) Leaf nutrient dynamics of two tree species and litter nutrient content in southern Bakundu Forest Reserve, Cameroon. *Journal of Tropical Ecology* 13, 1–15.

Spaccini, R., Piccolo, A., Mbagwu, J.S.C., Teshale, A.Z. and Igwe, C.A. (2002) Influence of the addition of organic residues on carbohydrate content and structural stability of some highland soils in Ethiopia. *Soil Use and Management* 18, 404–411.

Stromgaard, P. (1991) Soil nutrient accumulation under traditional African agriculture in the miombo woodland of Zambia. *Tropical Agriculture (Trin)* 68, 74–80.

Tolsma, D.J., Ernst, W.H.O. and Verwey, R.A. (1987) Nutrients in soil and vegetation around two artificial waterpoints in eastern Botswana. *Journal of Applied Ecology* 24, 991–1000.

Vaje, P.I., Singh, B.R. and Lal, R. (2005) Soil erosion and nutrient losses from a volcanic ash soil in Kilimanjaro Region, Tanzania. *Journal of Sustainable Agriculture* 26, 95–118.

Vitousek, P.M. (1984) Litterfall, nutrient cycling and nutrient limitation in tropical forests. *Ecology* 65, 285–298.

Vitousek, P.M. and Sanford Jr, R.L. (1986) Nutrient cycling in moist tropical forest. *Annual Review of Ecology and Systematics* 17, 137–167.

Vitousek, P.M., Gerrish, G., Turner, D.R., Walker, L.R. and Mueller–Dombois, D. (1995) Litterfall and nutrient cycling in four Hawaiian montane rainforests. *Journal of Tropical Ecology* 11, 189–203.

Young, A. (1997) *Agroforestry for Soil Management.* CAB International, Wallingford, UK.

Zarin, D.J., Duchesne, A.L. and Hiraoka, M. (1998) Shifting cultivation on the tidal floodplains of Amazonia: impacts on soil nutrient status. *Agroforestry Systems* 41, 307–311.

Zinke, P.J., Sabhasri, S. and Kunstader, P. (1978) Soil fertility aspects of the Lua forest fallow system of shifting cultivation. In: Kunstadter, P., Champman, E.C. and Sabhasri, S. (eds) *Farmers in the Forest*. University of Hawaii Press, Honolulu, Hawaii, pp. 134–159.

4 Soil Dynamics during the Fallow Period

The cropping phase of the shifting cultivation cycle was considered in Chapter 3, with emphasis on soil changes. The present chapter examines soil changes during the fallow period. It is axiomatic to say that soil changes that occur during the fallow period are closely related to those that take place in fallow vegetation. In fact, the pace and vigour of fallow vegetation regeneration largely determines the extent of soil fertility restoration. The changes fallow vegetation undergoes over time are the subject of Chapter 5. Suffice to say for now that the process of soil fertility restoration during shifting cultivation is closely related to and dependent on the regeneration and development of fallow vegetation. If fallow vegetation does not develop, or when the process of vegetation succession in fallow land stagnates as a result of the imposition of biotic factors such as regular burning or grazing, the process of restoration of soil fertility under bush fallow is retarded. It is imperative, therefore, that the process of natural re-generation of fallow vegetation be unhindered by farmers. In some cases, native farmers try to enhance the process of soil fertility restoration during the fallow period by selectively retaining certain trees, or by planting those perceived to have beneficial effects on soil restoration, in addition to planting those that provide fruits or are of economic importance. The management of fallow vegetation by shifting cultivators is discussed in Section 5.7.

The restoration of soil fertility during the fallow period involves the following processes: (i) organic matter build-up, especially in the topsoil; (ii) nutrient accumulation in the vegetation biomass and in the topsoil; and (iii) improvement of soil physical status. The restoration of soil biological status – especially an increase in the population and diversity of soil microbes – is also important. Soil organic matter dynamics are crucial and central to the restoration of soil fertility under bush fallow. Nutrient storage in plant biomass, a major function of the fallow, is discussed in Chapter 5.

4.1 Soil Organic Matter Dynamics

Soil organic matter, which declines during cultivation, usually builds up during the fallow period if the regenerating fallow vegetation produces adequate litter which more than offsets the rate of soil organic matter decomposition and mineralization. Organic matter build-up in the soil may not be instantaneous with the inception of the fallow period. In fact, the first few years of the fallow period may be char-acterized by a decline in the level of organic matter. This implies that the process of organic matter diminution during cropping as a result of leaching, erosion and low quantity of litter generated may continue into the first few years of the fallow period.

Aweto (1981a) observed that there was a decline in organic matter level in the topsoil of forest fallows in south-western Nigeria during the first 3 years of the fallow. Organic matter build-up occurred thereafter, and was evident in the 0–10 cm layer of 7-year and 10-year fallows. Nakano and Syahbuddin (1989) reviewed studies on organic matter and nutrient dynamics in forest fallows in South-east Asia, drawing mainly on data from Thailand. They observed that soil organic matter and nutrient levels were lowest in fallows of 3–4 years. Roder *et al.* (1997) studied slash-and-burn agriculture in northern Laos; their data clearly show a trend of decline in organic matter 3 years after cropping. Aweto (1981a, b) attributed the decline in organic matter in fallow topsoils to the dominance of fallow vegetation by *Chromolaena odorata* during the first 3–5 years of the fallow period. Litter production by the forb fallow of *C. odorata* appeared inadequate to offset the rate of organic matter decomposition. The fallows of 1–3 years in northern Laos, which experienced a steady decline in organic matter levels, were also dominated by *C. odorata*. In southern Nigeria, the decline was reversed when *C. odorata* forbs were replaced by tree perennials, and thereafter, a substantial accretion of organic matter occurred in the soil. This implies that the floristic and structural characteristics of fallow vegetation influence organic matter and nutrient dynamics in fallow soil. The relationships between fallow soil and vegetation are examined in Chapter 6.

The study of Szott and Palm (1996) in Yurimaguas, in the Amazon basin of Peru indicated a 20% decline in organic matter in the 0–45 cm layer of soils (ultisol) under natural bush fallow vegetation, within the first 8 months of the fallow period. The fallow plot, which was previously under 15-year-old secondary forest, had been cleared using the slash-and-burn method. It was subsequently fallowed after a single harvest of upland rice. Although a gradual accretion of organic matter occurred in the fallow soil within 2 years of fallow inception, organic matter levels were still slightly below the pre-fallow level after nearly 4.5 years.

Organic matter decline during the first few years of the fallow period, although quite common, is by no means a universal feature of shifting cultivation. Where many stumps and logs are left in the cultivated field, organic matter

addition to the soil from decaying plant roots and fallen logs may offset the rate of organic matter loss from the farmland topsoil. Also, where fields are established as small clearings within forests, leaf litter from the trees in the adjacent forest may be a supplementary source of plant matter that would be available for conversation into soil organic matter in the cultivated field. Furthermore, dense fallow vegetation dominated by trees will establish quickly – within a few months of the fallow period – in such fields (if they were cropped for a short period before fallowing), due to intensive coppicing of numerous tree stumps and the abundance of seed-bearing parent trees in the forest surrounding the cultivated field. This ensures that the process of organic matter and nutrient accretion in fallow soil commences within the first few months of the fallow period and would have progressed significantly by the second or third year. Bebwa and Lejoly (1993) observed that there was no decline in the organic matter level in the 0–20 cm layer of oxisols under regenerating fallows between the first and sixth years after clearing an equatorial rain forest in the Kisangani area of the Democratic Republic of the Congo. The study of Tiessen *et al.* (1992) in the semi-arid thorn *caatinga* forest of north-eastern Brazil also indicated that, although the build-up of soil organic matter was gradual, the first few years of the fallow period were not characterized by organic matter diminution.

Provided that the process of regeneration is not hindered, organic matter builds up relatively fast in fallow topsoils, especially in the 0–15 cm layer. Organic matter level in the topsoil under forest bush fallow usually approaches the equilibrium level under climax vegetation in about 10–15 years. In southern Nigeria, Aweto (1981a) observed that after 10 years of fallowing, organic matter in the 0–10 cm layer of natural bush fallows attained 76% of the equilibrium level under mature forest which is over 80 years old. Uhl and Jordan (1984) observed that following 5 years of successional development of vegetation after cutting and burning Amazonian rain forest near San Carlos, Venezuela, soil organic matter and nutrients attained the levels of the pre-burn forest. The burnt forest plot that they studied was not cultivated before it was allowed to revert to bush fallow vegetation. This implies that organic matter decline in the soil prior to fallowing may not have been pronounced. Furthermore, burning forest vegetation increases soil nutrient levels beyond the level in pre-burn forest soil. Since the plot was not cultivated, soil nutrient levels probably exceeded the pre-burn forest level at the inception of fallowing. Hence, the results obtained by Uhl and Jordan (1984) do not necessarily indicate an exceptionally fast rate of nutrient and organic matter build-up in fallow soil. Organic matter accumulation in fallows is usually considerably slower, and it takes much longer than 5 years for organic matter in soils under forest fallows to approach forest equilibrium level. The fact that the organic matter status of Uhl and Jordan's (1984) 5-year successional plot is slightly higher than that of the adjoining pre-burn forest presumably suggests that their initial organic matter levels were not quite the same. Site differences in respect of initial soil organic matter levels between the rainforest and the 5-year-old successional plot presumably accounted for the 'exceptionally rapid' rate of organic matter accretion under the 5-year fallow plot.

4.1.1 Organic matter equilibrium concept

It is important to examine the organic matter equilibrium concept, as it largely determines organic matter dynamics and whether nutrients will accumulate substantially in the soil under bush fallow over time. According to Nye and Greenland (1960), the organic matter or humus equilibrium level is the maximum level humus can accumulate under the fallow. They further pointed out that when soil organic matter is well below the equilibrium level at the inception of the fallow period, soil organic matter increment would be substantial, but negligible if the humus level is close to or at the equilibrium level at fallow inception. It is difficult to determine empirically whether soil organic matter level is below the maximum level that would accumulate under a particular kind of fallow at the beginning or middle of the fallow period. Hence, it is difficult to apply the concept and judge empirically whether humus would accumulate in the soil at the inception or middle of the fallow period. This situation is further compounded by the fact that the structure and floristic com-

position of fallow vegetation change over time during the course of its successional development. This calls for a re-examination and redefinition of the organic matter equilibrium concept to make it more relevant to an understanding of humus dynamics under fallow soil or any type of vegetation.

Soil organic matter or humus is dynamic, being in a state of continuous flux. Its behaviour or quantity in the soil is largely governed by the relationship between the rate of addition of plant litter and its subsequent decomposition to form humus/soil organic matter on the one hand, and the rate of organic matter loss from the soil through microbial decomposition and leaching and erosion on the other hand. The organic matter equilibrium concept is based on this dynamic relationship between the rate of humus addition and humus loss from the soil. When the rate of organic matter addition to the soil approximately equals the rate of humus/organic matter loss, the soil organic matter level will remain fairly stable over a period of several years, and it can be said that the soil organic matter has attained a state of equilibrium for that particular type of ecosystem – natural, semi-natural, or anthropogenic. Organic matter equilibrium levels under climax ecosystems such as rainforest or deciduous seasonal (monsoon) forest differ from those of successional communities derived from them. Within a major tropical biome such as rainforest, the soil humus equilibrium level varies regionally and locally, depending on variations in climate, topographic and soil conditions and especially on soil textural composition and ground drainage. Also, the equilibrium level of soil organic matter in savanna ecosystems varies spatially in response to changes in the structure of the vegetation (especially the density of trees and the capacity of the vegetation to produce litter and reduce soil erosion), regularity and intensity of burning and grazing, and also in relation to climatic, soil and topographic conditions. Under bush fallow vegetation the soil organic matter equilibrium level varies depending on whether the vegetation is woody or grass fallow. In addition, the interplay of climatic, topographic, soil and substrate conditions affects the equilibrium level of organic matter under bush fallow.

When soil organic matter has attained a state of equilibrium, it exhibits relatively minor fluctuations over a period of several years, but does not remain constant in absolute terms. In contrast, soil organic matter levels change rapidly (either increasing or decreasing appreciably) in soil that has not attained the equilibrium level. McDonald *et al.* (2002) observed that organic matter in the 0–10 cm layer of soil (inceptisols) under a secondary forest of about 12 years old in the Blue Mountains of Jamaica fluctuated within narrow limits of the mean value. In contrast, organic matter declined sharply under cleared plots used for slash-and-burn agriculture (Fig. 4.1). Clearly, organic matter in the topsoil of the secondary forest is at or is close to the equilibrium level, while the soil in the cultivated plot is not. Under a mature rainforest ecosystem, the soil organic matter level will fluctuate slightly over time, although it has attained the equilibrium level. For instance, the soil humus level under rainforest fluctuates slightly in response to changes in seasons which affect the rates of litter supply to the soil, litter decomposition, humus decomposition and mineralization in the soil and leaching losses of organic matter.

4.1.2 Organic matter equilibrium concept: an explanatory model

As pointed out in Section 4.1.1, the organic matter equilibrium model can be used to explain organic matter dynamics during the fallow period. The concept is also relevant to an understanding of soil organic matter and nutrient dynamics during cultivation. Figure 4.2 is an illustration of the organic matter equilibrium level under forest fallow. It is based on the relationship between the rate of litter production by fallow vegetation and its subsequent decomposition/humification, and the state of organic matter loss from the soil as a result of microbial decomposition, leaching and erosion. At the beginning of the fallow period and during cultivation, soil organic matter may decline because the rate of organic matter addition to the soil through litter production, decomposition and humification is less than the rate of organic matter loss from the soil through humus decomposition, leaching and erosion. As long as the rate of addition of organic matter to the soil is less than the rate of its loss from the soil, organic matter will not build up in the soil. Rather, it would continue to diminish over time. It is only when the rate of organic matter addition exceeds the rate of loss that organic matter builds up in the soil. This development is contingent on certain changes in fallow vegetation, especially an increase in vegetation cover which reduces soil organic matter loss from the soil through reductions in soil temperature and in soil leaching and erosion.

In addition, an increase in litter generation by fallow vegetation (through an increase in vegetation biomass) is required for organic matter accretion in the soil. The rate of organic matter accumulation in fallow soil largely depends on the magnitude of the difference

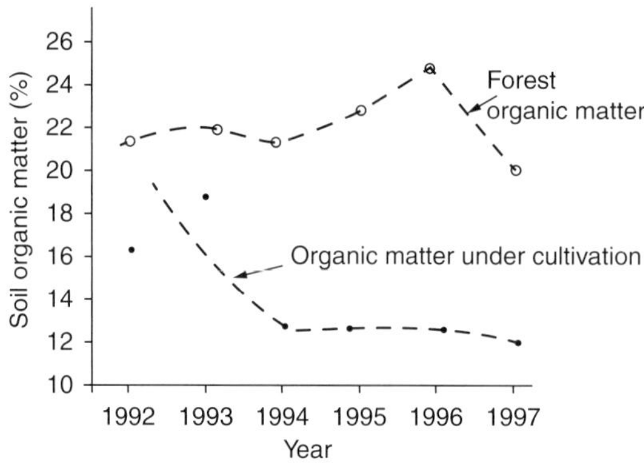

Fig. 4.1. Relative stability of soil organic matter under secondary forest in Jamaica contrasted with sharp decline under cultivation. (Based on data of McDonald *et al.*, 2002.)

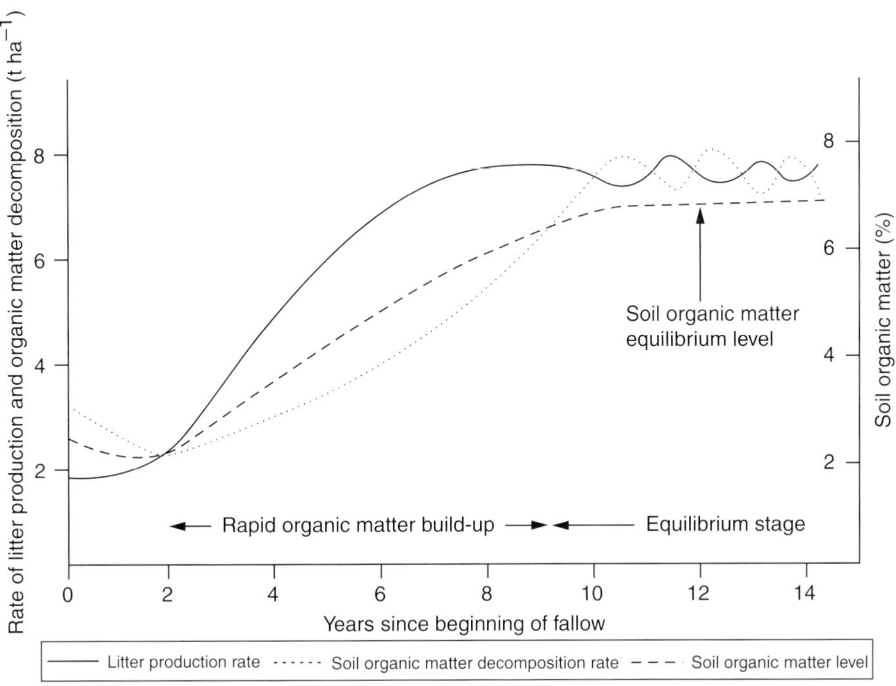

Fig. 4.2. A graphical (diagrammatic) illustration of soil organic matter equilibrium level under forest fallow.

between the rates of organic addition and loss from the soil. If the rate of organic matter addition greatly exceeds the rate of loss, organic matter will accumulate rapidly in the soil. On the contrary, if the rate of organic matter addition is only slightly greater than the rate of loss, organic matter will build up slowly in the soil over time. Organic matter will not continue to build up in the soil indefinitely. This is partly because the rate of litter supply to the soil by fallow vegetation, which increases over time, usually stabilizes after about 8–14 years following fallow inception, as the study of Ewel (1976) in Guatemala has shown. Ultimately, the rate of organic matter accumulation slows down considerably and organic matter attains an equilibrium level for that particular type of fallow vegetation.

It is important to note that when organic matter accumulates in the soil, there is an increase in the rate at which it is degraded. This is because organic matter is a source of food for soil micro-organisms that derive their energy from organic matter and its decomposition

products (Brady and Weil, 2002). Consequently an increase in soil organic matter leads to an increase in soil microbial populations and in macrofauna, resulting in enhanced rates of organic matter decomposition and mineralization in the soil. This has the effect of slowing down the rate at which organic matter accumulates in the soil.

4.1.3 Organic matter accretion in fallow soil in different ecological zones

The rates of organic matter accretion and the quantity of organic matter that accumulates in soil under bush fallow vegetation vary considerably, particularly in relation to the equator–tropics gradient of decreasing rainfall and increasing evapotranspiration. Generally, the rate of organic matter build-up in fallow soil is greatest in the humid tropics, especially in the rainforest environment where biological production, and hence the rate of plant growth, is highest. As the biomass of natural vegetation decreases from about 400 t ha^{-1} in the equatorial rainforest environment to less than 10 t ha^{-1} in the semi-arid savannas towards the poleward limit of the tropics, the amount of plant materials available for conversion into soil organic matter also decreases. In the savanna regions and the deciduous seasonal (monsoon) forests of the tropics, a marked seasonality in rainfall, coupled with biotic factors such as burning and grazing, combine to reduce the rate of growth of fallow vegetation and its efficacy in accumulating organic matter in the soil over time.

In south-west Niger, West Africa, the beneficial value of savanna fallows in restoring soil fertility has been seriously undermined because of their use for grazing livestock coupled with their short duration (Achard and Banoin, 2003). In fact, Gandah *et al.* (2003) observed that millet grain yields in land in western Niger fallowed consecutively for 2 years were an average of 27% of the yield obtained from manured land. This clearly indicates that savanna fallows, particularly in semi-arid regions, have a low capacity for rejuvenating soil fertility. This is especially so if the fallow period is short. In forest fallows in the rainforest zone of south-western Nigeria, soil organic matter accumulated relatively fast from an initial level of 2.5% at fallow inception and attained 4.2% in the top 10 cm of the soil (oxisols) after 10 years of fallowing (Aweto, 1981a). In the Ibarapa Division of south-western Nigeria, where the original forest has been degraded into derived savanna, the rate of organic matter accretion is much slower than in the rainforest. The data of Areola (1980) showed that in Ibarapa, after 30 years of fallowing, soil organic matter content in the 0–10 cm layer reached 1.7%. This was below the value of 2.2% obtained by Aweto (1981a) for the corresponding layer of soil under 3-year forest fallow, also in south-western Nigeria. In the Ibarapa savanna fallow, soil organic matter increased marginally from 1.1% in 5-year fallow to 1.7% in 30-year fallow. The organic matter level was not measured at fallow inception but was likely to be about 1%. The slower rate of organic matter accretion in the savanna soil is due to annual burning of savanna vegetation, destroying standing herbaceous plant biomass and litter, resulting in soil exposure which accelerates the rate of organic matter loss from the soil through leaching and erosion.

The deciduous seasonal forest ecosystems, especially those of Asia, have been subjected to natural and human-induced fires for several thousand years (Stott *et al.*, 1990), as have the savanna formations of Africa, where burning is pervasive (Trollope, 2011). This, coupled with the markedly seasonal nature of the rainfall, has somewhat reduced the rate of organic matter build-up in the deciduous seasonal forests of the tropics, compared with their rainforest counterparts. The data of Ramakrishnan and Toky (1981) indicated that for oxisols in Burnihat, Meghalaya, in north-east India, soil organic matter in the 0–7 cm layer built up slowly from a mean value of 4.2% at fallow inception to 5.0% after 15 years, and ultimately to 5.2% after 50 years. It should be pointed out, however, that the organic matter levels in the Indian bush fallows studied by Ramakrishnan and Toky were quite high and close to the equilibrium level under deciduous seasonal (monsoon) forest at fallow inception. This largely accounted for the slow build-up of organic matter in soil under the Meghalaya bush fallows referred to above. Sabhasri (1978), who studied soil dynamics under bush fallows of the Lua' people of Pa Pae, northern Thailand, observed no build-up of organic matter in the 0–15 cm layer of the soil after 7 years of leaving the land fallow. Again, organic matter in the surface layer of the bush fallow during the first year of the fallow was 6.1%. This is presumably an indication that the soil organic matter level in the bush fallow was at, or very close to, the equilibrium level at fallow inception.

In semi-arid ecosystems, the soil organic level is usually low and the rate of humus accumulation is also slow. The study of Tiessen *et al.* (1992) in semi-arid north-eastern Brazil corroborates this assertion. Their results indicated that the topsoil organic matter level increased from 1.6% during the first year of the fallow period to 2.2% at the end of the tenth year. The organic matter level under the climax *caatinga* vegetation of thorn savanna or thorn forest was also low, being 2.4%. This was similar to the value obtained by Aweto (1981a) for a 1-year forest fallow in south-western Nigeria.

4.2 Nutrient Dynamics

Sanchez (1994) observed that the fallow period (and hence, fallow vegetation) does not improve soil fertility per se, but merely accumulates nutrients in plant biomass. Such nutrients stored in fallow vegetation are subsequently made available to crops, prior to cultivation, when the dried slash of cleared vegetation is burnt. It is erroneous to think that fallow vegetation does not improve soil fertility, although this may sometimes be true of short fallow periods of 3 years or less. The beneficial value of short fallow periods may be essentially limited to accumulating nutrients in the standing biomass of fallow vegetation. Planted short-duration fallows of fast-growing legumes such as *Gliricidia sepium* may, however, have beneficial effects in accumulating nitrogen and organic matter in the soil. Long fallow periods, in most cases, result in build-up of nutrients in addition to restoring soil physical status. Such fallows, therefore, help to restore soil fertility, contrary to the observation of Sanchez (1994). In fact, Nye and Greenland (1960) observed that nutrient accumulation in the topsoil is a major function of bush fallow vegetation. The functions of bush fallow vegetation are discussed in Section 5.1.

It is important to observe that although nutrient storage in the standing plant biomass is one of the major functions of the bush fallow vegetation, more nutrients may accumulate in the soil than in fallow vegetation. The data published by Nye and Greenland (1960) for both forest and savanna fallows in southern Ghana indicated that the stocks of soil nutrients, especially total nitrogen, calcium and magnesium, were considerably higher in the top 30 cm of the soil than in the aerial parts of 20-year forest and savanna fallows. In addition, the data of Jaiyebo and Moore (1964) indicated that the quantities of nitrogen, calcium and magnesium in the 0–30 cm layer of soil under 6-year bush fallow vegetation in Ibadan, south-western Nigeria, were 10–17 times greater than the quantities immobilized in the above-ground component of the vegetation. The bush fallow vegetation, however, had a superior ability in accumulating phosphorus than the top 30 cm soil layer. Also, the data of Toky and

Ramakrishnan (1983) showed that for bush fallows ranging between 1 year and 20 years, the total stocks of nitrogen, calcium and magnesium in the 0–40 cm soil layer were greater than those in aerial parts of fallow vegetation and litter. These findings suggest that soil nutrient stocks are of pivotal importance to the sustainability of soil productivity under shifting cultivation, and that nutrients stored in fallow vegetation merely supplement the soil nutrient capital during cropping. It must be pointed out that fallow vegetation plays a crucial role in respect of phosphorus stocks in the soil and vegetation system.

The trend of nutrient dynamics in fallow soil over time closely follows that of organic matter build-up. This is understandable because soil organic matter is not only the major determinant of the cation exchange capacity (CEC) of most soils in shifting cultivation areas dominated by kaolinitic clays, but also because organic matter is the source and store of major plant nutrients. Whether or not nutrients will accumulate appreciably in fallow soil over time depends on soil organic matter dynamics, which, in turn, depend on the regeneration and the nature and successional development of fallow vegetation.

4.2.1 Forest fallows

In forest bush fallows in south-western Nigeria, Aweto (1981a, b) observed no increase in total nitrogen and exchangeable potassium, calcium and magnesium in the 0–10 cm soil layer between the first and third year of the fallow period when bush fallow vegetation was dominated by the forb, *Chromolaena odorata*. By the seventh year, when woody fallow vegetation had become established, exchangeable calcium, magnesium and potassium had accumulated in fallow soil, although this increase was largely restricted to the 0–10 cm layer (Fig. 4.3). However, soil-exchangeable calcium and magnesium levels declined between the seventh and tenth years of the fallow period, presumably due to nutrient uptake and storage in the vigorously regenerating woody fallow vegetation. This partly accounted for the trend of decline in soil pH levels over time. Total nitrogen build-up in

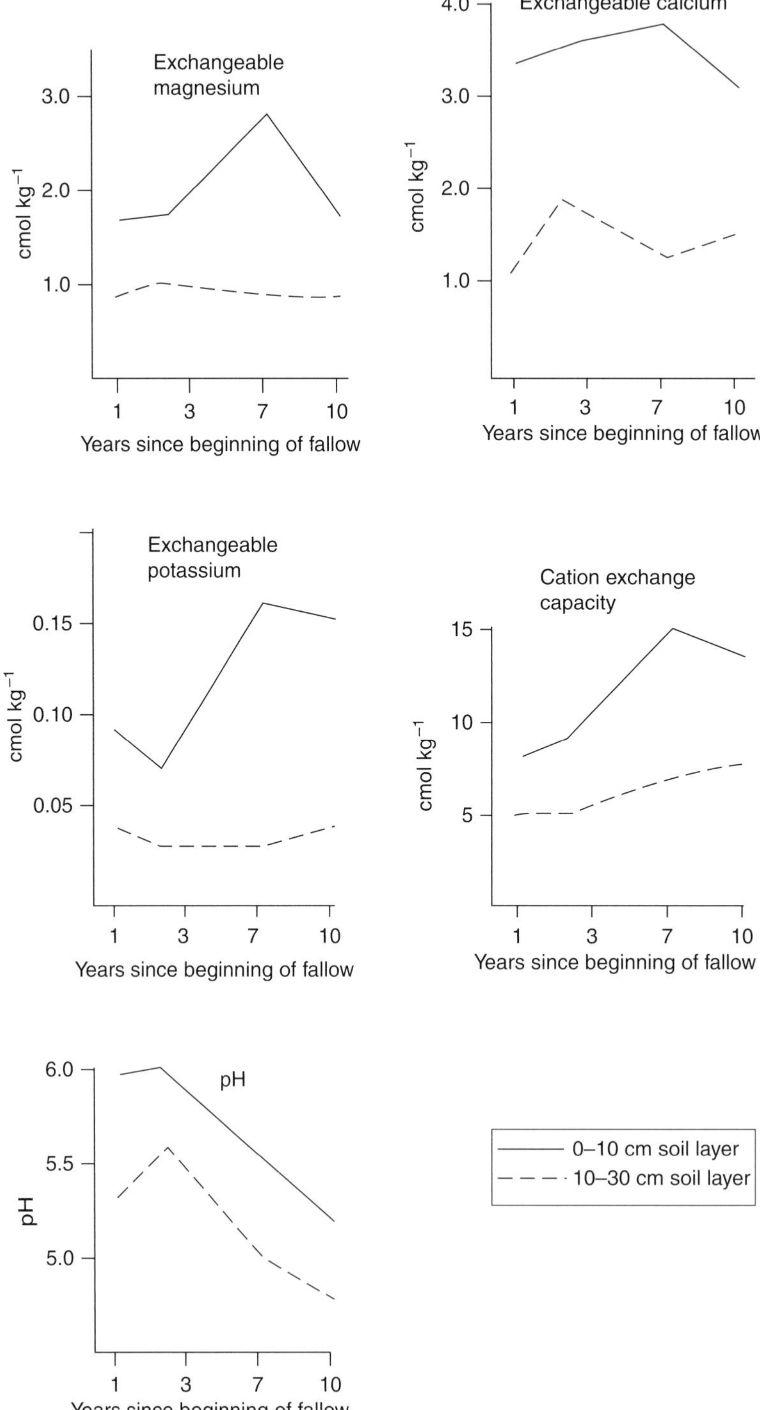

Fig. 4.3. Changes in the levels of soil exchangeable cations, cation exchange capacity, and pH in forest fallows in south-western Nigeria. (Based on data of Aweto, 1981a.)

forest fallow also occurred by the tenth year of the fallow, but was also restricted to the topsoil layer. CEC increased after the third year and, as with nutrient cations, this increase was restricted to the top 0–10 cm of the soil. It is evident from the foregoing that the build-up of nutrients and organic matter in the forest fallows coincided with the emergence and dominance of trees in fallow vegetation. In recognition of the importance of trees to soil fertility restoration, shifting cultivators integrate trees into their fields (a practice known as agroforestry), and these subsequently become a part of fallow vegetation. Such trees provide farmers with poles, income, fruits and medicine, and also help to increase crop yields. Farmers integrate the tree legume *Faidherbia* (*Acacia*) *albida* into their fields in countries such as Niger, Malawi, Zambia and Ethiopia to increase crop yield (Garrity and Stapleton, 2011), while in the fields of the eastern and central African highlands, farmers plant another tree legume, *Calliandra calothyrsus*, to provide feed for livestock, poles and to improve soil fertility (Mandal, 2011). Agroforestry is discussed further in Chapter 9.

Unless fallow vegetation that adequately protects the ground is quickly established after the inception of the fallow period, the process of nutrient diminution during cropping lasts into the first few years of the fallow period. In the forest region of Luang Prabang Province of northern Laos, Roder *et al.* (1997) observed that following the harvesting a single crop of rice in slash-and-burn fields, soil organic matter and nutrient diminution continued into the third year of the fallow period, resulting in loss of 29 t ha^{-1} of organic carbon, 1.1 t ha^{-1} of total nitrogen and 0.7 t ha^{-1} of extractable potassium from the top 100 cm of the soil. Furthermore, there was a sustained loss of exchangeable calcium and extractable phosphorus from the soils (ultisols) during the 3 years of bush fallow, this effect being more pronounced in the top 3 cm of the soil profile. Wadsworth *et al.* (1990) studied soil dynamics in shifting cultivation plots established on forest soil (ultisols) in the south-western Tobasco area of Mexico. The plots were initially intercropped with corn and manioc (cassava) for 1 year and this was followed by 3–4 years of cropping to pineapple before they were fallowed for up to 20 years. They

observed that, in line with the pattern of organic matter dynamics, total nitrogen in the 0–20 cm soil layer declined throughout the cropping period and continued to decrease during the fallow period, reaching the lowest level during the fifth year of the bush fallow. Thereafter, total nitrogen accumulated progressively in the soil until the 15th year of the fallow period, when soil nitrogen attained 85% of the level under 50-year bush fallow. The rate of nitrogen accumulation in fallow soil slowed down considerably after 15 years of bush fallow, indicating that soil organic matter had approached the forest equilibrium level. Exchangeable potassium and available phosphorus and sulphur levels did not increase appreciably over time, even after 20 years of bush fallowing. Similarly, exchangeable calcium and magnesium did not accumulate appreciably in the Mexican fallow soil over time, although the levels in soil under 5-year fallow were lower than those under 20-year bush fallow.

The first 3–5 years of forest fallows are not always marked by nutrient stagnation or nutrient decline in the topsoil. Where trees are quickly established in bush fallow vegetation within the first few months of fallow inception, there will be a build-up of organic matter and nutrients in the topsoil, even within the first 3 years of the fallow period. Brubacher *et al.* (1989) have shown that woody plants, especially trees, accounted for up to 80% of the plant biomass in the first year of forest fallows on mollisols in Belize, Central America, and that the first 3 years of the fallow period were characterized by a gradual increase in soil total nitrogen and exchangeable potassium, due to a rise in the level of soil organic matter. In the 0–15 cm soil layer, pH rose gradually from 7.4 in 1-year fallow to 7.7 in 3-year fallow, and this presumably indicates an increase in the soil base cation level.

Hughes *et al.* (1999) observed no significant accretion of organic carbon in the top 10 cm soil layer during 50 years of succession on volcanic soils under evergreen rainforest in the Los Tuxlas region of Mexico. They attributed the stagnation of organic carbon levels in the soil after 50 years of rainforest succession to the capacity of the young volcanic soil to sequester large quantities of organic matter. The various successional plots studied had different land-use histories. While some had been used for maize cultivation for periods ranging from 15 to 22 years, and were subsequently used as pastures for 8–15 years prior to fallowing, others were cropped to maize for only 1–2 years before the land was left fallow. Differences in land-use history, rather than the inherent capacity of the volcanic soils to sequester organic matter, are the major factors accounting for the absence of build-up of organic matter in the successional plots aged 45–50 years.

In fallows in the monsoon forest of north-eastern India, there was a gradual build-up of nitrogen in the 0–7 cm soil layer from 0.22% in 1-year bush fallow to 0.24% in a 50-year bush fallow (Ramakrishnan and Toky, 1981). Figure 4.4, which depicts changes in the levels of soil nutrients and pH during the first 15 years of monsoon forest succession in India, indicates that available phosphorus declined until the fifth year of the fallow, after which it accumulated progressively in the soil until the 15th year.

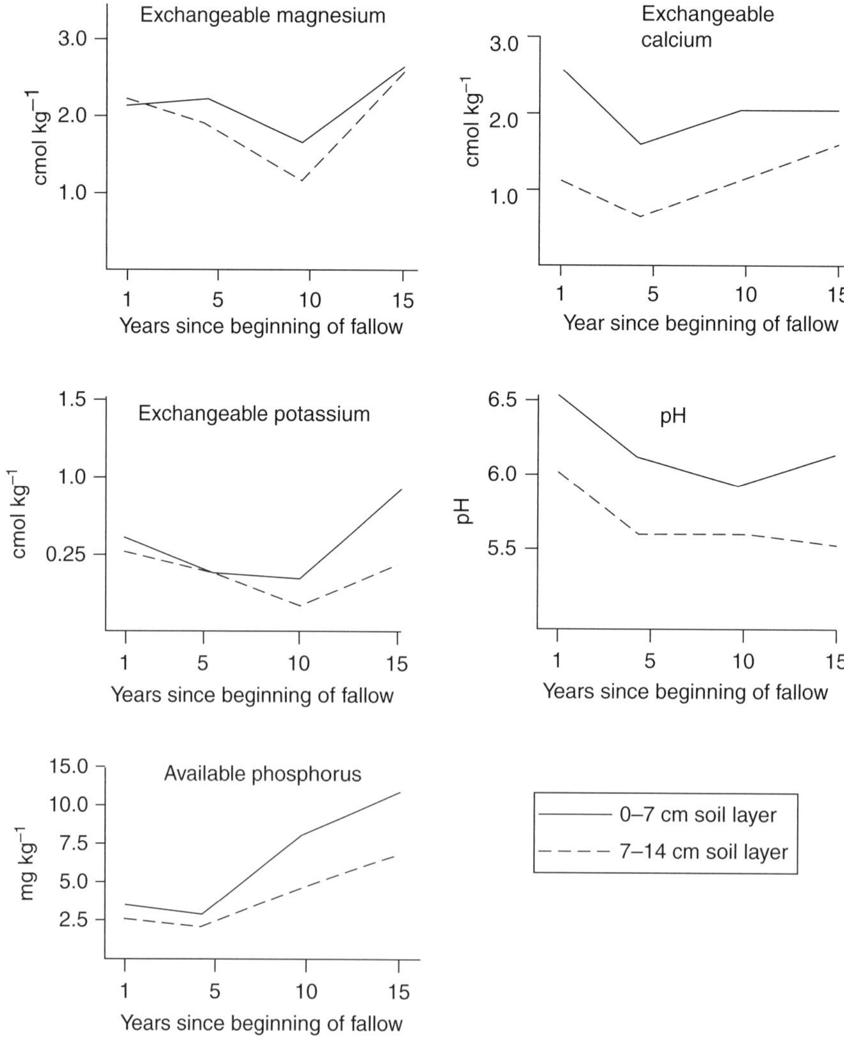

Fig. 4.4. Patterns of nutrient and pH dynamics during the first 15 years of the fallow period in the monsoon forest of north-eastern India. (Based on data of Ramakrishnan and Toky, 1981.)

Exchangeable calcium and magnesium declined throughout the 50-year fallow period due to immobilization in fallow vegetation. Consequently, soil pH declined over time, this decline being particularly marked during the first 10 years of the fallow period (Fig. 4.4). In contrast, the decline in exchangeable potassium was restricted to the first 10 years of the fallow period, after which potassium levels recovered and a slight build-up occurred by the 50th year of the fallow period.

4.2.2 Savanna fallows

Given the very slow rate of organic matter accumulation in savanna fallows due to burning and grazing, the process of nutrient accretion in savanna topsoils over time is also very slow, at least compared to forest fallows that are not subjected to such biotic influences. The data of Areola (1980) revealed that there was no build-up of available phosphorus in the 0–10 cm layer of sub-humid derived savanna fallows in south-western Nigeria between the fifth and 30th year of the fallow period. For the same period, mean topsoil nitrogen content increased slowly from 0.04 to 0.10%. It is pertinent to observe here that the mean total nitrogen content of 0.10% observed by Areola (1980) in the 0–10 cm layer of 30-year derived savanna fallow was below the value of 0.19% observed by Aweto (1981a) for the corresponding layer of soil under 1-year forest bush fallow, also in south-western Nigeria. This clearly shows that nutrient accretion under savanna fallows, such as those in West Africa, which are burnt annually and so subject to erosion, is minimal. Abubakar (1996), who studied a forest reserve in the drier Sudan savanna zone of northern Nigeria, observed that bush fallows which had been subjected to 'minimal disturbance' had a soil nitrogen increase in the 0–15 cm layer from 0.14% in a 2-year fallow to 0.25% in a 15-year fallow, but that there was no substantial increment in available phosphorus. He also observed a gradual build-up of exchangeable cations in the topsoil, especially calcium, magnesium and potassium, between the second and 15th year of the fallow period. This largely accounted for a slight rise in CEC from 5.24 to 6.74 cmol kg^{-1} for the same period.

The efficacy of semi-arid savannas in restoring soil nutrients appears to be limited by their low biomass, and hence their ability to add organic matter and nutrients to the soil, and by their low capacity to protect the soil against erosion. In the semi-arid thorn savanna of north-eastern Brazil, Tiessen *et al.* (1992) observed no significant increase in the levels of soil exchangeable calcium and magnesium, and only a slight increase in exchangeable potassium between the first and tenth years of the fallow period.

In areas close to the forest–savanna boundary, or where land for shifting cultivation is limited (as a result of which land may be subjected to prolonged cropping), forest fallows may be transformed into grass fallows which are relatively stable over time. Grasses usually invade forest fallows as a result of repeated burning during successive periods of cultivation. Burning kills fire-tender forest trees and encourages the dominance of grasses in fallows, and this will further enhance the occurrence of fire, leading to the persistence of grasses in such fallows. The transformation of forest fallows into grass fallows usually alters the process of successional development of vegetation, with an attendant slowing down or stagnation in the process of soil fertility restoration. Scott (1987) studied the changes in soil properties following the transformation of forest fallows dominated by trees into grass fallows in Gran Pajonal in the rainforest region of Peru. His results indicated that the soil nitrogen level remained fairly constant at 15 t ha^{-1} after a fallow period of 10–30 years, during which pioneer fallow communities dominated by the bracken fern *Pteridium aquilinum* and the grass *Imperata brasiliensis* were ultimately replaced by grass fallow communities dominated by *Andropogon*. Soil potassium and calcium increased during the first 3 years, and thereafter decreased till the tenth year before stabilizing, while the magnesium level increased slightly during the first 2 years of fallow, after which it remained stable until the tenth or 30th year of fallow. In Panama, fallow land in forest areas is invaded by the exotic grass *Saccharum spontaneum*, and this hinders the regeneration of native forest (Hooper *et al.*, 2005). The retardation of forest regeneration in such fallows leads to stagnation in organic matter and nutrient accretion in the soil.

4.3 Improvement in Soil Physical Status

Undoubtedly, the restoration of soil physical status is a major function of the fallow period. As pointed out in Section 3.5, soil physical status deteriorates during cropping, mainly due to site exposure and organic matter diminution. The bush fallow vegetation that regenerates on the abandoned farm at the cessation of cropping helps to improve soil physical condition for the crops that will be planted during a subsequent period of cropping. The extent of improvement in soil physical condition during the fallow period will largely depend on the degree to which organic matter accumulates in the soil. In common with research into soil dynamics during cropping, most studies on soil changes during the fallow period did not consider changes in soil physical properties (e.g. Brubacher *et al.*, 1989; Nakano and Syahbuddin, 1989; Tiessen *et al.*, 1992). This is because of the pervasive view that most tropical soils are nutrient-limited and that when adequate nutrients are present in the soil, productivity of the land is largely assured. Such a view does not take cognizance of the degradation of soil physical status during cropping, and of the fact that the maintenance of satisfactory yields during cropping partly depend on the restoration of soil physical condition during the preceding fallow period.

Fallows, natural or planted, have been demonstrated to improve soil physical status over time (Salako *et al.*, 1999). In forest oxisols in south-western Nigeria, Aweto (1981a) observed that in the 0–10 cm layer, soil bulk density decreased from 1.19 g cm^{-3} in 1-year bush fallow to 0.98 g cm^{-3} under 10-year bush fallow. Similarly, soil total porosity increased from 54.4% under 1-year bush fallow to 63.0% under 10-year bush fallow. Although soil organic matter declined slightly in the topsoils of 3-year bush fallows compared to 1-year fallows in the forest region of south-western Nigeria, there was a slight improvement in soil bulk density in the former compared with the latter. This presumably suggests that the roots of trees and other woody perennials in the 3-year bush fallow may have some beneficial effects in loosening the soil to improve soil macroporosity. In eastern Zambia, Torquebiau and Kwesiga (1996)

observed that the bulk density of the top 25 cm of soil under 1-year *Sesbania sesban* fallow was 1.18 g cm^{-3} compared to 1.32 g cm^{-3} under control plots used for maize cultivation. Since the organic matter contents of the *Sesbania* fallow and maize plot were similar, they attributed the better bulk density status of the fallow soil to the effects of the plant roots.

In the rainforest of northern Guatemala, Central America, Popenoe (1957) did not observe any significant decrease in soil bulk density following 3–5 years of bush fallowing. It is important to note that the rainforest soils studied by Popenoe had very high organic content matter and very low bulk density that was below 0.80 g cm^{-3}. Given such conditions, substantial decreases in soil bulk density may not result from a fallow period of 5 years or less.

In the forest fallows of south-western Nigeria, soil water-holding capacity in the top 10 cm increased from 36.4% under 1-year bush fallow to 46.5% in 10-year fallow (Aweto, 1981a). The increase in soil water-holding capacity followed a pattern closely similar to that of organic matter increment. This presumably suggests that an increase in soil water-holding capacity during the fallow period depends primarily on organic matter accretion. An increase in soil water-holding capacity over time is an important function of the bush fallow, particularly in coarse-textured savanna soils that are prone to drought due to irregularity in rainfall distribution.

Provided savanna fallow is not burnt and grazed intensively or too frequently, it normally has a salutary effect in improving soil physical status over time. Soil under semi-arid savanna fallow in Tiel, Senegal, had a bulk density of 0.92 g cm^{-3} in the 0–5 cm layer, compared to 1.55 g cm^{-3} in soil under cultivation for 4 years (Elberling *et al.*, 2003). The savanna vegetation at Tiel was exploited for fuelwood and also used for dry season grazing of livestock. Hence, the land was not left completely fallow. This notwithstanding, the savanna fallow soil had a much lower bulk density than the cultivated soil. This suggests that savanna fallows, even in semi-arid parts of the tropics, have considerable potential in improving soil physical status, especially if biotic influence is minimal. Savanna fallows can mitigate and even reverse the process of soil compaction which occurs during cropping. In fact, the study of Abubakar (1996) in the Sudan savanna of northern Nigeria indicated that soil bulk density of the 0–15 cm layer decreased from 1.34 g cm^{-3} in a 2-year fallow to 1.24 g cm^{-3} in a 15-year fallow. The proportion of water-stable aggregates also increased from 34.3% under the 2-year fallow to 56.3% under 15-year fallow. It is important to observe that the restoration of soil physical status is usually much slower in savanna fallows than in forest fallows. Figure 4.5 shows that the extent of reduction in soil bulk density (and hence the degree of soil compaction) and improvement in soil porosity in the Sudan savanna fallow soil in Nigeria are considerably lower than in forest soil which was fallowed for a shorter period. This is due to the slower rate of organic matter build-up in fallow soil, for reasons adduced earlier. The savanna fallow soil was sampled to a greater depth and this partly accounted for the slower rate of improvement in soil bulk density and porosity over time.

4.4 Soil Organic Matter and Nutrient Dynamics in High-altitude Fallows

It was pointed out earlier that shifting cultivation is not restricted to tropical lowlands. This subsection, therefore, examines soil dynamics under bush fallow vegetation in the tropics in areas characterized by high altitude. The literature on high-altitude bush fallows is very limited compared with their counterparts in tropical lowlands. The discussion that follows will be largely restricted to high-altitude fallows in South America. Bautista-Cruz and del Castillo (2005) examined the trend of changes in soil chemical properties for 100 years of succession, following cultivation of field crops, in the tropical montane cloud forest at 1850 m above mean sea level at El Rincón in Oaxaca state, Mexico. Their findings indicated that there was a progressive increase in the thickness of the O (organic) horizon during the 100 years of succession. The organic matter content of the soils was generally high due to the prevalent low temperatures and the attendant low rates of soil organic matter decomposition and mineralization. In Juquila, where soil organic matter was relatively low at fallow inception, soil organic matter increased from nearly 6% to about 26% in the 0–20 cm layer after 45 years of forest succession; thereafter, there was no further substantial increase until the 100th year. It would seem that owing to the lower rates of biological production and organic matter decomposition and mineralization, it takes a much longer time for soil organic matter to attain the equilibrium level in montane forests compared to lowland bush fallow vegetation. In contrast, at Tanetze, where soil organic matter

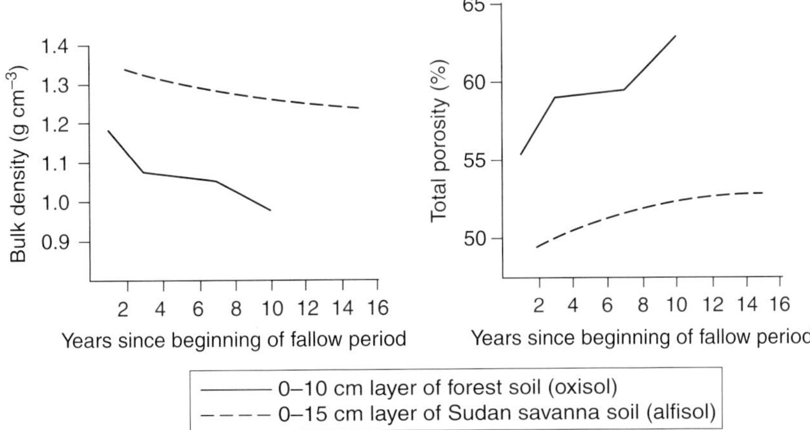

Fig. 4.5. Changes in soil bulk density and total porosity in forest and savanna fallows. (Based on data of Aweto, 1981a; Abubakar, 1996.)

was very high (18% in the 0–20 cm layer at fallow inception) they observed no build-up of organic matter during the first 45 years of the fallow period and thereafter. Clearly, the soil with very high organic matter levels at the beginning of the fallow period must have attained the equilibrium level under the montane forest. Hence, there was no accretion of soil organic matter, even after 45 years or more of fallowing. Soil pH, exchangeable calcium, magnesium, potassium, CEC and base saturation declined during the fallow period, particularly during the first 15 years. This suggests that the rate of nutrient uptake and storage in fallow vegetation during the first 15 years of the fallow period is greater than the rate of nutrient return to the soil, especially through litterfall, decomposition and mineralization.

Above the tree line, shifting cultivation is also practised at high elevations in tropical mountains. This is particularly true of the Andes in South America, where it is a sideline to livestock farming at high elevations, usually above 3000 m above mean sea level in inter-montane plateaux covered with grasses and rosettes, known as *paramo* ecosystems. Abadin *et al.* (2002) described the main features of soil dynamics during the fallow period in *paramo* ecosystems at an elevation of 3350–3700 m above mean sea level in Paramo de Gavidia, in the northern part of the Venezuelan Andes. The mean annual temperature is only 8°C, with very little seasonal departure from the mean value. Given the persistently low temperatures at such high elevations, one would naturally expect that organic matter and nutrient dynamics in the soil would differ from those of lowland bush fallows in the tropics. Their data for the 0–15 cm layer of a sandy inceptisol indicated that organic matter levels are high under bush fallow vegetation of 4 years or less, and even under cultivation. This does not imply considerable input of litter to the soil during the fallow or cropping period. Rather, the contrary is the case, as litter supply to fallow soil is relatively low, in view of the small biomass of fallow vegetation. The high organic matter levels are due to the prevalent and persistently low temperatures which retard microbial decomposition and mineralization of organic matter. Organic matter levels exhibit little fluctuation during the fallow period and the levels of soil organic matter

during cropping are similar to levels under the climax *paramo* vegetation (Fig. 4. 6). This would seem to suggest that soil organic matter levels did not significantly deviate from the equilibrium level during fallow periods of 8 years or more, and even during cultivation, and that there was no significant build-up of organic matter during the fallow period. This is a major difference between tropical bush fallows in lowlands and at high elevations in the tropics. The former are characterized by a rapid rate of organic matter depletion during cropping and a relatively fast rate of organic matter accretion in the soil during the fallow period, subject to the proviso that the regeneration of fallow vegetation is not hindered by biotic factors such as burning and grazing. In the latter, soil organic matter diminution is slow during cropping due to the prevalent low temperature, and the rate of organic build-up in fallow soil is slow due to the low rate of biological production and the prevalent low temperatures at high altitudes. Hence, soil organic matter levels tend to exhibit little fluctuation around the equilibrium level in soil under cultivation or fallow at high elevations in the tropics.

In spite of the very high levels of organic matter in soils under *paramo* bush fallow, their total nitrogen levels are comparatively low. The levels of total nitrogen in the 0–15 cm layer of 8-year *paramo* bush fallow and virgin *paramo* that had never been cultivated were 0.39% and 0.41%, respectively (Abadin *et al.*, 2002). These values are not much higher than the value of 0.31% obtained by Aweto (1981a) for the 0–10 cm layer of soil (oxisols) under 7-year bush fallow vegetation in the Ijebu-Ode area, south-western Nigeria, although the organic matter levels in the soil under the 8-year *paramo* fallow and virgin *paramo* ecosystem were up to four times the level in the south-western Nigeria bush fallow soil. This would seem to suggest that the *paramo* bush fallows, and even the virgin *paramo* ecosystem that Abadin *et al.* (2002) claimed had never been cultivated, are relatively poor in soil total nitrogen. This can be attributed to a number of factors:

1. The *paramo* bush fallows appear to be deficient in nitrogen-fixing species, unlike their counterparts in tropical lowlands that contain a significant proportion of legumes in their flora.

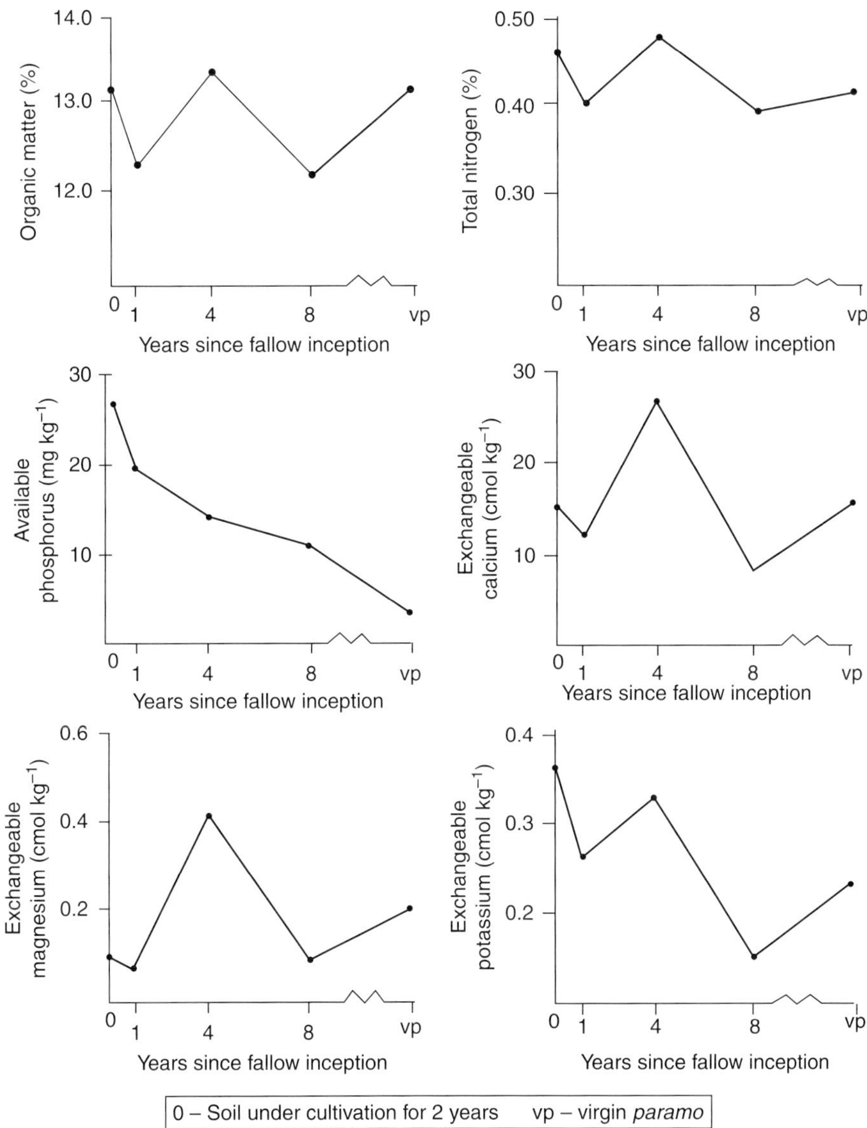

Fig. 4.6. Changes in soil properties during the fallow period in *paramo* ecosystems of northern Venezuela. (Based on data of Abadin *et al.*, 2002.)

2. Were the *paramo* fallows rich in nitrogen-fixing species, their soils would have relatively low amounts of total nitrogen, due to prevailing low temperatures which hinder the process of atmospheric nitrogen fixation by bacteria in root nodules.

3. The fixation of atmospheric nitrogen during lightning flashes will most probably be much slower than in tropical lowlands, as thunderstorms and lightning are less frequent and less intense in high elevations on mountains, where temperatures are low.

Plant available phosphorus declined throughout the fallow period (Abadin *et al.*, 2002), presumably due to immobilization in bush fallow vegetation that is not matched by the rate of phosphorus return to the soil through litterfall, decomposition and mineralization. Exchangeable phosphorus levels in the soil generally declined over time, particularly between the fourth and eighth year of the fallow period. After an initial build-up of exchangeable calcium and magnesium between the first and fourth year of the fallow period, there was a sharp decline between the fourth and eighth year (Fig. 4.6). On the whole, there was no significant accretion of soil exchangeable calcium, magnesium and potassium during the first 8 years of the fallow period and even under the climax *paramo* vegetation, relative to the levels in 1-year fallow.

Finally, it is pertinent to observe that although low temperatures at high elevations in the tropics have a salutary effect of conserving soil organic matter, the low temperatures also retard microbial decomposition of plant materials and the processes of humification and mineralization. Hence, plant nutrients may be locked up in organic forms in organic matter, and so mineral nutrients would not be readily available in the soil for plant use. Sarmiento (1995) observed that in spite of fertilizer application by native farmers in the Venezuelan Andes, the yield of potatoes declined to 27% of the initial level, following four consecutive years of cultivation. This is a clear pointer to the fact that the nutrients in mulch ploughed into the soil prior to cultivation are not quickly mineralized, and so do not become readily available for plant use.

4.5 Soil Erosion

Erosion is a major cause of deterioration in both the chemical and physical status of soil in shifting cultivation areas. This is particularly so after slash burning, as the field is largely exposed to the impact of raindrops and the attendant splash erosion and run-off during the first few months of cultivation. Soil erosion may assume considerable magnitude in highland regions such as the north-western highlands of Ethiopia where Bewket and Sterk (2003) estimated wet season erosion in cultivated fields in Chemoga watershed to be of the magnitude of 18 Mg ha^{-1} and 79 Mg ha^{-1} in Kechemo and Erene areas, respectively. The fallow period serves as a kind of erosion break, helping to check erosion in the temporarily abandoned field through the action of the regenerating fallow vegetation. The increment in soil total porosity under bush fallow over time, referred to in the preceding section, would lead to reduced run-off and an attendant reduction in soil erosion and degradation. Salako and Kirchhof (2003) observed that the perennial vegetation that characterizes long fallow systems enhances the development of a more continuous pore system which will improve soil permeability. In addition, fallow systems, including the bush fallow, have been found to enhance the stability of soil aggregates (Salako *et al.*, 1999), and this will further reduce the proneness of soil under fallow to erosion. In fact, bush fallow vegetation and other perennial fallow systems that provide adequate cover for the soil help to check the accelerated erosion that is initiated during cropping. Soil erosion is considerably reduced during the fallow phase of the shifting cultivation cycle in north-eastern India. In this area, sediment loss from farmlands under cultivation ranged between 22.5 and 30.1 t ha^{-1} per year compared to 1.13 and 0.76 t ha^{-1} per year observed in 5-year and 10-year bush fallows, respectively (Toky and Ramakrishnan, 1981). The 10-year bush fallow is more effective than the 5-year fallow in checking soil erosion. This implies that the extent of bush fallow regeneration is an important factor in reversing soil degradation resulting from erosion. The regeneration of fallow vegetation through the natural process of ecological succession is the theme of Chapter 5.

References

Abadin, J., Gonzalez-Priesto, S.J., Sarmiento, L., Villar, M.C. and Carballas, T. (2002) Successional dynamics of soil characteristics in a long fallow agricultural system of the high tropical Andes. *Soil Biology and Biochemistry* 34, 1739–1748.

Abubakar, S.M. (1996) Rehabilitation of degraded land by means of fallowing in a semi-arid area of northern Nigeria. *Land Degradation and Development* 7, 133–144.

Achard, F. and Banoin, M. (2003) Fallows, forage production and nutrient transfers by livestock in Niger. *Nutrient Cycling in Agroecosystems* 65, 183–189.

Areola, O. (1980) Some problems and issues in studying savanna fallows. *African Environment* 13, 51–62.

Aweto, A.O. (1981a) Organic matter build-up in fallow soil in a part of south-western Nigeria and its effects on soil properties. *Journal of Biogeography* 8, 76–74.

Aweto, A.O. (1981b) Secondary succession and soil fertility restoration in south-western Nigeria: II. Soil fertility restoration. *Journal of Ecology* 69, 609–614.

Bautista-Cruz, A. and del Castillo, R.F. (2005) Soil changes during secondary succession in a tropical montane cloud forest area. *Soil Science Society of America Journal* 69, 906–914.

Bebwa, B. and Lejoly, J. (1993) Soil dynamics and nutrients content in a traditional fallow system in Zaire. In: Mulongoy, K. and Merckx, R. (eds) *Soil Organic Matter Dynamics and Sustainability of Tropical Agriculture.* Wiley-Sayce, Chichester, UK, pp. 135–142.

Bewket, W. and Sterk, G. (2003) Assessment of soil erosion in cultivated fields using a survey methodology for rills in the Chemoga watershed, Ethiopia. *Agriculture, Ecosystems and Environment* 97, 81–93.

Brady, N.C. and Weil, R.R. (2002) *The Nature and Properties of Soils.* Prentice-Hall, Upper Saddle River, New Jersey.

Brubacher, D., Arnason, J.T. and Lambert, J.D.H. (1989) Woody species and nutrient accumulation during the fallow period of milpa farming in Belize, C.A. *Plant and Soil* 114, 165–172.

Elberling, B., Toure, A. and Rasmussen, K. (2003) Changes in soil organic matter following groundnut-millet cropping at three locations in semi-arid Senegal, West Africa. *Agriculture, Ecosystems and Environment* 96, 37–47.

Ewel, J.J. (1976) Litterfall and leaf decomposition in a tropical forest succession in eastern Guatemala. *Journal of Ecology* 64, 293–308.

Gandah, M., Brouwer, J. Hiernaux, P. and Van Duivenbooden, N. (2003) Fertility management and landscape position: farmers' use of nutrient sources in western Niger and possible improvements. *Nutrient Cycling in Agroecosystems* 67, 55–66.

Garrity, D. and Stapleton, P. (2011) More trees on farms. *Farming Matters* 6, 8–9.

Hooper, E., Legendre, P. and Condit, R. (2005) Barriers to forest regeneration of deforested and abandoned land in Panama. *Journal of Applied Ecology* 42, 1165–1174.

Hughes, R.F., Kaufman, J.B. and Jaramillo, V.J. (1999) Biomass, carbon, and nutrient dynamics of secondary forests in a humid tropical region of Mexico. *Ecology* 80, 1892–1907.

Jaiyebo, E.O. and Moore, A.W. (1964) Soil fertility and nutrient storage in different soil-vegetation systems in a tropical rain forest environment. *Tropical Agriculture (Trin)* 41, 129–139.

Mandal, T. (2011) A farmer friendly method. *Farming Matters* 6, 28–29.

McDonald, M.A., Healey, J.R. and Stevens, P.A. (2002) The effects of secondary forest clearance and subsequent land-use on erosion losses and soil properties in the Blue Mountains of Jamaica. *Agriculture, Ecosystems and Environment* 92, 1–19.

Nakano, K. and Syahbuddin (1989) Nutrient dynamics in forest fallows in South-East Asia. In: Proctor, J. (ed.) *Mineral Nutrients in Tropical Forest and Savanna Ecosystems.* Blackwell, Oxford, UK, pp. 325–336.

Nye, P.H. and Greenland, D.J. (1960) *The Soil under Shifting Cultivation.* Commonwealth Bureau of Soils, Harpenden, UK.

Popenoe, H. (1957) The influence of shifting cultivation cycle on soil properties in Central America. *Proceedings of the 9th Pacific Science Congress,* Bangkok, 7, 72–77.

Ramakrishnan, P.S. and Toky, O.P. (1981) Soil nutrient status of hill agro-ecosystems and recovery pattern after slash and burn agriculture (jhum) in north-east India. *Plant and Soil* 60, 41–63.

Roder, W., Phengchanh, S. and Maniphone, S. (1997) Dynamics of soil and vegetation during crop and fallow period in slash-and-burn fields of northern Laos. *Geoderma* 76, 131–144.

Sabhasri, S. (1978) Effects of forest fallow cultivation on forest production and soil. In: Kunstadter, P., Chapman, E.C. and Sabhasri, S. (eds), *Farmers and the Forest.* University Press of Hawaii, Honolulu, pp. 160–174.

Salako, F.K. and Kirchhof, G. (2003) Field hydraulic properties of an Alfisol under various fallow systems in southwestern Nigeria. *Soil Use and Management* 19, 340–346.

Salako, F.K., Babalola, O., Hauser, S. and Kang, B.T. (1999) Soil macroaggregate stability under different fallow management systems and cropping intensities in southwestern Nigeria. *Geoderma* 91, 103–123.

Sanchez, P.A. (1994) Alternatives to slash and burn: a pragmatic approach for mitigating tropical deforestation. In: Anderson, J.R. (ed.) *Agricultural Technology: Policy Issues for the International Community.* CAB International, Wallingford, UK, pp. 451–479.

Sarmiento, L. (1995) Restauration de la fertilite dans un systeme agricole jachere longue des hautes Andes du Venezuela. PhD thesis, Universite de Paris-Sud, Paris.

Scott, G.A.J. (1987) Shifting cultivation where land is limited. In: Jordan, C.F. (ed.) *Amazonian Rain Forests: Ecosystem Disturbance and Recovery.* Springer-Verlag, New York, pp. 34–45.

Stott, P., Goldammer, J.G. and Werner, W.L. (1990) The role of fire in the tropical lowland deciduous forests of Asia. In: Goldammer, J.G. (ed.), *Fire in the Tropical Biota: Ecosystem Proceesses and Global Challenges.* Ecological Studies 84, Springer, Berlin, pp. 32–44.

Szott, L.T. and Palm, C.A. (1996) Nutrient stocks in managed and natural humid tropical fallows. *Plant and Soil* 186, 293–309.

Tiessen, H., Salcedo, I.H. and Sampaio, E.V.S.B. (1992) Nutrient and soil organic matter dynamics under shifting cultivation in semi-arid northeastern Brazil. *Agriculture, Ecosystems and Environment* 38,139–151.

Toky, O.P. and Ramakrishnan, P.S. (1981) Run-off and infiltration losses related to shifting agriculture (jhum) in north-eastern India. *Environmental Conservation* 8, 313–321.

Toky, O.P. and Ramakrishnan, P.S. (1983) Secondary succession following slash and burn agriculture in north-eastern India. II. Nutrient cycling. *Journal of Ecology* 71, 747–757.

Torquebiau, E.F. and Kwesiga, F. (1996) Root development in a *Sesbania sesban* fallow-maize system in eastern Zambia. *Agroforestry Systems* 34, 193–211.

Trollope, W.S.W. (2011) Personal perspectives on commercial versus communal African fire paradigms when using fire to manage rangelands for domestic livestock and wildlife in southern and east African ecosystems. *Fire Ecology* 7, 57–73.

Uhl, C. and Jordan, C.F. (1984) Succession and nutrient dynamics following forest cutting and burning in Amazonia. *Ecology* 65, 1476–1490.

Wadsworth, G., Reisenauer, H.M., Gordon, D.R. and Singer, M.J. (1990) Effects of forest fallow on fertility dynamics in a Mexican ultisol. *Plant and Soil* 122, 151–156.

5 Fallow Vegetation Dynamics

Chapter 4 considered soil dynamics during the fallow period. The changes that take place in soil properties during the fallow period are related to those of fallow vegetation. The dynamics of the soil and vegetation components of bush fallow ecosystems represent the two sides of the same process of ecosystem development over time – ecological succession. Having outlined soil changes under bush fallow over time in Chapter 4, the current chapter focuses on vegetation dynamics, emphasizing changes in the floristic composition and structure of fallow vegetation over time. The uses and resource value of fallow vegetation to local communities, and nutrient storage in fallow vegetation, are also examined. In some cases, shifting cultivators influence or manipulate the floristic composition of fallow vegetation to enhance its utilitarian value or in order to improve its capacity to restore soil fertility under bush fallow. The management of fallow vegetation by farmers, although not a widespread practice, is also considered in this chapter.

5.1 Fallow Vegetation as a Resource

The cultivation phase of the shifting cultivation cycle usually brings more economic benefits to the farmer in terms of food production to meet family requirements and, in some cases, a little surplus for sale to earn cash. Although the cropping period yields more tangible benefits to the farmer, the fallow period is also beneficial as it complements the cropping period by restoring soil fertility which declines during cultivation. The fallow vegetation that colonizes a farm at the cessation of farming is undoubtedly a valuable resource that can be put to several uses, as listed in Table 5.1.

The benefits farmers derive from fallow vegetation may be tangible, as in the case of fruits, fuelwood, or medicine that can be obtained from the plants in fallow land. Other benefits are intangible and these include soil fertility restoration and suppression and elimination of weeds by fallow vegetation.

5.1.1 Intangible benefits

A major function of bush fallow vegetation is the restoration of soil fertility. This was given consideration in Chapter 4 and will not be given further treatment here. The significance of fallow vegetation in checking and controlling soil erosion was referred to in Section 4.5. Soil erosion loss under a 10-year bush fallow vegetation was reduced to less than 4% of the corresponding soil erosion loss from cultivated field in north-eastern India (Toky and Ramakrishnan, 1981). By substantially reducing the soil erosion rate, regenerating bush fallow vegetation halts or considerably slows

Table 5.1. Benefits farmers derive from fallow vegetation.

Intangible benefits	Tangible benefits
Soil fertility restoration	Fruits and food
Soil protection against erosion	Medicine
Reduction of weeds and pests	Poles and materials for building
Protection of surface and underground water	Fodder and forage for livestock
Wildlife conservation	Fuelwood
	Timber
	Game
	Fibres for making mats, hats, baskets, etc.
	Tannins and dyes

down the process of soil degradation during cropping. Furthermore, by accumulating organic matter and recycling nutrients to the topsoil, bush fallow vegetation can help rehabilitate and regenerate the fertility of degraded land. Abubakar (1996) observed that bush fallow vegetation in the semi-arid Sudan savanna zone of northern Nigeria can rehabilitate degraded land. The beneficial value of bush fallow vegetation in rehabilitating degraded land is not limited to the savanna environment. The significance of forest bush fallow in improving soil physical and chemical status, and hence its potential in rehabilitating degraded soil, was discussed in Sections 4.2 and 4.3. Nepstad *et al.* (1991) have described the use of forest regrowth vegetation for reclaiming and restoring the fertility of degraded pastures established on forest land in the Amazon basin of Brazil. This is a clear indication that regenerating secondary forest or natural fallow has a regenerative capacity of restoring degraded land.

Another function of bush fallow vegetation is the control and checking of the proliferation of weeds. When a field is cultivated for more than a few years, the growth and re-seeding of weeds become very rapid. Fallow vegetation helps to suppress the growth of weeds by shading them and preventing their re-seeding (De Rouw, 1995). In this way, the bush fallow helps the farmers save much time and energy that would have been expended on weeding the field. Also, by checking the growth of weeds during the subsequent period of cultivation, fallow vegetation helps to reduce economic losses they would have incurred as a result of proliferation

of weeds which reduce crop yields. Lambert and Arnason (1986) observed that weed biomass in shifting cultivation fields in Belize, Central America, increased several-fold from 560 kg ha^{-1} during the first year to 1939 kg ha^{-1} at the end of the third year. Although there was no significant decline in available nutrient levels in the cultivated site during the 3 years of cropping, maize grain yield declined dramatically from 2971 kg ha^{-1} during the first year to a mere 751 kg ha^{-1} during the third year. They attributed the drastic reduction in maize yield during the third year of cropping to competition with weeds for nutrients (particularly nitrogen), light and water.

5.1.2 Tangible benefits

In the preceding paragraphs, attention focused primarily on the intangible or indirect benefits of bush fallow vegetation. We shall now consider the tangible resources of fallow vegetation from which farmers benefit directly in terms of provision of food, fibre and other economic products.

Fruits

Among the most valuable economic products of the bush fallow, from the point of view of farmers, are fruits. In the rainforest zone of West Africa, fruits are obtained from trees such as *Elaeis guineensis* (the oil palm), *Spondias mombin* (the hog plum), *Irvingia gabonensis* and *Maesobotrya barteri*. These fruit trees are usually protected by farmers on farmlands during cropping, and consequently become integral and important elements of the flora of fallow vegetation. The oil palm is one of the most valuable economic trees in the forest zone of West Africa, being the main source of vegetable oil and palm wine (which may be distilled locally into gin), while the fronds are used for making brooms and baskets and for thatching houses. The kernels of oil palm are valuable exports, while their oil is used locally for manufacturing soap and margarine. The fruits of *Irvingia gabonensis* and *Chrysophyllum albidum*, apart from being valuable dietary supplements, are major items of trade and important contributors to the rural economy. In the rainforest of the Peruvian Amazon, the Bora Indians preserve important fruit trees such as *Theobroma bicolor* (the macambo tree) and peach palm in bush fallow vegetation, for fruit and food (Denevan *et al.*, 1984).

In the rural and agricultural economy of north-eastern Brazil, *Attalea* (*Orbignya*) *speciosa* (babassu palm) plays a vital role that is similar to that of the oil palm in West Africa. As with the oil palm, farmers selectively leave and protect babassu palm in cultivated fields, and they subsequently become a significant element in fallow vegetation. The babassu palm is an important source of edible oil, while the kernels provide milk and oil for soap-making (May *et al.*, 1985a). The leaves and fruits of *Adansonia digitata* (the baobab tree), widely distributed in the savanna lands of West, East and parts of Central and Southern Africa, are used for food and are vital dietary supplements. Other important fruit trees in savanna fallows in West Africa include *Vitellaria paradoxa* (shea butter) and *Parkia biglobosa* (locust bean), while *Sclerocarya birrea* is important for food and wine brewing in both West and Southern Africa. Some useful fruit trees in fallow vegetation, as well as those that provide other benefits to farmers in West Africa and the Amazon basin of South America, are shown in Tables 5.2 and 5.3, respectively.

Medicine

Although Western orthodox medicine is gaining ground in many parts of the tropics, especially in the urban centres, a significant proportion of the indigenous people in the rural areas still depend predominantly on traditional medicine. Bark, roots and leaves of trees and herbs in

Table 5.2. Some useful trees in fallow vegetation in West Africa.

Species	Common name	Ecological zone	Uses
Elaeis guineensis	Oil palm	Rainforest	Edible fruits, oil, wine; leaves used for thatch and making baskets
Irvingia gabonensis	Wild mango or Dika nut	Rainforest	Edible fruits
Chrysophyllum albidum	Golden berry	Rainforest	Edible fruits
Triplochiton scleroxylon	Obeche	Rainforest	Timber
Newbouldia laevis	Smooth newbould	Rainforest	Roots, bark and leaves are medicinal; stems used for roof rafters
Parkia biglobosa	Locust bean	Savanna	Edible fruits also provide forage; bark is used for dying and tanning
Vitellaria paradoxa	Shea butter	Savanna	Seeds yield fat/oil used for cooking, lamp oil and cosmetics.
Faidherbia (*Acacia*) *albida*	Apple-ring acacia	Savanna	Leaves and fruits are livestock fodder; improves soil fertility
Adansonia digitata	Baobab	Savanna	Leaves and fruits are edible; bark and seeds are medicinal; bark fibres used for making ropes
Hyphaene thebaica	Doum palm	Savanna	Leaves and roots provide fibres for mat-making; young seeds are edible and also provide dye
Gueira senegalensis	Gueira	Savanna	Leaves and fruits medicinal

Table 5.3. Some useful trees in fallow vegetation in the Amazon basin.

Species	Common name	Ecological zone	Uses
Bactris gasipaes	Peach palm	Dry deciduous and rainforest	Edible fruits
Bertholletia excelsa	Brazil nut	Rainforest	Edible seeds
Attalea (*Orbignya*) *speciosa*	Babassu palm	Rainforest and *cerrado* savanna	Fruits provide cooking oil, 'milk' and medicine; leaves are used for thatch and the stem for timber
Inga edulis	Ice cream bean	Savanna, dry deciduous and rainforest	Edible fruits; roots, leaves and barks are medicinal, plant improves soil fertility
Theobroma bicolor	Mocambo	Rainforest	Edible seeds
Bixa orellana	Achiote	Dry deciduous and rainforest	Fruit sap is medicinal; non-edible seeds provide dye
Pourouma cecropiifolia	Amazon tree grape	Rainforest	Fruit eaten raw or used for making wine
Anacardium occidentale	Cashew	Dry deciduous and rainforest	Edible fruits, roasted nuts edible; stem yields gums and resins; medicinal bark and leaves

fallow vegetation or forests are commonly used for preparing concoctions for treating various ailments and diseases. In north-eastern Brazil, the fruits of the babassu palm yield medicine for treating gastrointestinal problems and toothache (May *et al.*, 1985b). In the savanna lands of West Africa, the neem tree (*Azadirachta indica*) has been integrated into bush fallow vegetation and farming systems, and is widely used for treating diseases such as malaria. Trees in forest fallow vegetation in West Africa that are important medicinal plants include *Morinda lucida* and *Newbouldia laevis*. The latter has medicinal uses in childbirth, while the former is used for treating malaria which is endemic in the region.

Nigeria. *Senna* (*Cassia*) *siamea*, often planted in rows along the margins of cultivated fields, later become sources of poles in fallow vegetation for building houses or farm sheds and fence construction in southern Nigeria. *Oxytenanthera abyssinica* (bamboo) and *Newbouldia laevis* are also used for building houses in southern Ghana (Amanor, 1994), as they are in southern Nigeria.

In north-eastern India, bamboos feature prominently in fallow vegetation after about 10–20 years of successional development of vegetation in monsoon forest (Toky and Ramakrishnan, 1983). Bamboos and other suitable poles from trees in fallow vegetation are used for building purposes and for fences.

Poles and materials for building

Both forest and savanna fallows are sources of poles and other materials that are used for building houses. In the savanna lands of the tropics, suitable grasses are often woven together to form thatch for roofing buildings. In West and Southern Africa, grasses are used for roofing houses and silos in rural areas. In southern Nigeria, the leaves of the oil palm are used for thatching houses, while in north-eastern Brazil, the leaves of the babassu palm are used for the same purpose. The poles of *Newbouldia laevis* are often used for making roof rafters in southern

Fodder

In savanna areas, both the grasses and woody plants in fallow vegetation are a major source of fodder and forage for feeding livestock. Savanna fallows in the drier Sudano–Sahelian zone of West Africa are intensively utilized for livestock grazing, as livestock population is high due to the absence of tsetse flies. Trees in savanna fallows, especially *Faidherbia* (*Acacia*) *albida*, are a major supplementary source of fodder for livestock during the dry season when most of the herbage is eaten up by foraging livestock. Herdsmen cut branches of *Faidherbia* trees for feeding livestock. Loppings of other trees such as *Acacia tortilis* and *Grewia flava* are widely used for livestock feeding. In parts of Southern Africa, especially in Botswana, trees such as *Combretum apiculatum*, *Terminalia sericea* and *Dichrostachys cinerea* provide browse for feeding livestock. In Botswana, these trees are usually allowed to grow in savanna rangelands, as well as land under cultivation or fallow.

Owing to the relatively low livestock population and paucity of suitable browse and fodder species in tropical forest ecosystems, forest fallows are generally far less important than their savanna counterparts as sources of fodder for livestock. This notwithstanding, forest fallows may provide fodder for livestock. Amanor (1994) identified tree species important as browse in forest fallows in southern Ghana. In the Philippines, farmers actually plant trees in forest fallows to provide fodder for livestock (Calub, 2003). The exotic tree *Gliricidia sepium* is now integrated into farmlands and fallows in the Ibadan area of south-western Nigeria and is sometimes lopped by farmers and used for feeding livestock, especially small ruminants such as goats and sheep. *Panicum maximum* (Guinea grass), found occurring naturally in bush fallow vegetation of about 1 or 2 years' growth, is also harvested and used for feeding livestock, especially cattle transported from northern Nigeria for slaughter in abattoirs in southern Nigeria.

Forests and wooded fallow vegetation have traditionally been the main sources of energy for shifting cultivators, who depend predominantly on biomass for heating and cooking. With the accelerated disappearance of forests and woodlands, shifting cultivators and other rural dwellers in developing countries increasingly source their fuelwood supplies from fallow vegetation, where trees and other woody perennials are an important resource. Usually, not all the tree boles and larger branches are consumed by fire when dried slash of cut fallow vegetation is burnt, prior to cultivation. Most of the unburnt tree stems and branches are taken for use as fuelwood and this is a major drain on nutrients during the shifting cultivation cycle. A significant proportion of the nutrients immobilized in trees and shrubs in fallow vegetation is lost from farmland through the exploitation of fuelwood.

Most of the trees that are initially established in bush fallow vegetation are fast growing, attaining the height of several metres in a few years. Such trees including *Musanga cecropioides* in tropical Africa and *Cecropia* spp. in tropical America are of very little value as commercial lumber because of the unfavourable quality of their wood; the soft wood of trees such as *M. cecropioides*, for example, renders the species practically useless as timber. Some of the pioneer tree species in bush fallow vegetation, however, do yield timber and other commercial products. A noteworthy example is *Aucoumea klaineana* (okoumé), which grows extensively, often forming dense stands as a pioneer tree species on farmlands in Gabon at the cessation of cropping. This tree is much sought after for use in veneers (Richards, 1996). Some pioneer trees such as *Ceiba pentandra* and *Antiaris toxicaria*, which frequently establish themselves as pioneer tree species during rainforest succession in tropical Africa, persist until later stages of forest regeneration to yield lumber.

In tropical America, the fast-growing *Ochroma lagopus* (balsa) establishes itself readily as one of the pioneer tree species in abandoned farmlands. It yields valuable products including lumber, for which it is sometimes grown in plantations.

5.2 Rainforest Succession

Undoubtedly, secondary succession in rainforest ecosystems has received considerable attention from ecologists and pedologists, particularly during the last three decades, which have witnessed intensification of rainforest conversion into arable land or pastures and other uses. In the wake of accelerated loss of rainforest ecosystems, especially in the Amazon basin of South America, there is an urgent need to understand the process of rainforest regeneration after human disturbance, in order to ascertain the rate of regeneration and to characterize the structure and floristic composition of secondary forests which have largely replaced undisturbed primary forests. As Corlett (1995) aptly pointed out, with the widespread replacement of primary forests by secondary forests, the latter will now have to perform the functions performed by the former.

The functions of secondary forests were discussed in Section 5.1, which dealt with the functions of the bush fallow, including woody forest fallows (secondary forests) and savanna fallows. The functions of secondary forests will not be considered further, but suffice it to say that secondary forests, which are the product of secondary succession and even wooded savanna fallows, will play an increasing role in regulating global hydrological and biogeochemical cycles.

5.2.1 General features of rainforest secondary succession

This book deals primarily with shifting cultivation and the associated process of vegetation development (secondary succession) on sites previously subjected to slash-and-burn agriculture. Hence, successional development of vegetation following the abandonment of pastures, which is similar in some respect to succession in shifting cultivation sites, will not be considered here. The establishment of pastures following the clearance of mature rainforest or forest regrowths, slash-and-burn agriculture, and the establishment of mono-cultural tree plantations, are the main features of the development of the Brazilian Amazon basin. The pastures established on former forest lands become unproductive and are invaded by trees and shrubs. Consequently, they are abandoned after about 10 years and the process of forest re-establishment proceeds unhindered. The trends of succession and soil dynamics on abandoned pastures in the Amazon basin of South America have been documented by Buschbacher *et al.* (1988), Uhl *et al.* (1988) and Nepstad *et al.* (1991).

Rainforest succession on abandoned slash-and-burn sites is initiated when grasses, forbs and tree species invade a previously cultivated site at the cessation of cultivation. In most cases grasses and forbs overrun the cultivated field before crops are harvested, and these weeds subsequently become an important element of the flora of the fallow vegetation at the end of cropping. Coppicing from stumps of trees left on the farmland, which is checked by cutting coppice shoots during cropping to avoid shading of crops, goes on without hindrance during the fallow period. Many trees are re-established on the field as a result of sucker regrowth from tree roots that were not killed by fire prior to cultivation. Also, trees, shrubs, climbers and grasses are established from seeds in the soil that survived fire prior to cultivation. In many instances, seeds of plants are blown by the wind from areas surrounding or adjacent to the clearing, or may have been dispersed to the clearing by animals including birds, bats and mammals. Brubacher *et al.* (1989) studied secondary succession during the first 3 years of the fallow period in semi-evergreen seasonally dry forest in Belize, and observed that sprouting from existing roots in the field is a major means of plant recolonization of slash-and-burn fields at the end of cropping. In the wetter parts of the humid tropics, especially in equatorial latitudes where there is no pronounced dry season, plant recolonization of clearings occurs primarily as result of germination of seeds, unlike in the drier forests that occur towards the poleward limits of the humid tropics, where sprouting of roots in slash-and-burn sites is a major means of revegetation (Ewel, 1980).

The first plant community to be established on a site stripped bare of vegetation is called the pioneer stage or community. The pioneer community, as was pointed out earlier, is a mixture of grasses, trees, shrubs and climbers. The composition of the pioneer community varies depending on the length of the preceding cropping, the severity of the fire prior to cropping (which destroys plant seeds and roots in the soil), the size of the clearing, whether there are mature trees or secondary forests nearby that can serve as sources of seeds, and also the extent of land degradation prior to fallowing, among other factors. Tree species usually become dominant within a few years and shade out the grasses and other herbaceous plants on the site. The first trees to be established in a clearing or slash-and-burn field following its abandonment are known as secondary species (Swaine and Hall, 1983) or pioneer tree species (Adedeji, 1984; Uhl, 1987). The secondary tree species usually form a young secondary forest a few years or months after the initiation of the process of secondary succession. Early pioneer or secondary tree species include *Harugana madagascariensis*, *Trema orientalis* and *Musanga cecropioides* in Africa; *Cecropia* spp., *Ochroma lagopus* and *Vismia* spp. in South and Central

America; while those of Asiatic secondary forests include *Macaranga* spp., *Anthocephalus* spp. and *Endospermum* spp. Saldarriaga *et al.* (1988) maintained that in the Upper Rio Negro of Colombia and Venezuela, the early successional tree species are replaced by a second group of successional trees after about 30–40 years. The latter group includes *Alchornia* sp. and *Jacaranda copaia*, which persist and become dominant in the secondary forest for another 50 years. The pioneer or secondary tree species are short-lived and usually die after about 10–20 years, and are gradually replaced by tree species that characterize mature forest. The latter are known as primary tree species, and they differ from the pioneer or secondary tree species that characterize secondary forests. The characteristics of secondary tree species are discussed in Section 5.2.2.

A secondary forest may appear similar to a climax or primary forest in some respects. Brown and Lugo (1990) showed, for instance, that the leaf and root biomass of a 15-year-old secondary forest is similar to that of a 'mature' forest of 80 years or older, while Aweto (1981) observed that species diversity in a 10-year-old secondary forest in south-western Nigeria is similar to that of the mature or climax rainforest vegetation. In spite of such superficial similarities, it usually takes several decades for a young secondary forest to attain the status of the climax or primary rainforest in terms of structure and floristic composition. Saldarriaga *et al.* (1988) observed that it takes about 190 years for forest established on a previously cultivated site to attain the biomass and basal area of mature forest. Richards (1996) has also noted that fairly old secondary forests usually differ in floristic composition from primary forests.

Secondary succession initiated on previously cultivated sites does not always lead to forest regeneration. When sites are cultivated and burnt too frequently, trees in farmlands are killed and this retards the capacity of a secondary forest to regenerate. This leads to the invasion of such sites by grasses, especially *Andropogon* spp. and *Imperata cylindrica*, which persist on the cultivated land over time. The conversion of forest fallows into grasslands, owing to biotic interference with the process of forest regeneration, is known as deflected succession and is considered in Section 5.5 of this chapter.

5.2.2 Characteristics of secondary or successional tree species

Secondary tree species are those that characterize secondary forests, as distinct from primary species that occur in climax or primary forests that have been minimally influenced by humans. Specifically, secondary tree species include the pioneer tree species that are the first trees to invade, colonize and establish themselves on a bare, exposed site, as well as those that subsequently replace the pioneer and other seral species prior to the attainment of the climax stage. Seral species are relatively transient or short-lived species that replace one another in an area during the course of ecological succession; that is, the process of ecosystem development over time that eventually results in the establishment of a stable or climax community in a state of dynamic equilibrium with the environment. Secondary tree species are usually absent in primary or climax rainforest communities, except in gaps created in climax forests by the falling of large trees, where microsuccession may be taking place (Richards, 1996). An important characteristic of secondary trees (especially the pioneer tree species) is that their seeds can germinate in the open, unlike the seeds of primary forest species that require shade to germinate. Secondary species such as *Trema orientalis*, *Musanga cecropioides*, *Cecropia ficifolia* and *Macaranga* spp. are able to quickly establish themselves in clearings, slash-and-burn sites during and at the end of cropping and at the sides of newly constructed roads or those still under construction. It is the ability of their seeds to germinate, in spite of the high light intensity and high temperature in exposed sites, that enables the pioneer tree and other pioneer species to invade and establish themselves on a bare, exposed land surface. Interestingly, many pioneer tree species produce copious seeds that can be dispersed by the wind or transported by bats or frugivorous birds to clearings and devegetated surfaces to facilitate their revegetation. In their study of young forest regrowths in Ghana, Swaine and Hall (1983) reported that more than half of the tree species (64%) were dispersed by bats or birds, while another 17% were wind-dispersed.

Another interesting feature of secondary, especially pioneer, species is their ability to establish themselves as a result of sucker regrowth from roots present in the soil that were not killed during farming. As pointed out earlier, sucker regrowth from roots is a major mechanism of revegetation of cleared land in forests with a marked dry season.

Mainly on account of the fact that the seeds of most pioneer species require light to germinate and their saplings are intolerant of shade, they cannot regenerate once they are sufficiently numerous to form a closed or nearly closed overhead canopy. This implies that their seedlings – if their seeds germinate at all under their own shade – cannot survive and attain maturity. Hence, most pioneer tree species have a short lifespan compared to primary forest tree species. In African rainforests, the two secondary trees *Harungana madagascariensis* and *Trema orientalis* rarely live for more than 15 years in secondary forests. *Musanga cecropioides* usually lives longer and may persist in forest regrowth vegetation for about 15–20 years (Richards, 1996). In the Upper Rio Negro region of the Amazon basin of Venezuela, Uhl (1987) observed that in forest regrowth vegetation established on a slash-and-burn site after cropping, the pioneer tree species *Vismia* spp. began to die out after 5 years, and was being replaced by other successional trees belonging to the family Melastomataceae. Not all successional tree species die before they attain the age of 20 years. Some live much longer. Saldarriaga *et al.* (1988) observed that during forest succession in Colombia and Venezuela, successional tree species such as *Vochysia* sp. and *Alchornia* sp. come into dominance after about 30 years and persist in secondary forests for another 50 years or longer. The trees *Albizia zygia* and *Albizia adianthifolia*, which are early successional species during rainforest succession in West Africa, live much longer than 20 years and may persist in secondary for forests for up to 40 years or longer.

Secondary or pioneer tree species are remarkable for their exceptionally rapid rate of growth compared to primary forest trees. Unless a site is severely degraded and depleted of soil nutrients at the inception of the fallow period, or if there is marked absence of seed-bearing parent trees in the surrounding area due to deforestation, a young secondary forest of about 2–5 m in height is quickly established on an abandoned slash-and-burn site after 3–5 years of abandonment. The rapidity with which a secondary forest is established after human disturbance, usually less than 3 years in recent clearings made in forests, is a vivid testimony to the rapidity of growth of pioneer tree species. Uhl *et al.* (1981) observed that in the Upper Rio Negro region of Venezuela, near San Carlos, a secondary forest with trees of 3–8 m in height was established in a site just 22 months after forest cutting and burning the dried slash. The site they studied was not cultivated before it was abandoned for secondary succession to take place. The fertilizing effect of the ash released as a result of burning the slash of the cut forest, had most probably enhanced the rate of secondary forest re-establishment. The very rapid rate of tree growth in secondary forest regrowth in the Rio Negro region does not, however, belie the fast growth rate of secondary forest tree species. Swaine and Hall (1983) reported that in Ghana, *Musanga cecropioides* attained a height of 11 m in 3 years. Longman and Jenik (1987) observed that the mean rate of growth of plantation-grown *Ochroma lagopus*, a secondary forest tree species of tropical America, was 5.5 m per year. In Venezuela, Uhl *et al.* (1981) observed that *Cecropia* spp. attained a height of 5 m in under 2 years, while in Borneo, Riswan (1982) found that *Trema cannabina* had grown to 7.8 m in just 1.5 years.

Given the relatively short lifespan of secondary or successional tree species, they produce seeds at an early age. Corner (1988) observed that a common successional tree in Malaya, *Adinandra dumosa*, flowers and produces seeds when it attains a height of 2 m at the age of 2–3 years. In an upland evergreen rainforest in Ghana, Swaine and Hall (1983) observed that the pioneer tree, *Trema orientalis*, was fruiting at the age of 6 months. In contrast, it takes primary forest tree species much longer, usually several years or even decades, to produce seeds. The seeds of several successional species are small and light to facilitate their dispersal by birds, bats or the wind.

5.2.3 Changes in floristic composition of vegetation

The trend of secondary succession from the devegetation of a surface to the establishment of a secondary forest and ultimately a mature or primary forest is basically the same throughout the tropical rainforest biome, as discussed in Section 5.2.1. However, the details of plant recolonization, especially the floristic composition of the pioneer community, and the sequence of species populations that replace one another, vary regionally and even locally. This is due to a number of factors including the nature of the climate, soil conditions and cultural factors relating to the intensity and length of cropping and hence the extent of soil fertility impoverishment prior to fallowing. Another important factor influencing the rate of plant recolonization is the size of the clearing and whether it was made inside a mature forest or an area which has been long deforested as a result of decades of the practice of shifting cultivation. This sub-section discusses changes in the floristic composition of vegetation during secondary succession using examples from South America, West Africa and Malaysia.

Owing to the ability of grasses and forbs to re-establish themselves quickly after human disturbance, they are usually an important element of the flora of the pioneer community that initiates secondary succession in the rainforest region. Trees may regenerate as coppices from stumps left on the farm during cropping or establish themselves from seeds or root suckers not killed during burning. Such pioneer tree species ultimately shade out the grasses and forbs within a few months or years, depending on how fast the tree canopy closes up. In the Upper Rio Negro region of the Amazon basin of Venezuela, the grasses established after slash-and-burn agriculture include *Andropogon bicornis*, *Panicum pilosum* and *Paspalum decumbens*, which dominate fallow vegetation with forbs (especially *Eupatorium cerasifolium* and *Phyllanthus* sp.) during the first year following cessation of cropping (Uhl, 1987). By the second year of farm abandonment pioneer trees belonging to the genus *Vismia* had formed a nearly closed canopy at the height of 2 m and the herbaceous understorey of grasses had been largely replaced by shrubs belonging to the family Melastomaceae. Another important pioneer tree genus in the Upper Rio Negro region of Venezuela is *Cecropia*. Uhl *et al.* (1981) found that *C. ficifolia*, which was established as a pioneer tree within 3 months of site abandon-

ment, became dominant and formed a loose canopy at the height of 5 m after 22 months. Eight years after the inception of secondary succession in the seasonally dry forest in the southern Amazon Basin, d'Oliveira *et al.* (2011) observed that *Cecropia* spp. were still the dominant species in the regenerating secondary forest. Pioneer tree species belonging to the genera *Cecropia*, *Vismia* and *Miconia* dominate young secondary forests during the first 10–20 years of Amazonian rainforest succession. Most of these trees die within 20 years and are subsequently replaced by other fast-growing trees such as *Jacaranda copaia*, *Vochysia* sp. and *Alchornia* sp., which remain in the secondary forest for the next 50 years or more (Saldarriaga *et al.*, 1988). Ultimately this latter group of successional species die out after about 60–80 years and mature forest species such as *Mourirri uncitheca* and *Couma utilis* gradually establish themselves in gaps created by the falling of dead successional tree species.

Detailed studies of West African rainforest succession include those of Ross (1954), Aweto (1981), Swaine and Hall (1983) and Adedeji (1984). The dynamics of species populations during secondary succession differ between intensively farmed areas that have been deforested for a long period and sites cleared recently within matured forests.

The rate of tree species colonization is much slower, and tree species population much lower, in intensively farmed areas that have been deforested for a long time, than in recent clearings made within mature forests. In the intensively farmed Ijebu-Ode area of south-western Nigeria, Aweto (1981) observed that tree density was relatively low in forest fallows after 10 years, when it attained an average of 2670 trees ha^{-1}. In contrast, Swaine and Hall (1983) observed that in a clearing made in the Atewa Range Forest Reserve in southern Ghana, tree density attained 25,000 trees ha^{-1} after 1 year of abandonment. This figure is about nine times higher than the tree density observed by Aweto (1981) in 10-year bush fallows in an intensively farmed area. Due to the paucity of seed-bearing parent trees and the progressive killing of stumps, seeds and root suckers in the soil, the regeneration of trees is considerably slower in fallows in intensively farmed areas. As will be pointed out in Section 5.5, the regeneration of trees and forests may be retarded as a result of frequent burning, leading to a deflected succession in which grasslands or savanna vegetation replace secondary forest. Ross (1954) described the trend of succession in an area near a forest reserve in south-western Nigeria where farming activities were minimal due to the setting up of the forest reserve. He observed a clearing not subjected to shifting cultivation, where trees, especially *Trema orientalis*, invaded the site to form a dense stand a few months after human disturbance. Also in Ghana, Swaine and Hall (1983) observed that *T. orientalis* was among the early pioneer tree species during secondary succession in a clearing in the Atewa Range Forest Reserve. Other early pioneer trees included *Musanga cecropioides*, *Rauvolfia vomitoria*, *Terminalia ivorensis* and *Harungana madagascariensis*. By the fifth year of rainforest succession in south-western Nigeria, Ross (1954) observed that *Musanga cecropioides* had become dominant and formed a closed canopy at a height of 10 m. Below the *Musanga* canopy was a lower tree layer at the height of about 5 m above the ground, in which trees such as *Macaranga barteri* and *Fagara macrophylla* were present. After 14 years, *Musanga cecropioides* was still dominant in secondary forest, forming a canopy at the height of 23 m. However, by the 17th year of the fallow, *M. cecropioides* had begun to die out from the upper canopy of the secondary forest and was being replaced by trees such as *Fagara macrophylla*, *Macaranga barteri* and *Anthocleista* sp.

Secondary succession in the intensively farmed area of south-western Nigeria studied by Aweto (1981) differs from that described by Ross (1954) for the lightly farmed area around the Shasha (Omo) forest reserve. Forbs, especially *Chromolaena odorata* and grasses such as *Panicum brevifolium*, *Oplismenus burmannii* and *Andropogon tectorum* dominate bush fallow vegetation during the first 1–3 years of their abandonment. Although other forbs such as *Waltheria indica*, *Aspilia africana* and *Triumfetta cordifolia* are present in fallow vegetation a few months after the cessation of cropping, *Chromolaena odorata* is usually overwhelmingly dominant and usually accounts for up to 80% of the basal cover of fallow vegetation during the first 3–5 years of rainforest succession. *Chromolaena odorata* and other weeds (including the recent invader *Tithonia diversifolia*) usually colonize farmlands before cassava is harvested. Since the weeds do not pose any major threat to a mature crop of cassava, the main staple crop, the farmer usually leaves the farm unweeded. Hence, *C. odorata* and other weeds are usually well established in fallow vegetation at the inception of the fallow period. Pioneer tree species such as *Musanga cecropioides*, *Trema guineensis* and *Harungana madagascariensis*, which Adedeji (1984) observed during secondary succession in forest clearings in the Ife area of south-western Nigeria, were absent in fallow plots during the first 3 years of succession in the intensively farmed Ijebu-Ode area. The absence of such pioneer tree species in fallows in the Ijebu-Ode area may be indicative of progressive deterioration in soil fertility and the killing of the seeds of the pioneer tree species in the soil as a result of repeated burning associated with cultivation. While *C. odorata* dominates fallow vegetation during the 1–3 years following cessation of cropping, trees are slowly regenerating from stumps left on the farm during cultivation. After about 5–7 years, the regenerating trees form a closed or nearly closed canopy which completely or partly shades out the forbs and grasses that initiated the process of secondary succession. The trees that form a canopy at a height of 5–6 m after about 7 years of fallow in the heavily farmed Ijebu-Ode area of south-western Nigeria include *Allophyllus africanus*, *Anthonotha marophylla*, *Ficus exasperata*, *Macaranga barteri*, *Phyllanthus discoideus* and *Rauvolfia vomitoria* (Aweto, 1981). Trees such as *Antiaris toxicaria*, *Albizia adianthifolia* and *Sterculia tragacantha* that are present in young (7–10-year-old) secondary forests usually persist until a mature secondary forest is established.

Secondary succession in rainforests in Asia, following site clearance and slash-and-burn agriculture has been described by Whitmore (1984) for Malaysia, Kochummen and Ng (1977) for Malaysia, and Sillitoe (1995) for Papua New Guinea, among others. The study of Kochummen and Ng (1977) deserves special mention because it was a long-term study of a permanent sample plot, the most accurate and reliable way of studying changes in the plant community during succession. Their study was based on a 0.36 ha plot in Kepong, western Malaysia, which had been subjected to several cycles of farming, before it was finally abandoned in 1945 and natural succession proceeded uninterrupted. Two years after abandonment, regular observations found revealed it was dominated by pioneer shrubs and trees, especially *Melastoma melabathricum*, the most frequent woody species accounting for 79.3% of the stems. Other species which are elements of the pioneer tree flora of the plot included *Mallotus paniculatus*, *Vitex pubescens* and *Grewia tomentosa*. At that time the dominant pioneer *Melastoma melabathricum* had attained the height of about 1.5–1.8 m. Presumably, when routine monitoring of the plot started 2 years after abandonment, these pioneer trees and shrubs had replaced the herbaceous plants, including grasses and climbers, which are usually among the first plants to be established on a devegetated site in the Malayan peninsula. Four years following site abandonment, *Melastoma* trees had begun to die out and their population had decreased to less than 40% of their initial population at 2 years. After about 10 years following site abandonment, the pioneer *Melastoma* trees had almost died out and was largely replaced by a dense understorey of the fern *Dicranopteris linearis*. Thirty years after site abandonment, a closed tree canopy had formed at 6–9 m, with a number of trees towering above the level of the general tree canopy and attaining a height of 18 m. At this stage, woody climbers had become an important feature of the vegetation and the fern *Dicranopteris* had been eliminated except in a few unshaded gaps. The rate of invasion of primary tree species that characterize the climax forest was very slow in the Kepong permanent sample plot. After 30 years, only one primary tree species, representing nearly 2% of the 51 woody species present in the plot, had invaded the secondary forest.

In contrast, the data of Swaine and Hall (1983) indicated that the rate of colonization of a clearing in a rainforest in Ghana was very rapid. After only 5 years of succession, primary forest tree species accounted for up to 42% of the 60 tree species observed on a 800 m² transect in Atewa Range Forest Reserve, Ghana. The study of Uhl *et al.* (1981) showed that the rate of invasion of a cleared and burnt plot in a Venezuelan forest near San Carlos was also rapid. They observed that after 5 weeks of site clearance and abandonment, primary tree species accounted for 41% of all the individuals in a plot, although this proportion declined considerably to 26% due to rapid increase in the population of pioneer tree species. The rather slow rate of secondary succession in the Kepong permanent sample plot, particularly the rate of primary forest tree species invasion and establishment, is most likely due to the chequered and perhaps prolonged history of cultivation in the past which impoverished the soil and depleted the soil seed bank (Kochummen and Ng, 1977). Another possible contributory factor is the considerable distance of the site from seed-bearing parent forest trees, as the plot was surrounded by plantations of young forest trees at the inception of the experiment.

5.2.4 Changes in number of species and species diversity

Contemporaneous with the change in the floristic composition of forest fallow communities over time is an increase in the number of species, particularly tree species, and in species diversity, as secondary succession progresses towards the state of the climax forest. Early seral or successional communities are usually characterized by low species diversity and a relatively low number of species, as a few species tend to be overwhelmingly dominant. In 1-year-old bush fallow vegetation in the heavily farmed Ijebu-Ode area of south-western Nigeria, a single species, *Chromolaena odorata*, is usually dominant, often accounting for up to 80% of the cover of the vegetation (Aweto, 1981). In the San Carlos area of Venezuela where the flora of the pioneer fallow community is more diversified, individuals belonging to the same genus may be dominant. Uhl (1987) observed that individuals belonging to the genus *Vismia* accounted for the majority of the woody individuals 2–5 years following fallow inception in the San Carlos area.

There is a continuous flux in species populations during the course of secondary succession, particularly during the first 10–15 years of forest regrowth. Herbaceous species and forbs that initiate the process of succession on devegetated surfaces are replaced by woody pioneer species in a few months or years. Early pioneer or successional tree species are in turn replaced by later successional tree species in a matter of a few years. In spite of this flux in plant species and species populations, there is a gradual or sometimes dramatic increase in the number of species of vascular plants and in species diversity over a period of a few years.

There was a substantial increase in the number of all species and tree species during 5 years of secondary succession in a clearing in the Atewa Range Forest Reserve in Ghana. The number of all species in a 0.08 ha plot increased from 100 at 1 year to over 150 during the fifth year (Swaine and Hall, 1983). In an area subjected to intense shifting cultivation in south-western Nigeria, Aweto (1981) recorded fewer species in successional plots of 0.09 ha aged between 1 and 10 years. He observed that the average number of species increased from 39 in 1-year fallow vegetation to 54 in 10-year fallow vegetation characterized by a secondary forest. The number of species and the tree density in Atewa Range Reserve plot, 1 year after the beginning of succession, were greater than those recorded for 10-year fallow plots in the Ijebu-Ode area, suggesting a faster rate of succession in the former due to its location inside a forest reserve. The findings of Mwampamba and Schwartz (2011) on the Eastern Arc Mountain of Tanzania also showed that the rate of forest regeneration is much faster in fallows located within forests. The Kepong permanent sample plot in western Malaysia yielded data on woody species dynamics during 31 years of rainforest succession. At the inception of observation, when the 0.36 ha plot was 2 years old, it contained 21 species of woody plants. Four years after cessation of cropping, the number of woody species had increased to 26. Eleven years later, that is, 15 years after the initiation of forest succession, there was no further increase in the number of woody species which had declined slightly to 25. Thereafter, the number of woody species doubled to 51 after another 16 years, that is, 31 years after the abandonment of the plot. In the Upper Rio Negro region of Colombia and Venezuela, Saldarriaga *et al.* (1988) observed that the number of tree species in a 0.03 ha plot increased from 55 in a 9-year secondary forest to an average of 95 in 80-year old forest regrowth, while mature forest of more than 200 years had an average of 99 species. The data of Saldarriaga

et al. (1988) would seem to suggest that the number of tree species increases in forest regrowths gradually until they are 80 years old, and even beyond. In contrast, species diversity appears to stabilize early during succession. Simpson's (1949) index of species diversity of 14-year forest regrowth of 0.94 was comparable with the average diversity for mature forest of more than 200 years in the Upper Rio Negro region (Fig. 5.1). In the Ijebu-Ode area of south-western Nigeria, species diversity based on Simpson's index stabilized after 7–10 years of forest succession when it attained the level of diversity in mature rainforest. Species diversity in successional forest communities in the Upper Rio Negro region of Venezuela was higher than that of successional communities in the Ijebu-Ode area of south-western Nigeria. Species diversity values obtained using Simpson's index, for forest fallows aged 9–15 years in Venezuela, are higher than those of forest 80 years old in south-western Nigeria (Fig 5.1). West African rainforests have been more intensively influenced by humans, and probably for a considerably longer period than forests in the Amazon basin of South America, on account of the much higher population density and longer history of human settlement in the former. Consequently, West African rainforests and successional communities are characterized by lower species diversity than their counterparts in South America, such as those that occur in Venezuela.

5.2.5 Changes in vegetation structure

One of the major changes occurring during forest succession that can be readily observed is the change in vegetation structure over time. The changes in species population and floristic composition in successional communities discussed in Sections 5.2.3 and 5.2.4 result in marked changes in vegetation structure as fallow vegetation develops towards the status of mature forest. This subsection focuses on changes over time in three elements of fallow vegetation structure, namely: vertical profile, tree density and life form composition. As plants, especially trees in fallow vegetation, grow over time and plant species replace one another, the profile of the vegetation changes, usually

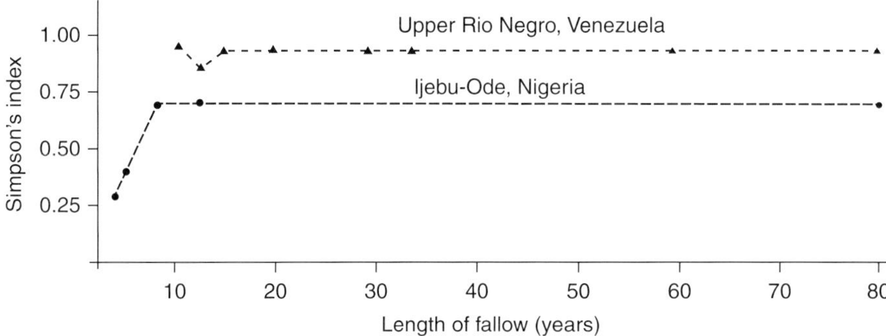

Fig. 5.1. Species diversity dynamics during rainforest succession in Venezuela and south-western Nigeria. (Based on data of Aweto, 1981; Saldarriaga *et al.*, 1988.)

becoming more complex as the community is transformed from a single-layered to a multi-layered community. A few weeks or months after the inception of succession on a devegetated surface, the vegetation profile is relatively simple, consisting of a single layer of pioneer trees and shrubs or a layer of pioneer trees or forbs interspersed with grasses. After a few years, a recognizable canopy is formed by the pioneer or seral tree species with or without an understorey of woody perennials.

Uhl (1987) reported that 2–3 years after slash-and-burn agriculture in the Upper Rio Negro region of Venezuela, the regenerating fallow vegetation was characterized by a closed canopy at a height of 8 m (formed by *Vismia* pioneer trees), below which was an understorey of shrubs that did not exceed 2 m. In Panama and Costa Rica in Central America, Budowski (1970) observed that successional communities not more than 15 years old, and on sites previously cultivated, were characterized by one or two clearly defined strata, while older stages of forest succession of 20–50 years were usually marked by three strata of plants. In old secondary forests that exceed 50 years in age, it is difficult to distinguish the three strata of trees as they tend to merge with one another.

Tree density dynamics during forest succession, especially during the first 5–10 years following site abandonment, depend largely on the intensity and duration of prior human activity such as the length of cropping, whether or not the site was burnt, and whether or not

there are stands of mature forest nearby to serve as a source of seeds for tree seedling establishment on the devegetated site. Tree density generally increases over time until a maximum level is attained, after which it declines as a result of increase in size of individuals and attendant competitive elimination of certain individuals. Maximum tree density is quickly attained in clearings made inside forests that were not burnt or farmed. The data of Swaine and Hall (1983) revealed that in a clearing in Atewa Range Forest Reserve, Ghana, which was not burnt or cultivated prior to abandonment, tree density reached a maximum level of 25,000 trees ha^{-1} after 1 year. Thereafter, tree density declined progressively to 7500 trees ha^{-1} by the fifth year following initiation of secondary succession, when observation was discontinued. Uhl (1987) noted in the San Carlos area of the Upper Rio Negro region of Venezuela, that in plots previously subjected to slash-and-burn agriculture, tree density increased till the fourth year following initiation of secondary succession, after which a decline occurred in the fifth year. In the Ijebu-Ode area of south-western Nigeria, which has been subjected to shifting cultivation for several decades, tree density is very low in fallow vegetation at the inception of secondary succession, being less than 100 trees ha^{-1}. Tree density increased progressively during the first 10 years of rainforest succession, this increase being particularly marked between the third and seventh years. It would seem that when clearings are made inside forests, tree density quickly attains the maximum level after a few years, usually 1–3 years, after which a process of thinning-down begins. In contrast, in heavily farmed areas, the process of increase in trees may continue for up to about 10 years before thinning of trees sets in (Fig. 5.2).

There are relatively few studies which quantitatively characterized changes in plant life form composition during forest succession following slash-and-burn agriculture. Aweto (1981) used life form spectra, based on Raunkiaer's (1937) plant life forms, to evaluate the changes in plant life form composition during the first 10 years of rainforest succession in the Ijebu-Ode area of south-western Nigeria. His results indicated that microphanerophytes (represented mainly by forbs such as *Chromolaena odorata* and climbers) and therophytes (annuals represented by *Tridax procumbens* and grasses such as *Oplismenus burmannii* and *Panicum brevifolium*) predominated in 1-year fallow vegetation. Megaphanerophytes (mainly trees that are more than 30 m tall) and mesophanerophytes (trees that are 8–30 m tall) were also present at the inception of succession as coppices regenerating from stumps of trees that were not killed during cropping. After the third year of the fallow there was a gradual elimination of therophytes, which were shaded out by trees (mesophanerophytes and megaphanerophytes) that had become more numerous 10 years after the inception of

succession (Fig. 5.3). In the mature forest stage, therophytes have been completely eliminated and there was a corresponding decrease in the proportion of microphanerophytes as the proportion of mesophanerophytes and megaphanerophytes increased. The substantial rise in the numbers of the two latter life forms in the mature forest resulted in increase in vertical development of the vegetation.

5.3 Succession in Deciduous Seasonal (Monsoon) Forest

Given the rather limited geographic extent of monsoon forests and the fact that the system of shifting cultivation is more widespread in rainforest areas, it is hardly surprising that the study of monsoon forest succession has received less attention from ecologists and pedologists than rainforest succession. Vegetation dynamics in monsoon forest ecosystems, following the cropping phase of the shifting cultivation cycle, have been extensively studied in north-eastern India (Toky and Ramakrishnan, 1983; Swamy, 1986; Swamy and Ramakrishnan 1987a). The nature of the pioneer community depends partly on the length and intensity of cropping and on the length of the fallow phase of the shifting cultivation cycle. In most areas, the length of the fallow period is 5 years and, in such areas, the regenerating fallow vegetation is dominated by *Chromolaena odorata* and *Imperata cyclindrica*

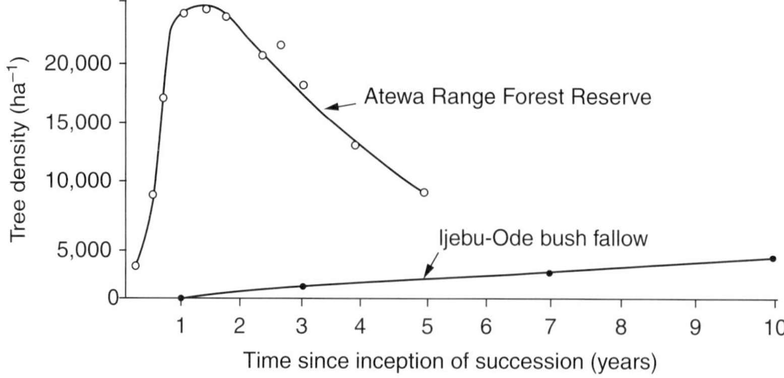

Fig. 5.2. Tree density dynamics during secondary succession in a clearing in the Atewa Range Forest Reserve, Ghana, and in the intensively farmed Ijebu-Ode area of south-western Nigeria. (After Aweto, 1981; Swaine and Hall, 1983.)

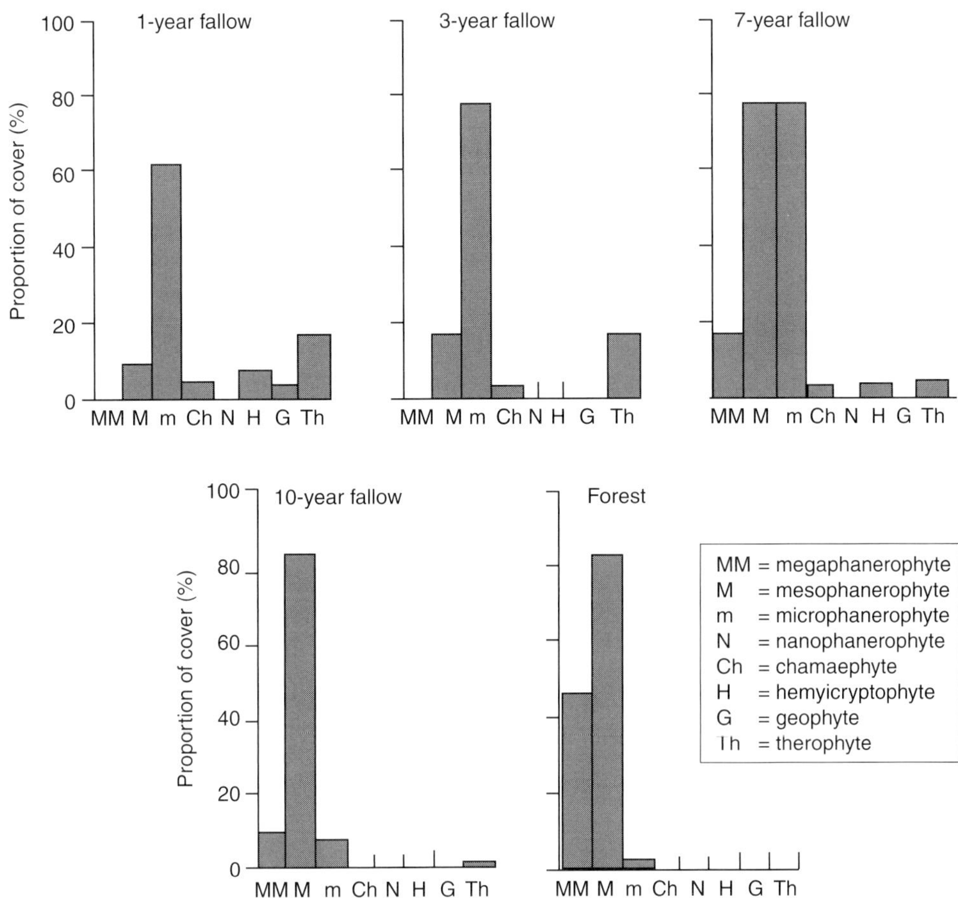

Fig. 5.3. Life form spectra, based on cover of forest successional communities in south-western Nigeria (After Aweto, 1981).

(speargrass), but in areas where the fallow period is up to 10 years, the exotic weed *Mikania micrantha* dominates fallow vegetation during the first few years following cessation of cropping (Toky and Ramakrishnan, 1983). Other herbaceous plants in monsoon forest fallows of north-eastern India include *Ageratum conyzoides*, *Panicum maximum* and *Borreria hispida*. *Mikania micrantha* usually attains maximum biomass in 4-year fallows, after which its biomass diminishes considerably by the 11th year of the fallow period (Swamy and Ramakrishnan, 1987a). The plant immobilizes substantial amounts of potassium in its standing biomass, which it subsequently recycles to the soil through litterfall (Swamy and Ramakrishnan, 1987b). It is therefore of immense value in conserving the potassium status of young fallows, especially in the soil, during the first few years following the inception of secondary succession. After about 5 years following the cessation of cropping, the early seral species, especially *C. odorata*, *I. cylindrica* and *M. micrantha*, are replaced by a woody fallow in which the bamboo, *Dendrocalamus hamiltonii*, features prominently. The latter species usually predominates in fallows of 10–20 years, although other trees that regenerated from stumps, roots and seeds in the soil that were not killed during farming, are also present (Toky and

Ramakrishnan, 1983). Such trees include *Cedrela toona, Vitex glabrata* and *Dillenia indica.* Figure 5.4 shows the frequency of herbaceous and woody species during 20 years of secondary succession in a monsoon forest in north-eastern India. Herbaceous species such as the grass *Panicum maximum* and forbs such as *Chromolaena odorata,* which initiate the process of succession, dominate fallow vegetation of 5 years or younger. However, they decline in dominance with time and are eliminated from fallow vegetation before the tenth year, being replaced by tree species such as *Vitex glabrata, Dillenia indica* and *Careya arborea.* These tree species usually develop from seeds, roots and stumps that survive the cropping period and usually increase in frequency after the tenth year.

As with rainforest ecosystems, the structure of monsoon forest fallow is simple at the inception of succession, usually consisting of a layer of grasses with small trees. However, after about 15–20 years, the structure of the secondary forest is more complex with a profusion of trees, often forming two strata. Usually, after about 10 years, the herbs and forbs that initiated the process of succession have been shaded out by trees which have become numerous and well established in the secondary forest. Species diversity, which was low at the inception of succession, increased throughout the 20 years of succession in the monsoon forest of north-eastern India (Toky and Ramakrishnan, 1983). In contrast, Aweto's (1981) data for rainforest successional communities in south-western Nigeria indicated that species diversity stabilized after 7 years following cessation of cropping. Also, Saldarriaga *et al.* (1988) observed that a 14-year secondary forest in the Upper Rio Negro region of Colombia and Venezuela had a Simpson's diversity index of 0.94. For the same area, they observed that an 80-year secondary forest plot had the same species diversity index (0.94) as the 14-year secondary forest. Their finding would seem to lend credence to the observation of Aweto (1981) that species diversity stabilizes, usually within the first decade, if rainforest succession is not arrested. One can infer from the foregoing that while species diversity may stabilize after a decade or so of rainforest succession, it may not stabilize after two decades of monsoon forest succession. This is presumably as a result of the slower pace of succession in monsoon forests, due to the distinct seasonality of the rainfall regime.

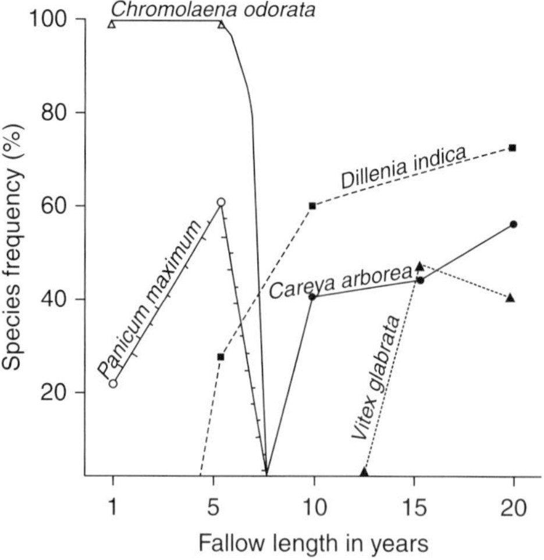

Fig. 5.4. Species frequency dynamics during monsoon forest succession in north-eastern India. (Based on data of Toky and Ramakrishnan, 1983.)

5.4 Succession in Savanna Ecosystems

Savanna ecosystems are far more extensive in the tropics than tropical rainforests. This notwithstanding, secondary succession has not been adequately studied in savanna regions, unlike in the rainforest region. Consequently, little is known about savanna succession, as Hopkins (1983) lamented. Although shifting cultivation is an important feature of the agricultural economy of savanna ecosystems, especially of humid and sub-humid savannas, secondary succession in tropical savanna lands has received less attention from ecologists than has rainforest succession. The comparatively little attention given to the study of savanna secondary succession is due to a number of factors.

First, unlike forest fallows, bush fallow vegetation in the savanna regions is usually burnt annually. Savanna vegetation is burnt in order to facilitate hunting, as part of land preparation for farming and to promote sprouting of grasses for livestock towards the end of the dry season when fresh grazing is very scarce (Hopkins, 1974). Burning helps to control pests such as ticks, and also crop diseases and weeds. Burning eliminates the grass/sedge layer of the savanna ecosystem, exposes the soil to erosion and gradually reduces tree species population if burning is severe. The savanna lands of the tropics are usually burnt annually, and Stott (1991) observed that fire is an important ecological factor in savanna eco-systems. Burning retards and, if severe, may reverse the trend of successional development in savanna ecosystems. This makes the study of savanna succession problematic, as savanna succession is usually 'arrested' and the vegetation assumes a dynamic equilibrium with a biotic factor – burning – instead of progressing towards the 'climatic climax' status. Burning is so pervasive in the savanna ecosystems of the tropics that it is often difficult to decipher the trend of succession in the absence of fire. This usually necessitates the delimitation of plots that are protected against burning for years or decades while the trend of succession is monitored in the permanent plots established.

Second, the problem of studying succession in savanna fallows is further exacerbated by another biotic factor, grazing. The herbage and forage resources of the savanna lands of the tropics are utilized by wildlife and domesticated animals such as cattle, sheep and goats. The dynamics of domestic animal population are an important ecological factor in savanna suc-cession. A couple of successive years with above-average rainfall encourages herdsmen to increase the population of their livestock, which in turn leads to overgrazing and soil and vegetation degradation. This slows down the rate, or even reverses the rate of vegetation succession. A natural increase in the population of large wild herbivores has been reported to adversely affect the structure of savanna vegetation and subsequently savanna vegetation dynamics (Laws *et al.*, 1975).

Finally, it is important to emphasize that the 'climax' vegetation of most areas now characterized by savanna vegetation is not known with certainty. Hence, it is difficult to ascertain the successional status of savanna vegetation and the rate at which various seral communities approach the climax status. The climatic climax vegetation of rainforests that have been degraded into 'derived savanna' is rainforest, as such areas can still support forest vegetation by virtue of their climate. In fact, patches of forest vegetation may coexist with savanna vegetation on dry land that has not been farmed for several decades in the derived savanna or 'forest–savanna mosaic' zone. Apart from the derived savanna, very little is known with certainty about the structure and floristic composition of the climatic climax communities that ultimately develop from other savanna formations. Some authorities (e.g. Keay, 1959; Chachu, 1982) consider dry forests such as those dominated by *Diospyros mespiliformis* and *Kigelia africana* to be the climax vegetation of the sub-humid Guinea savanna vegetation in Nigeria and, by extension, other parts of West Africa where Guinea savanna vegetation occurs. Other researchers (e.g. Ajayi and Hall, 1975) regard such patches of dry forests to be relics of more extensive forests that existed in the Guinea savanna during more moist periods in the past, and not necessarily the climax stage of present-day successional processes taking place in the savanna vegetation. The controversy regarding the status of patches of dry forest in the Guinea savanna zone of West Africa is a pointer to the problem of determining the successional status of savanna vegetation.

5.4.1 Temporal dynamics of savanna vegetation

The trend of savanna succession following cessation of cropping in farmlands largely depends on whether or not the site is burnt, and on the intensity and regularity of burning. Given the pervasive occurrence of fire in most savanna lands of the tropics, the frequency and intensity of burning are key factors determining the temporal dynamics of savanna vegetation during secondary succession. In some instances, secondary succession in savanna vegetation has been studied in plots protected from burning. Clearly, the trend of succession in savanna plots protected against burning differs strikingly from the temporal dynamics of savanna vegetation subjected to burning. When savanna vegetation is burnt periodically, succession results from the interplay of both allogenic and autogenic factors. Allogenic factors are external to the plant community or vegetation, and they include burning and grazing by domesticated and wild herbivores that partly influence the trend of vegetation dynamics. In contrast, where succession occurs in savanna plots protected against external biotic influences such as burning and grazing, vegetation dynamics result primarily from autogenic factors, that is, factors that result from the plant community or vegetation itself. Such autogenic factors include competition between the plant individuals in the vegetation and the associated dynamics of species populations, and also the modification of the site (including soil and microclimate) by the vegetation, through the addition of organic matter and nutrients to the soil and also by shading the ground surface. In Sections 5.4.2 and 5.4.3, the trend of savanna succession in both burnt and protected plots is discussed.

5.4.2 Succession in savanna vegetation subjected to burning

In most parts of the tropics, savanna bush fallow vegetation is usually burnt. The reasons for burning have already been referred to in Section 5.4. The trend of succession of savanna vegetation subjected to burning, discussed here, is typical of most savanna areas of the tropics where shifting cultivation is practised. At or prior to the inception of the fallow period, those remaining shoots or stumps that survived the cropping period begin to coppice vigorously, as the farmer no longer checks the sprouting. Contemporaneous with this development, sucker regrowth begins from roots of trees and shrubs, and the seeds of annual and tufted perennial grasses begin to germinate and revegetate the farm at the cessation of cropping. After a couple of months, a herb layer with grasses and sedges is established and after a few years the coppice shoots and sucker regrowths, together with live trees retained by the farmer on the cultivated field during cropping, constitute the tree and shrub layers of the regenerating fallow vegetation. Sucker regrowth from rootstocks in the soil is an effective means of tree species re-establishment in previously cultivated sites in the savanna lands and in the drier forests of the tropics. After a few years of savanna vegetation succession, basically the same tree species that were on the site prior to cropping would be re-established (Hopkins, 1983). The rate of regeneration of trees and shrubs varies considerably, usually being much faster in the humid and sub-humid savannas adjoining the forest zone than in semi-arid savannas fringing hot deserts. Owing to considerable regional and even local disparities in the rates of fallow vegetation re-establishment, it is often difficult to make reliable generalizations concerning the rate of tree growth and colonization of fallow vegetation. Over much of the sub-humid Guinea savanna and the drier Sudan savanna of West Africa, trees and shrubs are usually well established in fallow vegetation after about 4–8 years following farm abandonment. Usually at this stage, fallow vegetation is a shrub or tree savanna, but it may ultimately be transformed into a savanna woodland after a few or several decades, depending, among other things, on the intensity of site disturbance prior to fallowing and the regularity and intensity of site burning. Menaut (1977) observed that some measure of fire protection is essential for the transformation of less wooded shrub or tree savanna into a savanna woodland that is more densely stocked with trees. This is because fire tends to inhibit tree establishment (Staver *et al.*, 2011). The obvious corollary is that a savanna bush fallow may not be altered into a heavily wooded savanna woodland if it is regularly exposed to intense burning.

Chachu (1982) has inferred the sequence of succession in Kainji National Park in the northern Guinea savanna zone of Nigeria, which leads to the establishment of dry forests of *Polysphaeria* sp. and *Kigelia africana*, which he considered to be the climatic climax vegetation of the area. He made his deductions from an ecological study of the current types of plant communities in the area. He proposed two lines of succession that lead to the establishment of the same climax dry forest. In one line of succession (Fig. 5.5), savanna communities of *Burkea africana* and *Terminalia avicennioides* are established on shallow and poorly weathered plinthite soil and these are subsequently replaced by *Detarium microcarpum* savanna. *Detarium microcarpum* trees help to weather the plinthite substratum, making it possible for seedlings of *Isoberlinia* spp. to invade and subsequently establish an *Isoberlinia* woodland. Other trees such as *Tamarindus indica* and *Diospyros mespiliformis* may subsequently invade and become important floristic elements of the *Isoberlinia* woodland. The woodland of *Isoberlinia* and other trees usually affords the ground and herb layer some measure of fire protection, thereby allowing a dry forest, in which tree species such as *Polysphaeria* sp., *Kigelia africana* and *Diospyros mespiliformis* feature prominently, to be established as the climax community. The other line of savanna succession according to Chachu (1982) is associated with deep sandy or loamy sand soils on which tree savanna of *Afzelia africana* is well established. This second line of succession is characteristically associated with farmlands as well as sites with deep, well-drained soils that are usually cultivated. At the cessation of

cropping, *A. africana* and *Isoberlinia* tree savanna is established on previously farmed sites. The increasing invasion of *Isoberlinia* seedlings leads to the establishment of *Isoberlinia* woodland, which in turn, will lead to the establishment of dry forest as described above.

5.4.3 Succession in fire-protected savanna vegetation

When savanna plots are protected against burning for a long period, they progressively become more woody over time. This is provided that other biotic influences such as grazing and fuelwood exploitation are also excluded, and that the savanna soil is not severely degraded during the preceding period of cropping or the soil characterized by impenetrable crusts and hard pans near the soil surface, inhibiting tree seedling establishment. Under effective and prolonged fire protection, trees become more numerous and a grass or tree savanna may be ultimately transformed into savanna woodland, or woodland that is densely stocked with trees. In some cases, when savanna vegetation is transformed into a denser arboreal community, tree crowns touch one another to form an overhead canopy which may partially or completely eliminate the herb layer to give rise to patches of dry forest. The transformation of savanna vegetation into more densely wooded vegetation, under fire-protection, occurs in both humid savannas and semi-arid/arid savanna. In parts of southern Africa, especially in Botswana, savanna vegetation has become increasingly woody, with an attendant decline in grass cover, over the past few decades. This phenomenon – known in southern Africa as bush encroachment (van Vegten, 1983; Moleele, 1999) – has been attributed mainly to an increase in cattle density and hence to increased grazing pressure from cattle (Moleele and Perkins, 1998; Moleele, 1999). While cattle undoubtedly play an important role in the spatial dispersal of seeds of the invading thorny woody species, the exclusion of fire from southern African savannas appears crucial to their becoming more densely wooded. In the Sudano–Sahelian zone of West Africa, the cattle population has increased substantially since the

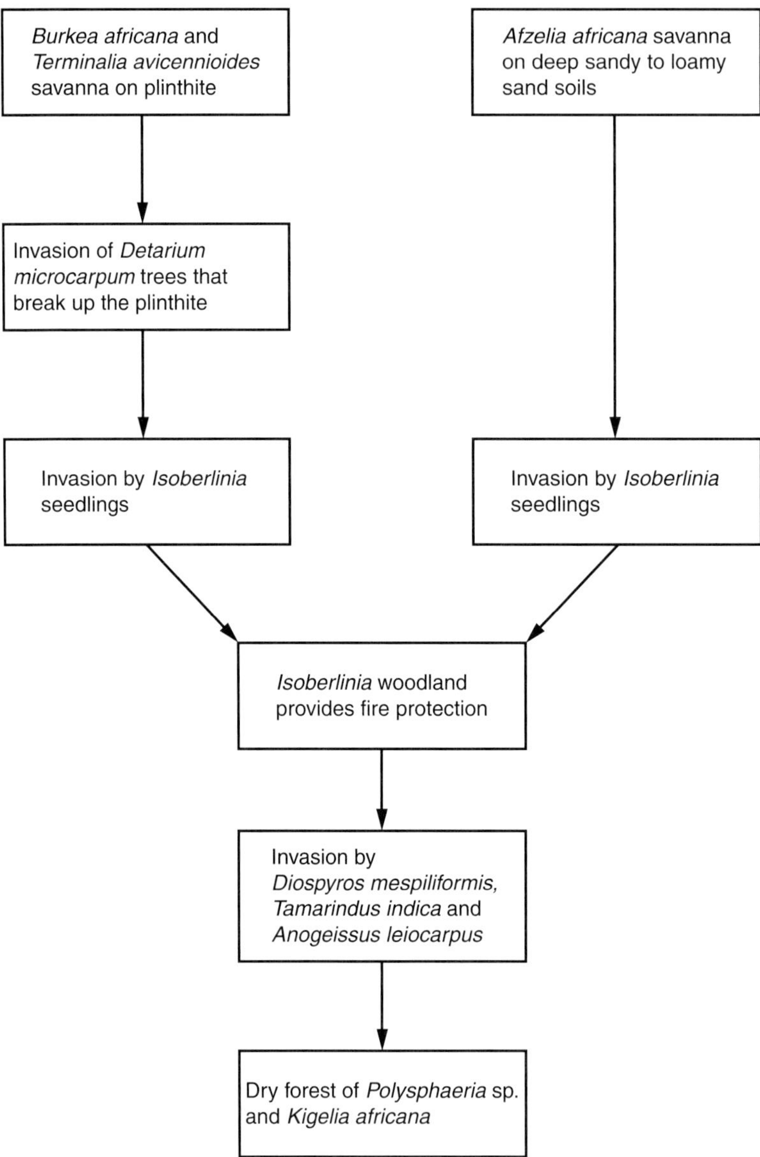

Fig. 5.5. Diagrammatic representation of savanna succession in the northern Guinea savanna zone that leads to the formation of a dry forest. (After Chachu, 1982.)

devastating drought of the early 1970s, yet there is no indication that the savanna vegetation, which is burnt annually, is progressively becoming more woody. The example of the Sahel and Sudan savannas seems to suggest that an increase in cattle density does not necessarily make savanna vegetation more woody, as inferred by Moleele and Perkins (1998). The study of Aweto and Adejumobi (1991) has shown that sedentarization and increased intensity of cattle grazing in the sub-humid Guinea savanna in southern Nigeria led to degradation of vegetation, reduced tree density and deterioration in soil fertility, as burning was adopted as a strategy of rangeland management by the herdsmen. The findings of the study of Aweto and Adejumobi (1991) would seem to suggest that increased grazing intensity by cattle does not make savanna vegetation a more densely wooded community, and that the contrary can be the case unless savanna vegetation is protected against burning.

The subtropical grasslands and savannas of the Rio Grande plains of Texas, USA, have become increasingly woody over recent decades and have ultimately been transformed into densely wooded thorn woodland (Archer *et al.*, 1988). The process of savannazation of the grasslands in southern Texas is initiated when they are invaded by lone stands of mesquite (*Prosopis glandulosa*). Each stand ultimately forms a cluster of *P. glandulosa* plants, as the seeds of the tree are dispersed and tend to be established around the parent plant. The cluster of woody plants or wood clumps expand over time and adjacent clusters coalesce to form larger wood clumps, and over a period of several decades an initial grassland or grass savanna is transformed into a woodland. The factors that facilitated the transformation of the grasslands into woodland vegetation include biotic factors such as overgrazing, drought and reduced occurrence of fire (Archer *et al.*, 1988). As with southern African savannas, partial or complete elimination of intense fire appears pivotal to the transformation of the grasslands and savannas in southern Texas into woodland vegetation. In Sections 5.4.4 and 5.4.5, succession in fire-protected experimental plots in West Africa and South America will be examined.

5.4.4 Succession in fire-protected savanna plots in West Africa

Autogenic succession in savanna vegetation has been studied in West Africa in the humid derived savanna (forest–savanna mosaic) zone that abuts on the rainforest zone, and in the less humid Guinea savanna zone. The former was originally a part of the rainforest zone and had become degraded into savanna vegetation following centuries of cultivation involving

burning (Keay, 1959). The latter, in contrast, adjoins the semi-arid Sudan savanna which is xeromorphic, due to the prolonged dry season and the intense evaporation in the zone. Given the differences in climate between the derived savanna and the Guinea savanna zones, one would naturally expect considerable differences between them in respect of the pathway of succession and in the nature of the climax communities.

A long-term experiment to evaluate the effects of fire exclusion and varying intensity of burning on savanna vegetation was conducted at Olokemeji, near Abeokuta, at the forest–savanna boundary in south-western Nigeria. The site, close to the Olokemeji forest reserve, was initially cleared of trees, burnt and then divided into three plots. One plot was burnt early, around the inception of the dry season in December when the savanna grass layer had not thoroughly dried up. This plot was subjected to light burning annually, as usually only about 50% of the grass layer is burnt at the beginning of the dry season. The second plot was burnt towards the end of the dry season in March. This second plot was subjected to intense burning, as the grass layer dries up completely and burns more intensely during the latter part of the dry season. The third plot was protected against burning. After 28 years, Charter and Keay (1960) evaluated the results of the Olokemeji fire experiment. Their results indicated that the late-burnt plot was characterized by a tree savanna with a well-developed herb layer. It contained trees, none of which was a fire-tender forest species (Fig. 5.6). In contrast, the early-burnt plot had been transformed into a savanna woodland with two wood clumps containing a significant number of invading fire-tender forest species. The fire-protected plot became a forest, with no grass layer and numerous fire-tender forest species, 28 years after the inception of the experiment. The fire-tender forest species included *Hildegardia barteri*, *Sterculia tragacantha*, *Phyllanthus discoideus* and *Albizia zygia*, which had invaded the fire-protected plot from the nearby forest reserve. The number of trees in the fire-protected plot was considerably higher than in the late-burnt and early-burnt plots, and this indicated that the rate of tree and species

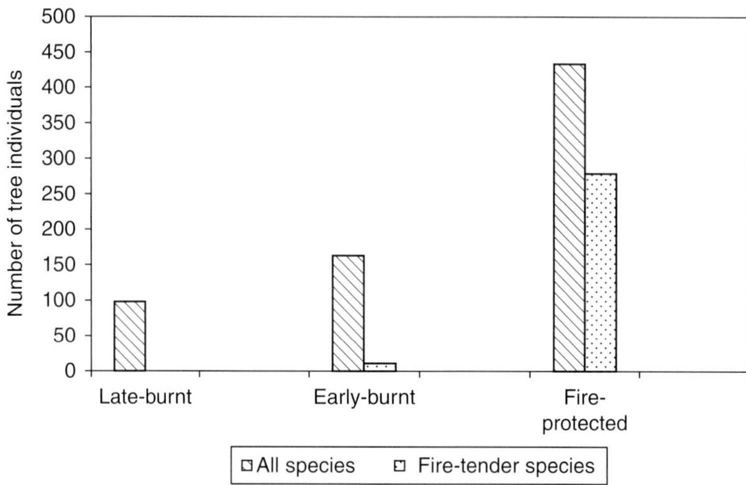

Fig. 5.6. Effects of fire protection and burning on derived savanna vegetation at Olokemeji, south-western Nigeria. (After Charter and Keay, 1960.)

invasion of savanna vegetation during secondary succession is considerably enhanced by fire exclusion. A major conclusion that can be drawn from the Olokemeji fire experiment is that succession in derived savanna vegetation would ultimately lead to the establishment of forest vegetation, provided the period of fire-protection is long enough.

In the Red Volta River forest reserve in the northern Guinea savanna zone of northern Ghana, an experiment that is similar to the Olokemeji fire control experiment was initiated in 1950. Three plots in the Guinea savanna zone were delimited and the trees enclosed by the plots were counted. Thereafter, the plots were cleared and one plot subjected to early burning during the inception of the dry season; the second plot was subjected to late burning towards the end of the dry season; and the third was protected against burning. After 27 years, the number of trees in the late-burnt, early-burnt and fire-protected plots had increased by 11, 91 and 239%, respectively, relative to the initial number of trees in the plots prior to the inception of the experiment. These data suggest that the rate of tree invasion of savanna vegetation over time increased substantially when fire was excluded, and that early burning did not completely halt the process of tree invasion and the conversion of tree savanna into a savanna woodland, but merely slowed it down. Among the tree species invading the fire-protected plot were late arrivals such as *Anogeissus leiocarpus*, *Afzelia africana* and *Diospyros mespiliformis*, characteristic of dry forest and savanna woodland (Brookman-Amissah *et al.*, 1980). This presumably suggests that with prolonged fire protection, the Guinea savanna may ultimately be transformed into a dry forest as postulated by Keay (1959), although this has not yet occurred in the Red Volta River fire-protected plot after 27 years. It is also important to note that in both of the savanna-burning experiments discussed above, the rate of tree invasion – and hence of savanna succession – was slowest in the late-burnt plots, in which burning was more intense. This is because savanna trees are vulnerable to fire and they are killed if burning is intense (Cochard and Edwards, 2011; Ryan and Williams, 2011) and, ultimately, intense fire prevents establishment of savanna trees.

5.4.5 Succession in fire-protected savanna in South America

As in West Africa, fire protection of savanna vegetation that has also been protected against grazing resulted in the remarkable transformation of the grass savanna of the Llanos of Venezuela into a more densely wooded savanna vegetation. In Calabozo in the Venezuelan Llanos, the study of San Jose and Farinas (1983) indicated that following 16 years of fire protection of an ungrazed savanna dominated by *Trachypogon* spp., the number of trees increased dramatically 15-fold in the grass savanna and fourfold in groves in the 3 ha permanent sample plot. Similarly, the number of species increased from three to ten in the open savanna, while in the groves dotting the grass savanna, the number of species increased from four to 16 over the 16-year period. Among the invading tree species was *Cochlospermum vitifolium*, a tree that appears to be fire-tender. It multiplied rapidly in isolated stands and surpassed the number of isolated stems of such fire-tolerant species as *Curatella americana* and *Byrsonima crassifolia* that were present in the plot at the inception of the experiment. Other invading tree species included *Casearia hirsuta* and *Genipa caruto* (San Jose and Farinas, 1983). It is important to note that the increase in tree species population was considerably higher in the open savanna than in the groves. This is largely due to competition between tree individuals and tree species in the groves in the permanent sample plot. The number of isolated stands of *Curatella americana* and *B. crassifolia* increased 20-fold and 13-fold, respectively, over the 16-year period of fire protection. In the groves or wood clumps, the increase in the number of each of the two species was not more than double the initial number of tree individuals at the inception of the experiment. Fire and soil moisture are important factors determining the rate of invasion of neotropical grasslands by trees (Garcia-Nunez *et al.*, 2011), although soil depth may assume increased significance in sites where lateritic crusts occur at or near the ground surface.

5.5 Deflected Succession

Biotic factors such as frequent cultivation, burning, grazing and fuelwood exploitation may prevent secondary succession from running its normal course of vegetation development that leads to the establishment of the climatic climax vegetation. In many instances, the interposition or operation of these biotic factors may alter the trend or pathway of secondary succession, leading to the establishment of subclimaxes that are in a state of equilibrium with biotic factors such as repeated burning or grazing, rather than with the prevailing regional climate. When the normal course or trend of ecological succession is altered as a result of the interposition of biotic factors and influences, vegetation succession is said to be deflected. Towards the drier northern margin of the rainforest zone in West Africa, repeated cultivation which involves land clearing by burning has encouraged the invasion of farmland and fallow land by grasses. Such grasses usually dry up during the dry season and the predominance of grasses in early successional communities has reinforced their proneness to fire. The frequent occurrence of burning in such fallow vegetation has hindered the invasion of fire-sensitive rainforest trees and also encouraged the preponderance of grasses and the invasion of fire-tolerant savanna trees. For this reason, secondary succession at the northern margin of the rainforest zone in West Africa leads to the formation of savanna vegetation known as derived savanna (Keay, 1959) rather than to the formation of secondary forest. This is a classic example of deflected succession. The derived savanna vegetation is stable and is in a state of equilibrium with fire, and is therefore a fire subclimax. Fire is the repressive ecological factor that prevents derived savanna vegetation from reverting to rainforest vegetation (Hopkins, 1974). As the results of the Okokemeji fire-protection experiment in Nigeria clearly showed, derived savanna vegetation reverts to forest when fire is excluded for a long period. Scott (1987) also pointed out that anthropogenic savanna in the Grand Pajonal of Peru, formed when Campa Indians cultivated land for about 1–4 years and subsequently subjected the fallow vegetation to annual burning on farm abandonment, reverted to forest vegetation when vegetation burning was discontinued for 18 years.

Deflected succession is closely related to arrested succession. In fact, the former usually leads to and terminates in arrested succession.

The successional development of vegetation is said to be arrested when deflected succession leads to the establishment of a stable subclimax community that is in a state of equilibrium with a biotic factor – such as fire – which prevents the vegetation from developing further to attain the status of the climax. In West Africa, as with most parts of the tropics, regular annual fires have hindered further successional development of savanna vegetation and hence maintained them in subclimax status. Hence, most present-day savanna vegetation is the product of arrested succession. In the subsections that follow, examples of deflected succession in Africa, tropical America and Asia will be discussed briefly.

5.5.1 Deflected succession in Africa

The formation of derived savanna at the drier northern margin of the rainforest zone in West Africa as a result of deflected succession was referred to in the preceding section and will not be discussed further here. Suffice it to say that the derived savanna is floristically similar to the southern part of the Guinea savanna zone, as most of the species that characterize the former actually originated and invaded from the latter zone. It is also important to stress that the derived savanna is physiognomically similar to the southern Guinea savanna and not to the rainforest from which it was derived.

Apart from the derived savanna which forms a distinct vegetation zone in West Africa, other examples of deflected succession abound in Africa, especially where the fallow period has been reduced to about 3 years and fallow vegetation is burnt annually. Usually when land is cultivated or burnt too frequently, *Imperata cyclindrica* (spear grass) tends to become dominant and forest regeneration is hindered. Spear grass invades over-cultivated land and land subjected to frequent burning in West Africa and in other parts of tropical Africa such as the Congo basin. In several instances, though, *Pteridium aquilinum* (bracken fern) becomes dominant and persists over time, as it can resist fire by means of its rhizomes and thereby hinder forest re-establishment (Richards, 1996).

5.5.2 Deflected succession in tropical Asia

Most savanna vegetation in tropical Asia is secondary in nature and is the product of deflected succession in tropical deciduous seasonal (monsoon) forest and dipterocarp rainforest. In the Indian subcontinent, for instance, the original sub-humid and dry deciduous (monsoon) forest has been replaced by savanna vegetation as a result of several millennia of human occupancy (Mistra, 1983). Monsoon forests are characterized by a herb layer that becomes flammable during the long dry season when the herb layer dries up. The regular burning of the monsoon forest eco-systems leads to the death and elimination of fire-sensitive trees and shrubs, and consequently to the dominance of grasses and fire-resistant trees such as *Tectona grandis* (teak). This process ultimately leads to the formation of savanna vegetation in which trees such as *T. grandis* and *Acacia* spp. feature prominently in the tree flora, the former tree species being a relict of the original monsoon forest. Obviously, such savanna vegetation is a fire subclimax and is maintained in its current status by annual burning and is, therefore, the product of deflected succession.

In many parts of Asia, especially in the Philippines and Indonesia, deflected succession has resulted in the replacement of forest vegetation by dense stands of *Imperata cylindrica* (speargrass). When rainforest is cleared and the site farmed, a secondary forest would become established if the regenerating fallow vegetation is not burnt repeatedly. However, when young fallow vegetation is burnt frequently, secondary succession does not progress towards the establishment of the climax dipterocarp forest. Instead, *I. cylindrica*, which can survive repeated burning, becomes established and dominant in fallow vegetation. Stands of *I. cylindrica* will persist unless the area is protected against burning for several years to permit forest re-establishment. Beets (1990) estimated that in Indonesia about 20 million ha of land are covered by *I. cylindrica*. In Papua New Guinea, Eden (1993) has reported the occurrence of secondary savanna which is burnt periodically. The savanna vegetation is dominated by *I. cylindrica* and *Ischaemum polystachyum*, and is the product of deflected succession associated with shifting cultivation in the wetland areas and forests of south-western Papua New Guinea.

Yoshida (2009) has reported that the successional pathway on the north-western subtropical Pacific island of Haha-jima has altered as a result of the invasion of the oceanic island by an exotic tree legume, *Leucaena leucocephala*. He observed that the invasion of the relatively species-deficient island by *L. leucocephala* has completely changed the course of secondary succession, as it hinders the invasion of secondary forests by native tree species such as *Macaranga tanarius* and *Machilus thunbergii*. The invading tree legume is an early successional species during secondary succession in the islands of the north-western Pacific Ocean and, when the tree dies, it is replaced by more aggressive exotics such as *Morus australis*, thereby leading to a deflected secondary succession (Yoshida and Oka, 2004). A major finding of Yoshida (2009) is that *L. leucocephala* does not alter the course of secondary succession in continental islands in the north-western Pacific, which are richer in species, implying that ecosystems with high species diversity are less prone to deflected succession than those which are deficient in native species.

5.5.3 Deflected succession in tropical America

In tropical America, as in Africa and tropical Asia, frequent cultivation and shortened fallow periods coupled with frequent burning of the regenerating fallow vegetation result in deflected succession. This usually culminates in subclimax savanna and grassland communities and those dominated by *Pteridium aquilinum* (bracken fern).

These subclimax communities are maintained by fire and would be invaded by trees and transformed into secondary forests if they are protected against burning for a couple of decades. Scott (1987) described deflected succession in the Gran Pajonal area of eastern Peru where the Campa Indians, who practise shifting cultivation, often burn fallow vegetation, to prevent it from being transformed into impenetrable secondary forests. The Gran Pajonal area is an uplifted plateau of an average elevation of 1000 m above sea level. The plateau is deeply dissected by numerous streams and land appropriate for cultivation is limited. The scarcity of suitable arable land encourages frequent cultivation of suitable areas and this, coupled with repeated burning, tends to trigger off the process of deflected succession. Scott (1987), who studied the formation of grassland vegetation from forest in this region, described the stages in the savannazation of the forest ecosystem as follows:

1. Primary or secondary forest is cleared and the land used for cultivation of field crops, especially cassava, for 1–4 years. At the cessation of cropping, weeds such as *Pteridium aquilinum* (bracken fern) and *Imperata brasiliensis* become dominant in the abandoned farmland.
2. When the young fallow vegetation is burnt annually, bracken fern tends to predominate. Vegetation burning also encourages erosion.
3. *Imperata brasiliensis* replaces bracken fern as the dominant weed with sustained annual burning, which further impoverishes soil nutrient status through erosion.
4. *Andropogon* spp. ultimately replace *Imperata brasiliensis* in old grasslands and the former grass species persists over time for as long as the site is subjected to annual burning. However, if burning is discontinued for several years, the process of tree invasion and forest re-establishment will be initiated.

5.6 Succession in Areas of High Altitude on Tropical Mountains

One interesting feature of the tropics is that enclaves of temperate and even polar climates occur at high elevations on mountains such as the Andes and even on Mount Kilimanjaro, which is located very near the equator. As pointed out in Chapter 4, shifting cultivation is practised at high altitudes of about 3000 m or above on the slopes and inter-mountain plateaux of the Andes mountains in South America. Although the slash of cut fallow vegetation is not usually burnt prior to cultivation, but ploughed into the soil, the farming system is essentially a type of shifting cultivation as it involves short periods of cultivation of about 1–4 years that alternate with longer periods of fallow of 7 to more than 20 years (Sarmiento *et al.*, 1993). The process of vegetation succession that takes place during the fallow period following cultivation in areas of high altitude differs from that in the lowland tropics, due to differences in climatic and soil conditions resulting from the prevalent low temperatures at high altitudes in tropical mountains. Sarmiento *et al.* (2003) have studied secondary succession during the first 12 years of the fallow period following shifting cultivation, in Paramo de Gavidia, in the State of Merida in the northern Andes of Venezuela. The fallow plots and virgin *paramo* vegetation studied by Sarmiento *et al.* (2003) were located at an elevation of 3200–3800 m above mean sea level. Their findings are briefly described here, and the main differences between secondary succession following cultivation in tropical lowlands and highlands are highlighted.

The pioneer species that initiate the process of secondary succession following the cessation of cropping were mainly introduced exotics, especially the forbs *Rumex acetosella* and *Erodium cicutarium*. The former were usually more abundant than the latter. The introduced exotic pioneer species were abundant in fallow vegetation and usually accounted for up to 70% of the vegetation in the first year of succession. Thereafter, the pioneer herbs and forbs declined progressively in abundance until the 12th year, when they accounted for a mere 5.5% of the vegetation (Sarmiento *et al.*, 2003). Contemporaneous with the decline in the abundance of the pioneer species, native forbs such as *Lupinus meridanus* and grasses including *Vulpia myuros* increase in abundance, and they can be regarded as mid-successional species. It is important to stress that although they attain a peak abundance in about the fifth year of the fallow period, they persist until the climax stage represented by the virgin *paramo* ecosystem. Over time the mid-successional species decline in relative abundance as late successional species increase and ultimately become the dominant species in fallow vegetation of 9–12 years and in the climax *paramo* ecosystem. Such late successional species include shrubs such as

Baccharis prunifolia and *Hesperomeles obtusifolia* and rosettes including *Espeletia schultzii* and *Ruilopezia floccose*. Grasses such as *Calamagrostis effusa* and *Agrostis subpatens* are also important elements of the flora of the older fallows of 9–12 years and the climax *paramo* vegetation.

The process of secondary succession in areas of high altitude in the Venezuelan Andes differs from that in the lowland tropics in some respects. As Sarmiento *et al.* (2003) have pointed out, there is no true species turnover during secondary succession in high elevations in tropical mountains. A species merely declines in abundance and is not completely replaced by another. The species that are dominant in the old fallow vegetation of 9–12 years are usually present in 1-year fallow vegetation, where they account for a relatively small proportion of the vegetation in terms of abundance. Over time they increase in abundance and so overshadow those species such as *Rumex acetosella* that are dominant during the first few years following the inception of succession in the Venezuelan *paramo* ecosystem. In contrast, there is usually a rapid turnover of species population during succession in tropical lowlands, at least within the first 10 years following the inception of succession. In forest fallows of south-western Nigeria and in the Upper Rio Negro region of Venezuela, as well in the monsoon forest fallows of north-eastern India in tropical lowlands, the turnover of species is rapid and several plant species replaced one another within a few months or years of inception of succession (Aweto, 1981; Toky and Ramakrishnan, 1983; Uhl, 1987) as was discussed in Sections 5.2.3 and 5.3. In extreme environments such as deserts, tundra and very high altitudes in tropical mountains that are characterized by persistently low temperatures, as with the *paramo* ecosystems of the Venezuelan Andes, the process of vegetation succession is characterized by a gradual and progressive colonization of the climax species, without a clearly defined replacement of plant species by one another (Svoboda and Henry, 1987).

The rate of succession in areas of very high elevation in the tropics is very slow, and this is due to the very low temperatures in high altitudes in the tropics. It is important to observe that the atmosphere is heated by conduction from the earth's surface, and this largely explains why ground surface temperatures decrease with increasing elevation. Sarmiento *et al.* (2003) concluded from their study of succession in the *paramo* ecosystems of northern Venezuela that the rate of succession is fast because vegetation physiognomy was restored within 12 years of succession. It should be pointed out that the high-altitude vegetation of the Andes is characterized by a comparatively simple structure, as trees are totally absent and the vegetation is dominated by grasses, forbs, rosettes and low shrubs. The restoration of vegetation structure within 12 years is, therefore, not necessarily indicative of a fast rate of succession as the vegetation is of low vertical development and simple in terms of structure. Perhaps a more reliable index of assessing whether the rate of succession is fast would be to determine the rate of biomass accumulation in the fallow vegetation during succession, and compare the biomass of seral stages with the climax *paramo* ecosystem. Similarly, the life form structure of seral fallow communities can be characterized and the rate at which they approach the climax quantitatively assessed. It is important to observe that alpine zones of mountains are characterized by harsh environmental conditions such as very low temperatures, soil moisture stress resulting from frequent frosts and freezing of soil moisture, and frequently by soil nutrient deficiency due to poor mineralization of soil organic matter. Given these environmental conditions, plants regenerate slowly after site disturbance, resulting in a slow rate of succession. The findings of Sarmiento *et al.* (2003) indicated that the dominant plant life forms in the alpine zones of the Venezuelan Andes, rosettes and sclerophilous shrubs, are present at the inception of succession and they gradually increase in abundance and subsequently become dominant in late successional stages. This is clearly indicative of a slow rate of succession.

Trees play an important role in succession in the lowland tropics. In both forest and savanna ecosystems of the lowland tropics, trees increase in size, number and diversity as succession progresses over time. This results in the elimination of grasses in rainforest fallows, while in savanna fallows, grasses diminish in abundance in later successional stages. In humid and sub-humid savanna ecosystems in the lowland tropics, secondary succession may lead to the formation of dry forests in which a grass layer is eliminated, if burning is excluded for a long time. Trees serve as foci of soil fertility regeneration as well as sources of seeds and suckers that facilitate regeneration of forest and savanna bush fallows in the lowland tropics. Their absence in alpine ecosystems, such as in the *paramo* of the Andes, would seem to suggest that vegetation structure is not only simpler, but that the process of vegetation recolonization is different from that of the lowland tropics. The absence of trees (which play an important role in secondary succession in the lowland tropics) from the alpine *paramo* ecosystem of the Andes is due to the effects of high altitude. Altitude may neutralize the effects of latitude (Prof. S.I. Okafor, Ibadan, 2011, personal communication). This observation clearly explains why trees are absent from the alpine *paramo* ecosystems of Venezuela, Colombia and Ecuador, in spite of their location within the tropics, and why the process of succession in the high-altitude *paramo* ecosystem differs strikingly from that of the lowland tropics.

Finally, due to the very low rate of humification and mineralization of organic matter in alpine environments, the soil in areas of high elevation in tropical mountains is often deficient in plant mineral nutrients. In other words, plant nutrients are not readily available in the soil, in spite of the very high levels of organic matter in the soils. The low levels and availability of nutrients in *paramo* ecosystems may further slow down the rate of succession. Ganade and Brown (2002) have shown that certain fallow species respond to nutrient availability and their rate of regeneration is enhanced by increasing soil nutrient levels through fertilization.

Sarmiento *et al.* (2003) made an interesting point concerning the rate of secondary succession in the alpine *paramo* ecosystem of the Venezuelan Andes, which they claimed to be constant. Except when secondary succession is initiated on degraded or impoverished sites with poor, nutrient-deficient soils, the process of succession is relatively fast during the first few years or decades following the inception of succession. Thereafter, the rate of succession slows down as the ecosystem matures and approaches the climax stage. If the observation of Sarmiento *et al.* (2003) that the rate of succession is constant is true, this will be a major way in which succession in alpine ecosystems in the tropics differs from that in their lowland counterparts. However, their proposition of constant rate of succession appears to rest on a thin empirical base, as their data on plant species richness dynamics during succession negate their proposition. Their data clearly showed that the number of species per plot doubled between the first and fourth years of succession, and thereafter the number of species stabilized until about the 12th year following the inception of succession. Clearly, the rate of increase of species in the plots they studied was faster during the first 4 years after cultivated fields were abandoned, than in the later years of succession. Hence, their proposition of a constant rate of succession is not sustained.

5.7 Management of Fallow Vegetation

Shifting cultivators do not usually manipulate the floristic composition of fallow vegetation in order to enhance the process of soil fertility restoration during the fallow period. The regenerating bush fallow vegetation is usually allowed to develop with minimal interference by farmers, and in fact most shifting cultivators are reluctant to embrace the use of planted fallows of leguminous trees or other soil-improving trees as an alternative to the natural bush fallow vegetation, or integrate such trees into bush fallow vegetation. They consider it a waste of time and resources to expend time and energy planting trees or herbaceous plants in fallow land, if such plants do not produce tangible benefits, such as fruits that can help meet their food requirements or be sold for cash. However, shifting cultivators usually leave the stumps of trees or even live trees on the farm. They may also leave pruned trees which may have been selectively retained for training climbing crops such cowpea or yams. By leaving tree stumps or selectively retaining certain trees on the farm during cropping, the farmer inadvertently influences the floristic composition of the fallow vegetation that subsequently develops after cultivation. In some cases, trees planted in lines along the boundaries of fields to demarcate their areal extent later become important elements of fallow vegetation.

Trees retained in farms by shifting cultivators in south-western Nigeria include *Newbouldia laevis, Phyllanthus discoideus, Alchornea cordifolia, Anthocleista vogellii* and *Elaeis guineensis* (Aweto, 2001). These and other trees such as *Baphia nitida* and *Anthonotha macrophylla* that are retained on farms in southern Nigeria, later become well established in bush fallow vegetation at the cessation of cropping, when the land is fallowed. *Attalea (Orbignya) speciosa* (babassu palm), which is well integrated on farms in north-eastern Brazil, has already been referred to. Trees such as *Faidherbia (Acacia) albida, Cordia abyssinica* and *Ziziphus* spp. are retained in farms in the Jebel Marra highlands of Sudan (Miehe, 1986). These multi-purpose trees which provide food, wood and fodder subsequently become important elements of fallow vegetation.

In a few cases, shifting cultivators deliberately enhance the utilitarian value of fallow vegetation by planting trees that yield useful products in fallow vegetation. The Bora Indians in the Peruvian Amazon plant bitter manioc intercropped with pineapple and fruit trees on their farms. They also plant *Bactris gasipaes* (the peach palm) and *Theobroma bicolor* (mocambo) trees or protect them in fallow vegetation to yield fruits (Denevan *et al.*, 1984). In this regard, the bush fallow vegetation supplements cultivated fields and gardens as a source of food. The Bora people also plant and protect tropical cedar seedlings in their fallows, which they may allow to grow for about 30 years to yield valuable timber for themselves or their children. Other useful products the Bora people obtain from fallow vegetation include liana fibres, tree barks and dyes (Denevan *et al.*, 1984).

The Wola people of Nipa province in Papua New Guinea also plant trees that yield useful products on farms during the period of cultivation. Such trees, which include *Casuarina oligodon* (casuarina), *Pandanus julianettii* (pandan) and *Cordyline fruticosa* (cordyline), are usually left in the field at the cessation of cropping and so become well established in fallow vegetation

(Sillitoe, 1995). Casuarina trees are of special importance because of their acknowledged ability to improve soil nutrients and to yield timber and fuelwood. The roots of casuarina trees harbour the nitrogen-fixing actinomycete *Frankia* sp., which enhances nitrogen accretion in bush fallow vegetation and soil.

The deliberate manipulation of the floristic composition of fallow vegetation in order to enhance the process of soil fertility restoration is not totally alien to the system of shifting cultivation, although the practice is rare. As a result of the practice of shifting cultivation over several millennia and indigenous knowledge handed down from one generation to another, swiddeners or native farmers of the tropics are aware of the fact that certain trees improve soil fertility. They sometimes plant trees which improve soil fertility in fallow vegetation or on farms, prior to onset of the fallow period. The Baduy people of western Java, Indonesia, regard shifting cultivation as central to their cultural identity and continue to practise this system of agriculture, even in spite of rising populations and increasing problems of land shortage (Iskandar and Ellen, 2000). In response to the problem of land shortage, the Baduy people have embraced agroforestry as a means of intensifying shifting cultivation. They grow crops, especially rice, in between rows of a commercially valuable tree legume, *Paraserianthes* (*Albizia*) *falcataria*, which helps to improve soil fertility. After cropping for about 2 years, a woody fallow of *Paraserianthes* trees is established from the trees left on the farm during cultivation. After about 3 or 4 years of fallow, the tree legume will have helped to rejuvenate soil fertility, and consequently the trees are felled by farmers for sale as timber, while their branches are used as fuelwood (Iskandar and Ellen, 2000).

In northern Thailand, farmers have integrated *Macaranga denticulata* (pada) trees into rice fields to help maintain or even improve soil fertility. At the inception of cropping when rice is sown in cultivated fields, pada seedlings germinate in the farm but they are not weeded out by the farmer, who may transplant them from where their density is considered too high to parts of the field where they are sparse (Yimyam *et al.*, 2003). At the end of cultivation, pada trees become an integral component of fallow vegetation and play an important role in restoring soil fertility.

5.8 Nutrient Storage in Fallow Vegetation

Nutrient storage in fallow vegetation is an important function of the bush fallow, although this subject has received comparatively little attention from students of shifting cultivation. Over time, fallow vegetation accumulates nutrients in its standing biomass and such nutrients are protected against leaching from the soil–vegetation system. Nutrients immobilized in the standing crop of fallow vegetation are released to fertilize the soil, except nitrogen and sulfur that are largely volatilized when fallow vegetation is cut, allowed to dry and burnt prior to cultivation. The shifting cultivator, therefore, depends not only on the nutrient capital that accumulates in the soil during the fallow period, but also on the nutrients immobilized in bush fallow vegetation, in order to grow crops for a few years before leaving the land fallow. It is important to stress that where farmers remove felled trees from cleared fallow vegetation for use as fuelwood, a sizeable proportion of the nutrient capital stored in fallow vegetation is removed from the cleared site, and is therefore not available to fertilize the site being prepared for farming.

A pioneering study on the dynamics of nutrient storage in forest fallows over time was carried out by Bartholomew *et al.* (1953). They analysed the nutrient content of fallow vegetation aged between 2 and 18 years in Yangambi, Democratic Republic of the Congo, and their results indicated that nutrient storage in fallow vegetation is usually fastest during the first few years, usually within 5 years of the fallow period. This is when bush fallow foliage, which is rich in nutrients, attains its maximum development. Thereafter, nutrient storage takes place mainly in the woody stems and branches, which are not only lower in nutrient content, but also grow slowly over time compared with vegetation foliage. In the forest successional communities developed on the soil of low base status (ultisols) studied by Bartholomew *et al.*

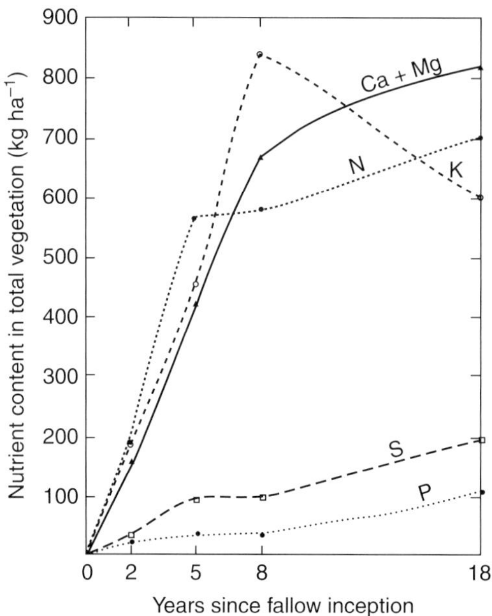

Fig. 5.7. Nutrient storage in the above-ground biomass of bush fallow vegetation in Yangambi, DR Congo. (After Bartholomew *et al.*, 1953; from Sanchez 1976.)

(1953), potassium storage in forest fallows peaked at 8 years following fallow inception, and thereafter declined until the 18th year. The rates of storage of phosphorus, sulfur, nitrogen, calcium and magnesium increased marginally between the eighth and 18th years of the fallow period (Fig. 5.7).

Szott and Palm (1996) evaluated nutrient storage in natural and planted fallows in Yurimaguas, Peru, during the first 53 months following the inception of the fallow period. They observed that nitrogen, calcium, magnesium, potassium and phosphorus storage in natural fallow and planted fallows, including that of a fast-growing indigenous tree, *Inga edulis*, increased over time and that the planted fallow of this native legume accumulated nitrogen and calcium faster than a natural bush fallow of comparable age (Fig. 5.8). After nearly 4.5 years of fallow, natural bush fallow vegetation accumulated more phosphorus, potassium, calcium and magnesium than planted fallows of the herbaceous legume *Pueraria phaseoloides* and of the leguminous shrub, *Cajanus cajan*. The results of Szott and Palm (1996) further indicated that the natural bush fallow accumulated more nutrients in its standing biomass than planted fallows of a woody perennial, *Desmodium ovalifolium*.

Nutrient storage in fallow vegetation impacts on soil nutrient status, as most of the nutrients immobilized in vegetation are derived from the soil. The quantities and rates of nutrient uptake from the soil depend on the type, floristic composition and stage of development of fallow vegetation. Soil characteristics, especially soil nutrient and physical status, in turn influence the development of fallow vegetation. Fallow soil and vegetation exert reciprocal effects on one another, and this is the subject of Chapter 6.

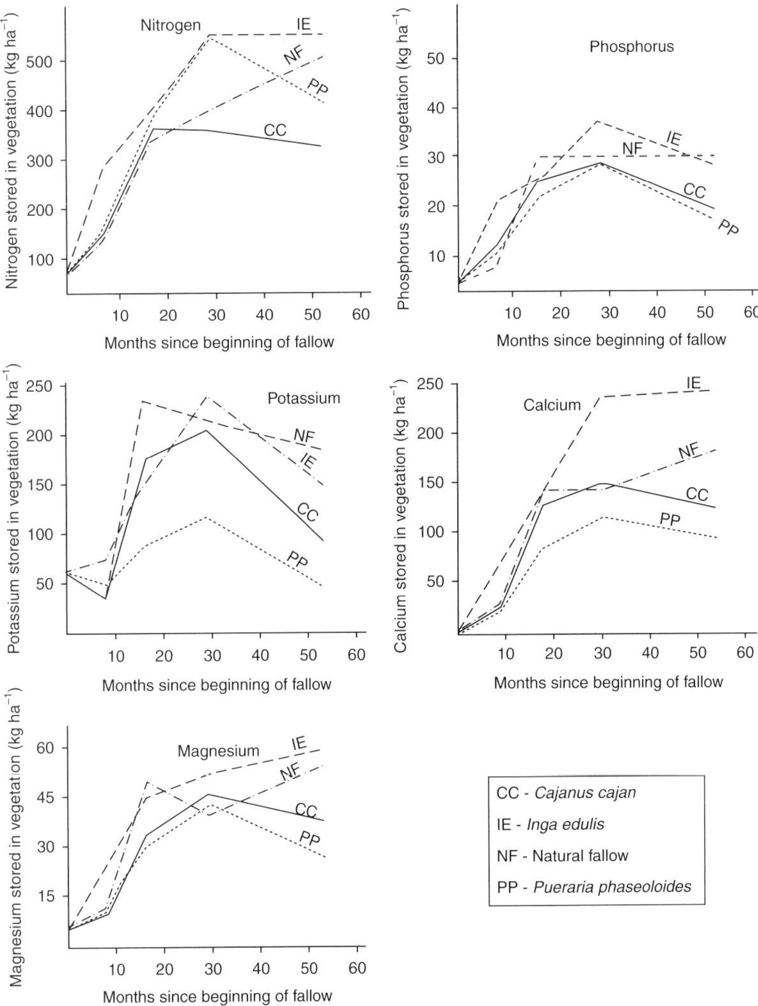

Fig. 5.8. Patterns of nutrient storage in natural and planted fallows in Yurimaguas, Peru. (Based on data in Szott and Palm, 1996.)

References

Abubakar, S.M. (1996) Rehabilitation of degraded land by means of fallowing in a semi-arid area of northern Nigeria. *Land Degradation and Development* 7, 133–144.

Adedeji, F.O. (1984) Nutrient cycles and successional changes following shifting cultivation practice in moist semi-deciduous forests in Nigeria. *Forest Ecology and Management* 9, 87–99.

Ajayi, S.O. and Hall, J.B. (1975) *Management Plan for Borgu Game Reserve.* Report to the Ministry of Agriculture and Natural Resources, Ilorin, Kwara State, Nigeria.

Amanor, K.S. (1994) *The New Frontier, Farmers Responses to Land Degradation: A West African Study.* Zed Books, London.

Archer, S., Scifres, C., Bassham, C.R. and Maggio, R. (1988) Autogenic succession in a subtropical savanna: conversion of grassland to thorn woodland. *Ecological Monographs* 58, 111–127.

Aweto, A.O. (1981) Secondary succession and soil fertility restoration in south western Nigeria: I, Succession. *Journal of Ecology* 69, 601–607.

Aweto, A.O. (2001) Trees in shifting and continuous cultivation farms in Ibadan Area, Southwestern Nigeria. *Landscape and Urban Planning* 53, 163–171.

Aweto, A.O. and Adejumobi, D.O. (1991) Impact of grazing on soil in the Southern Guinea Savanna zone of Nigeria. *Environmentalist* 11, 27–32.

Bartholomew, W.V., Meyer, I. and Laudelot, H. (1953) Mineral nutrient immobilization under forest and grass fallow in the Yangambi (Belgian Congo) region. *INEAC Ser. Sc.* 57, 1–27.

Beets, W.C. (1990) *Raising and sustaining productivity of Smallholder Farming Systems in the Tropics.* AgBe Publishing, Alkmaar, The Netherlands.

Brookman-Amissah, J., Hall, J.B., Swaine, M.D. and Attakorah, J.Y. (1980) A re-assessment of fire protection experiment in north-eastern Ghana savanna. *Journal of Applied Ecology* 17, 85–99.

Brown, S. and Lugo, A.E. (1990) Tropical secondary forests. *Journal of Tropical Ecology* 6, 1–32.

Brubacher, D., Arnason, J.T. and Lambert J.D.H. (1989) Woody species and nutrient accumulation during the fallow period of milpa farming in Belize, C.A. *Plant and Soil* 114, 165–172.

Budowski, G. (1970) The distinction between old secondary and climax species in tropical central American lowland forests. *Tropical Ecology* 11, 44–48.

Buschbacher, R., Uhl, C. and Serrao, E.A.S. (1988) Abandoned pastures in eastern Amazonia. II. Nutrient stocks in the soil and vegetation. *Journal of Ecology* 76, 682–699.

Calub, B.M. (2003) Indigenous fodder trees for rehabilitation. *LEISA* 19, 22–23.

Chachu, R.E.O. (1982) A tentative hypothesis of plant succession in Kainji Lake National Park. In: Sanford, W.W., Yesufu, H.M. and Ayeni, J.S.O. (eds) *Nigerian Savanna. Proceedings MAB State-of-Knowledge Workshop*, Kainji, Nigeria, pp. 406–417.

Charter, J.R. and Keay, R.W.J. (1960) Assessment of the Olokemeji fire control experiment (investigation 254) 28 years after institution. *Nigerian Forestry Information Bulletin* 3, 32 pp.

Cochard, R. and Edwards, P.J. (2011) Tree dieback and regeneration in secondary *Acacia zanzibarica* woodlands on an abandoned cattle ranch in coastal Tanzania. *Journal of Vegetation Science* 22, 490–502.

Corlett, R.T. (1995) Tropical secondary forests. *Progress in Physical Geography* 19,159–172.

Corner, E.J.H. (1988) *Wayside Trees of Malaya.* Malayan Nature Society, Kuala Lumpur, Malaysia.

Denevan, W.M., Treacy J.M., Alcorn J.B., Padoch, C., Denslow, J. and Paitan, S.F. (1984) Indigenous agroforestry in the Peruvian Amazon: Bora Indian management of swidden fallows. *Interciencia* 9, 346–357.

De Rouw, A. (1995) The fallow period as a wind-break in shifting cultivation (tropical wet forests). *Agriculture, Ecosystems and Environment* 54, 31–43.

d'Oliveira, M.V.N., Alvarado, E.C., Santos, J.C. and Carvalho Jr, J.A. (2011) Forest natural regeneration and biomass production after slash and burn in a seasonally dry forest in the southern Brazilian Amazon. *Forest Ecology and Management* 261, 1490–1498.

Eden, M.J. (1993) Swidden cultivation in forest and savanna in lowland southwest Papua New Guinea. *Human Ecology* 21,145–166.

Ewel, J. (1980) Tropical succession: Manifold routes to maturity. *Biotropica* 12 (suppl.), 2–7.

Ganade, G. and Brown, V.K. (2002) Succession in old pastures of central Amazonia: role of soil fertility and plant litter. *Ecology* 83, 743–754.

Garcia-Nunez, C., Azocar, A. and Rada, F. (2011) Seasonal gas exchange and water relations in juveniles of two evergreen Neotropical savanna tree species with contrasting regeneration strategies. *Trees: Structure and Function* 25, 473–483.

Hopkins, B. (1974) *Forest and Savanna.* Heinemann, Ibadan, Nigeria.

Hopkins, B. (1983) Successional processes. In: Bourliere, F. (ed.) *Tropical Savannas.* Elsevier, Amsterdam, pp. 605–616.

Iskandar, J. and Ellen, R.F. (2000) The contribution of *Paraserianthes (Albizia) falcataria* to sustainable swidden management practices among the Baduy of West Java. *Human Ecology* 28, 1–17.

Keay, R.W.J. (1959) *An Outline of Nigerian Vegetation.* Government Printer, Lagos, Nigeria.

Kochummen, K.M. and Ng, F.S.P. (1977) Natural plant succession after farming in Kepong. *Malayan Forester* 40, 61–78.

Lambert, J.D.H. and Arnason, J.T. (1986) Nutrient dynamics in milpa agriculture and the role of weeds in the initial stages of secondary succession in Belize, C.A. *Plant and Soil* 93, 303–322.

Laws, R.M., Parker, I.S.C. and Johnstone, R.C.B. (1975) *Elephants and Their Habitats: the Ecology of Elephants in North Bunyoro*, Uganda. Oxford University Press, Oxford, UK.

Longman, K.A. and Jenik, J. (1987) *Tropical Forest and its Environment.* Longman Scientific and Technical, Harlow, UK.

May, P.H., Anderson, A.B., Frazao, J.M.F. and Balick, M.J. (1985a) Babassu palm in the agroforestry systems in Brazil mid-north region. *Agroforestry Systems* 3, 275–295.

May, P.H., Anderson, A.B., Balick, M.J. and Frazao, M.F. (1985b) Subsistence benefits from the babassu palm (*Orbignya martiana*). *Economic Botany* 39, 113–129.

Menaut, J.C. (1977) Evolution of plots protected from fire since 13 years in a Guinea savanna of Ivory Coast. *Proceedings 4th International Symposium on Tropical Ecology*, Instituto National de Cultura, Panama, pp. 541–558.

Miehe, S. (1986) *Acacia albida* and other multipurpose trees on the farmlands in the Jebel Marra highlands, Western Darfur, Sudan. *Agroforestry Systems* 4, 89–119.

Mistra, R. (1983) Indian savannas. In: Bourliere, F. (ed.) *Tropical Savannas.* Elsevier, Amsterdam, pp. 151–166.

Moleele, N.M. (1999) Bush enchroachment and the role of browse in cattle production. PhD thesis, Department of Physical Geography, Stockholm University, Stockholm.

Moleele, N.M. and Perkins, J.S. (1998) Encroaching woody plant species and boreholds: is cattle density the main driving factor in the Olifants Drift communal grazing lands, south-eastern Botswana. *Journal of Arid Environments* 40, 245–253.

Mwampamba, H. and Schwartz, M.W. (2011) The effects of cultivation history on forest recovery in fallows in the Eastern Arc Mountain, Tanzania. *Forest Ecology & Management* 261, 1042–1052.

Nepstad, D., Uhl, C. and Serrao, E.A.S. (1991) Recuperation of a degraded Amazonian Landscape; forest recovery and agricultural restoration. *Ambio* 20, 248–255.

Raunkiaer, C. (1937) *Plant Life Forms.* Oxford University Press, Oxford, UK.

Richards, P.W. (1996) *The Tropical Rain Forest.* Cambridge University Press, Cambridge, UK.

Riswan, S. (1982) Ecological studies on primary, secondary and experimentally cleared mixed dipterocarp forest and Kerangas Forest in East Kalimantan, Indonesia. PhD thesis, University of Aberdeen, Aberdeen, UK.

Ross, R. (1954) Ecological studies on the rain forest of southern Nigeria. III. Secondary succession in the Shasha Forest Reserve. *Journal of Ecology* 42, 259–282.

Ryan, C.M. and Williams, M. (2011) How does fire intensity and frequency affect miombo tree populations and biomass? *Ecological Applications* 21, 48–60.

Saldarriaga, J.G., West, D.C., Tharp, M.L. and Uhl, C. (1988) Long-term chronosequence of forest succession in the Upper Rio Negro of Columbia and Venezuela. *Journal of Ecology* 76, 938–958.

Sanchez, P.A. (1976) *Properties and Management of Soils in the Tropics.* John Wiley, New York.

San Jose, J.J. and Farinas, M.R. (1983) Changes in tree density and species composition in a protected Trachypogon savanna, Venezuela. *Ecology* 64, 447–453.

Sarmiento, L., Monasterio, M. and Montilla, M. (1993) Ecological bases, sustainability, and current trends in traditional agriculture in the Venezuelan high Andes. *Mountain Research and Development* 13, 167–176.

Sarmiento, L., Llambi, L.D., Escalona, A. and Marquez, N. (2003) Vegetation patterns, regeneration rates and divergence in an old-field succession of the high tropical Andes. *Plant Ecology* 166, 63–74.

Scott, G.A.J. (1987) Shifting cultivation where land is limited. Campa Indian agriculture in the Gran Pajonal of Peru. In Jordan, C.F. (ed.) *Amazonian Rain Forests: Ecosystem Disturbance and Recovery.* Springer-Verlag, New York, pp. 34–44.

Sillitoe, P. (1995) Fallow and fertility under subsistence cultivation in the Papua New Guinea Highlands: I. Fallow successions. *Singapore Journal of Tropical Geography* 16, 82–100.

Simpson, E.H. (1949) Measurement of diversity. *Nature* 163, 688.

Staver, A.C., Archibald, S. and Levin, S. (2011) Tree cover in sub-Saharan Africa: rainfall and fire constrain forest and savanna as alternative stable states. *Ecology* 92, 1063–1072.

Stott, P. (1991) Recent trends in the ecology and management of the world's savanna formation. *Progress in Physical Geography* 15, 18–28.

Svoboda, J. and Henry, G.H.R. (1987) Succession in marginal arctic environments. *Arctic and Alpine Research* 19, 373–384.

Swaine, M.D. and Hall, J.B. (1983) Early succession on cleared land in Ghana. *Journal of Ecology* 71, 501–627.

Swamy, P.S. (1986) Eco-physiological and demographic studies of weeds of secondary successional environments after slash and burn agriculture in northeastern India. PhD thesis, North-Eastern Hill University, Shillong, India.

Swamy, P.S. and Ramakrishnan, P.S. (1987a) Contribution of *Mikania micrantha* during secondary succession following slash-and-burn agriculture (Jhum) in north-east India: I. Biomass, litterfall and productivity. *Forest Ecology and Management* 22, 220–237.

Swamy, P.S. and Ramakrishnan, P.S. (1987b) Contribution of *Mikania micrantha* during secondary sucession following slash-and-burn agriculture (Jhum) in north-east India. II. Nutrient cycling. *Forest Ecology and Management* 22, 239–249.

Szott, L.T. and Palm, C.A. (1996) Nutrient stocks in managed and natural humid tropical fallows. *Plant and Soil* 186, 293–309.

Toky, O.P. and Ramakrishnan, P.S. (1981) Run-off and infilteration losses related to shifting agriculture (Jhum) in north-eastern India. *Environmental Conservation* 8, 313–321.

Toky, O.P and Ramakrishnan, P.S. (1983) Secondary succession following slash and burn agriculture in north-eastern India. 1. Biomass, litterfall and productivity. *Journal of Ecology* 71, 735–745.

Uhl, C. (1987) Factors controlling succession following slash-and-burn agriculture in Amazonia. *Journal of Ecology* 75, 377–407.

Uhl, C., Clark, K. Clark, H. and Murphy, P. (1981) Early plant succession after cutting and burning in the Upper Rio Negro region of the Amazon basin. *Journal of Ecology* 69, 631–649.

Uhl, C. Buschbacher, R. and Serrao, E.A.S. (1988) Abandoned pastures in eastern Amazonia. 1. Patterns of plant succession. *Journal of Ecology* 76, 663–681.

Van Vegten, J.A. (1983) Thornbush invasion in a savanna ecosystem in eastern Botswana. *Vegetatio* 56, 3–7.

Whitmore, T.C. (1984) *Tropical Rain Forests of the Far East.* Clarendon Press, Oxford, UK.

Yimyam, N., Rerkasem, R. and Rerkasem B. (2003) Fallow enrichment with pada (*Macaranga denticulata* (Bl.) Muell. Arg.) trees in rotational shifting cultivation in northern Thailand *Agroforestry Systems* 57, 79–86.

Yoshida, K. (2009) Alteration of secondary-successional pathways on northwestern Pacific islands by the invasion of *Leucaena leucocephala*. *Geographical Reports of the Tokyo Metropolitan University* 44, 37–45.

Yoshida, K. and Oka, S. (2004) Invasion of *Leucaena leucocephala* and its impact on the native plant community in the Ogasawara (Bonin) Islands. *Weed Technology* 18, 1371–1375.

6 Relationships between Fallow Soil and Vegetation

It is axiomatic to say that the soil and vegetation components of fallow ecosystems are inter-related and exert reciprocal effects on one another. A single tree or grass is rooted in the soil, which not only provides it with anchorage, but also with the water and nutrients that the plant requires to survive. The plant, in turn, modifies the soil over time by adding organic matter to the soil through litter fall and decomposition and by trapping dust, raindrops and other aerosols from the atmosphere. In addition, it also modifies the microclimate of the immediate surroundings in which it grows by shading the ground and absorbing part of the incoming solar radiation and through gaseous exchange with the atmosphere. Just as an individual plant modifies and is influenced by the environment, an entire plant community or stand of vegetation such as fallow vegetation is intimately associated with the environment, and especially with the soil medium on which it develops. The nature and vigour of the regenerating fallow vegetation partly depends on soil fertility status at the inception of the fallow period, and hence on the extent of soil degradation resulting from the intensity of the previous land use (Uhl *et al.*, 1988; Aide *et al.*, 1995; Lafon *et al.*, 2000; Tasser *et al.*, 2007).

The study of Congdon and Herbohn (1993) has shown that initial soil conditions influence the trend of succession. They observed that rainforest plots disturbed by selective logging 25 years ago in Birthday Creek, north Queensland, Australia, had soils with a lower nutrient status, higher bulk density and a changed floristic composition compared with undisturbed forest plots. Their findings suggest that forest regeneration is partly dependent on soil fertility and the extent of soil degradation at the inception of secondary succession.

The extent of development of fallow vegetation – that is, its successional status – largely depends on the extent to which soil fertility is restored during the fallow period. Hence, farmers in the tropics who practise shifting cultivation usually use the characteristics of fallow vegetation for judging the extent to which soil fertility is rejuvenated before clearing the land to cultivate. The rate of soil fertility restoration under bush fallow varies considerably in the tropics depending on the climate, the nature of the soils and cultural practices, among other factors. Undoubtedly, within the same climatic region, soil variation is an important factor influencing the rate of vegetation development. In the Amazon basin of South America, Moran *et al.* (2000) have shown that inherent soil fertility is a major factor influencing the rate of fallow vegetation regeneration and that the rate of growth of trees is higher on soils of higher fertility status (alfisols) than on soils of lower fertility (ultisols and oxisols) during the first 5–15 years of rainforest succession. Similarly, on coarse-textured soils (oxisols) in south-western Nigeria, Aweto (1981a) observed a positive correlation between tree biomass parameters such as tree height and diameter on the one hand, and soil clay content on the other, implying that trees grow bigger, and presumably faster, on the more clayey sites. In a forest community in the Luquillo Mountains of Puerto Rico, Johnston (1992) observed that soil moisture, exchangeable base nutrients and pH were important determinants of the dominant tree species that occur in the forest; while in Ghana, Swaine (1996) observed that soil fertility is an important factor influencing the distribution of species in rainforest communities. The findings of Golley (1986) indicated that nutrient content of rainforest developed on fertile soil was strongly correlated with soil nutrient levels. In the Sudano–Sahelian zone of northern Cameroon, Donfack *et al.* (1995) observed that the floristic composition of savanna successional communities varied with soil types. The plant life forms, especially trees, in fallow vegetation have been reported by Aweto (1981d) to enhance nutrient and organic matter accretion in fallow soil. Also, some plant species in fallow vegetation, especially leguminous trees, have been reported to improve soil nutrient status (particularly total nitrogen levels) and to substantially increase crop yield (Yimyam *et al.*, 2003).

The findings of the studies briefly reviewed above clearly show that the floristic composition of forest and savanna vegetation, as well as the various successional communities associated with them, not only influences the soil, but that the soil in turn influences the regeneration of fallow vegetation. As with other terrestrial eco-systems, the process of nutrient cycling intimately links the soil and vegetation components of the bush fallow ecosystem to become functional and integrated ecological entities. One cannot fully comprehend the changes or processes that take place in bush fallow soil without considering changes that occur in the vegetation. Moss (1969), recognizing the inter-dependence of soil and vegetation, especially in shifting cultivation systems, observed that both soil and vegetation operate as strongly dependent open systems. He further posited that to neglect the study of the soil is to fail to recognize the part of the plant–soil system that would remain for cultivation after the vegetation has been cleared away. On the other hand, to ignore the vegetation in studies of shifting or rotational agriculture makes it impossible for one to recognize the role of vegetation in soil fertility restoration, and also those attributes of the plant community which are influenced by the soil (Moss, 1969). In view of the reciprocal relationships between fallow soil and vegetation, Aweto (1981a, b, c, d) used the soil–vegetation system approach to study forest fallow ecosystems in different successional stages in south-western Nigeria. The study was undertaken in order to quantitatively char-acterize the relationships between the soil and vegetation components, with a view to identifying the properties of fallow that enhance the process of soil fertility restoration. Such properties of fallow vegetation should be taken into consideration when designing measures to make shifting agriculture more sustainable and intensive, or when attempting to replace shifting cultivation with more intensive systems of agriculture. Also, characterizing the soil–plant community interrelationship is of pivotal importance in elucidating the effects of the soil on the regeneration of fallow vegetation.

This chapter draws heavily on the author's work on forest fallows in south-western Nigeria and studies carried out in the Amazon basin of South America in characterizing soil–vegetation interrelationships in bush fallow ecosystems. Reference is, however, made to work done elsewhere in the tropics. The focus is primarily on the relationships between the structural and floristic aspects of fallow vegetation on the one hand, and the physical and chemical properties of fallow soil on the other. The structural aspects of fallow vegetation considered here include plant life form composition; tree height, diameter and density; and vegetation cover, while the floristic aspects include total number of species, number of tree species and species diversity. Simple correlations were used to highlight the nature and strength of the relationships between pairs of soil and vegetation variables. They cannot, however, be used to analyse the relative effects of several variables on the same parameter. Fallow soil is influenced by several attributes of fallow vegetation, which in turn is influenced by the whole gamut of soil variables. Soil organic matter, for instance, is influenced by several attributes of fallow vegetation including tree density and biomass, vegetation cover, the proportion of different plant life forms and the quantity of litter generated by the vegetation. Stepwise regression was employed to elucidate the relative importance of vegetation attributes that influence soil attributes and vice versa.

6.1 Fallow Soil–Vegetation Interrelationships: Correlation Analysis

In the subsections that follow, the correlations between pairs of fallow soil and vegetation parameters will be examined. Pearson's correlation will be used to examine the nature and strengths of the relationships between pairs of soil and vegetation variables. In all cases, only correlation coefficients that are significant at the 1% confidence level are discussed.

6.1.1 Relationships between plant life forms and soil properties

Different plant life forms come into dominance at different periods during vegetation succession. The various plant life forms exert differential effects on soil properties because they vary in biomass (and hence in their ability to protect the soil), and their rates of nutrient uptake and recycling to the soil also vary. Some plant life forms such as therophytes (annuals) are short-lived and their effects on the soil are transient compared to mesophanerophytes and megaphanerophytes (trees), some of which persist in fallow vegetation for several decades until secondary forests attain the climax status. Obviously, the plant life forms that persist longer in fallow vegetation, such as mesophanerophytes and megaphanerophytes, have a more profound and lasting effect on the process of soil fertility restoration under bush fallow than the smaller and short-lived plant life forms (such as therophytes) that precede them during the course of vegetation succession.

Table 6.1 shows the correlations between plant life forms in forest successional stages in south-western Nigeria and the chemical properties of the 0–10 cm soil layer. Non-phanerophytes, represented mainly by thero-

phytes such as *Tridax procumbens*, *Erigeron floribunda* and *Ageratum conyzoides*, and also by hemicryptophytes such as *Andropogon tectorum*, exert relatively little effect on soil properties during rainforest succession, as revealed by the low and insignificant correlations between the properties of the 0–10 cm layer of the soil and non-phanerophytes. This presumably suggests that the lower plant life forms play a relatively small role in the process of soil fertility restoration under forest fallow vegetation. This should be expected, since, as pointed out earlier, non-phanerophytes are short-lived and they are shaded out of fallow vegetation by the longer-lived plant forms (especially trees) a few months or years after the inception of succession. Microphanerophytes, represented mainly by the forb *Chromolaena odorata*, are negatively correlated with soil exchangeable potassium, sodium and cation exchange capacity (CEC). *C. odorata* is usually present in fallow vegetation at the beginning of the fallow period and generally dominates the vegetation in 1-year and 3-year forest fallows. However, when soil organic matter and nutrient levels improve, it is usually replaced by trees after about 4–6 years following fallow inception. Besides, *C. odorata* is a fast-growing forb, and so rapidly immobilizes soil nutrients in its standing biomass. This partly explains why microphanerophytes are negatively correlated with soil nutrients and CEC. The study of Roder *et al.* (1997) in Laos has shown that soil nutrient levels declined during a 2-year fallow period when fallow vegetation was dominated by *C. odorata*. This is presumably because of the rapid rate of nutrient immobilization by *Chromolaena* forbs, which was not matched by the rate of nutrient recycling back to the soil (referred to above).

Although grasses are relatively insignificant in forest fallows in terms of soil fertility restoration, they have a salutary effect on soil fertility in savanna fallows, in which grasses persist for several years. This is particularly true of perennial grasses in areas where savanna fallows are protected against burning. The data of Elberling *et al.* (2003) indicated that soil organic matter was 72% higher and soil bulk density lower in the 0–10 cm layer in a grass savanna fallow than in a field cultivated for 5 years, in Dahra, Senegal. This should be expected, as grasses add litter to the soil and also

Table 6.1. Correlations between plant life forms and soil properties of the 0–10 cm layer.

	Vegetation characteristics				
Soil characteristics	Mega	Meso	Micro	Nano	Non-phan
Organic matter	0.73	0.48	−0.68	−0.35	−0.36
Total nitrogen	0.69	0.41	−0.58	−0.36	−0.26
Exchangeable potassium	0.69	0.61	−0.76	−0.40	−0.25
Exchangeable sodium	0.52	0.74	−0.79	−0.39	−0.21
Exchangeable calcium	0.45	−0.05	−0.23	−0.15	−0.04
Exchangeable magnesium	0.36	0.19	−0.43	−0.24	0.18
Cation exchange capacity	0.59	0.54	−0.63	−0.40	−0.24

Mega, megaphanerophytes; meso, mesophanerophytes; micro, microphanerophytes; nano, nanophanerophytes; non-phan, non-phanerophytes. Correlations that exceed the absolute value of 0.32 are significant at the 1% confidence level.

afford the ground some protection against leaching and erosion.

Megaphanerophytes are strongly and positively correlated with soil organic matter, total nitrogen and exchangeable potassium of the 0–10 cm soil layer, with each of the correlations exceeding 0.60. The correlations between mesophanerophytes on the one hand, and soil organic matter, exchangeable potassium, sodium and CEC on the other were also positive and significant. These positive correlations indicate a reciprocal relationship between mesophanerophytes and megaphanerophytes and soil chemical properties. Increases in soil organic matter and nutrient status allow mesophanerophytes and megaphanerophytes not only to invade fallow vegetation, but also to increase in population over time. At the same time, the positive association between mesophanerophytes/megaphanerophytes and soil organic matter/nutrient status implies that trees enhance the process of soil fertility rejuvenation. In fact, underneath tree canopies in both forest and savanna regions and even in deserts, distinct 'islands of soil fertility' have been observed by various researchers (e.g. Dean *et al.*, 1999; Wezel *et al.*, 2000; Aweto and Dikinya, 2003; Schade and Hobbie, 2005). The soil in such fertile islands under tree canopies is characterized by much higher levels of organic matter and nutrients than soil outside the canopy. Trees have large biomass, and so provide adequate ground cover against soil erosion and leaching (Aweto, 1981d), and this is a precondition for soil fertility restoration under bush fallow. In addition, trees enhance organic matter and nutrient build-up in the soil by adding litter to it and by helping to

recycle nutrients leached to the subsoil during cultivation back to the topsoil.

The correlations between plant life form composition and the chemical properties of the immediate subsoil layer (10–30 cm) are generally weaker than those of the 0–10 cm layer. Only ten correlations, representing 28.6% of the correlations for the subsoil, were significant at the 1% confidence level (Table 6.2). This situation sharply contrasted with that of the topsoil layer, where 68.7% of the total correlations were significant. Furthermore, only one of the correlations for the subsoil exceeded 0.60, unlike in the topsoil where nine correlation coefficients exceeded 0.60. This clearly shows that fallow vegetation exerts reduced effects on the subsoil layer, and that fallow vegetation in turn is less influenced by the subsoil than by the topsoil layer. Consequently, subsequent discussions on correlations between fallow vegetation and soil will be restricted to the topsoil layer.

Soil physical properties also influence fallow vegetation, and are in turn influenced by fallow vegetation (Table 6.3). Non-phanerophytes in forest fallow in south-western Nigeria are positively correlated with soil bulk density (0.55), implying that the lower plant life forms, especially annuals with low ground cover and which usually die off during the dry season leaving the ground surface exposed, would lead to soil compaction and deterioration in soil physical status. Besides, non-phanerophytes include annuals which spring up quickly following cultivation, when the soil bulk density is still high due to the effects of the preceding cultivation. They are replaced by higher plant

Table 6.2. Correlations between plant life forms of stages of forest succession in south-western Nigeria and soil chemical properties of the 10–30 cm layer.

Soil properties	Vegetation characteristics				
	Mega	Meso	Micro	Nano	Non-phan
Organic matter	0.17	0.60	−0.51	−0.25	−0.11
Total nitrogen	0.29	0.36	−0.37	−0.24	−0.08
Exchangeable potassium	0.51	0.10	−0.30	−0.34	−0.29
Exchangeable sodium	0.14	−0.56	0.18	0.29	0.37
Exchangeable calcium	0.07	−0.03	−0.06	0.09	−0.10
Exchangeable magnesium	0.20	−0.24	0.11	−0.08	0.03
Cation exchange capacity	0.29	0.55	−0.51	−0.22	−0.14

Mega, megaphanerophytes; meso, mesophanerophytes; micro, microphanerophytes; nano, nanophanerophytes; Non-phan, non-phanerophytes. Correlations that exceed the absolute value of 0.32 are significant at the 1% confidence level.

Table 6.3. Correlations between plant life forms of vegetation stages of forest succession in south-western Nigeria and physical properties of the 0–10 cm layer of the soil.

Soil variables	Vegetation characteristics				
	Mega	Meso	Micro	Nano	Non-phan
Sand (%)	−0.55	−0.06	0.41	0.07	−0.12
Silt (%)	0.36	0.19	−0.44	−0.08	0.48
Clay (%)	0.49	−0.26	−0.18	0.09	0.21
Water-holding capacity (%)	0.64	0.55	−0.70	−0.34	−0.19
Total porosity (%)	0.49	0.56	−0.53	−0.44	−0.54
Bulk density (g cm^{-3})	−0.49	−0.56	0.53	0.43	0.55

Mega, megaphanerophytes; meso, mesophanerohytes; micro, microphanerophytes; nano, nanophanerophytes; non-phan, non-phanerophytes. Correlations that exceed the absolute value of 0.32 are significant at the 1% confidence level.

life forms, especially trees, when soil bulk density improves over time. Microphanerophytes are negatively correlated with soil water-holding capacity (−0.70) and total porosity (−0.53), but positively correlated with soil bulk density (0.50). As pointed out earlier, microphanerophytes in forest fallows in south-western Nigeria are represented mainly by the forb *Chromolaena odorata*, and as soil water-holding capacity and total porosity improve over time, the higher life forms – mesophanerophytes and megaphanerophytes – invade fallow vegetation and replace microphanerophytes. Hence, microphanerophytes are negatively correlated with soil water-holding capacity and total porosity.

Mesophanerophytes have positive correlations with soil water-holding capacity (0.55) and total porosity (0.56) and a negative correlation with soil bulk density (−0.56). Similarly, megaphanerophytes are positively correlated with soil water-holding capacity (0.64) and total porosity (0.49). Both mesophanerophytes and megaphanerophytes are trees, and they enhance organic matter build-up in fallow soil. Consequently, an increase in the proportion of both mesophanerophytes and megaphanerophytes in fallow vegetation would lead to organic matter accretion in the soil. This would, in turn, lead to an improvement in soil water-holding capacity and porosity, as organic matter enhances soil water-holding capacity and also increases porosity by promoting aggregation of soil granules (Brady and Weil, 2002). The improvement in soil physical properties also enhances the growth of trees in fallow vegetation, as will be pointed out in a subsequent section.

6.1.2 Relationships between vegetation structural/floristic characteristics and soil chemical properties

Table 6.4 shows that fallow vegetation structural attributes, such as ground cover, tree density, tree diameter and tree height are strongly and positively correlated with soil organic matter. With the exception of tree density (which had a correlation of 0.64 with organic matter), the correlations between the other three structural vegetation characteristics and soil organic matter exceeded 0.80. Similarly, the correlations between the four structural attributes of fallow vegetation listed above and soil total nitrogen, exchangeable potassium, sodium, CEC and available phosphorus generally exceeded 0.70, indicating a strong positive relationship between the two sets of fallow ecosystem characteristics. As might be expected, soil organic matter is strongly and positively correlated with fallow vegetation cover, tree density, tree height and diameter.

An increase in vegetation cover in fallow vegetation enhances the capacity of the vegetation to protect the soil against organic matter losses through leaching and erosion; this is a precondition for reversing the process of organic matter diminution, which usually characterizes the first few years of the fallow period. An increase in the sizes of trees in fallow vegetation improves vegetation ground cover and enhances the capacity of the fallow community to protect the soil against leaching and erosion.

The results obtained by Vasquez-Mendez *et al.* (2010) in Santo Domingo Ranch, Cadereyta, in semi-arid central Mexico, indicated that an increase in vegetation canopy and ground cover decreased runoff and soil erosion by 87–97%, and this effect is marked under trees such as *Acacia farnesiana* and *Prosopis laevigata*. In addition, an increase in tree density has the effect of increasing litter generation and supply to the soil, as well as reducing soil temperatures and the rate of thermally induced organic matter decomposition in the soil. Hence, an increase in the density and sizes of trees in fallow vegetation usually leads to organic matter build-up in the soil, which in turn leads to an increase in the levels of total nitrogen and mineral nutrients, organic matter being a source and reservoir of nutrients in the soil. Increased availability of nutrients, consequent upon organic accretion, in turn leads to an increase in the size of the trees and hence in vegetation cover, indicating that soil fertility influences the process of plant succession. Mesquita *et al.* (2001) have shown that the nature of the pioneer community and the subsequent pathway of secondary succession in Amazonian forests largely depend on soil fertility at the inception of vegetation recolonization. Similarly, in the Mediterranean forests of northern Italy and eastern Austria, the findings of Tasser *et al.* (2007) indicated that the rate of tree regeneration is inversely related to intensity and length of the land use prior to its abandonment. In the Amazon basin, Moran *et al.* (2000) observed that the rate of vegetation regeneration

Table 6.4. Correlations between the structural/floristic characteristics of vegetation stages of forest succession and chemical properties of the 0–10 cm layer of the soil.

	Vegetation characteristics					
Soil characteristics	Ground cover	Tree density	Tree diameter	Tree height	Species diversity	Number of species
Organic matter	0.85	0.64	0.86	0.87	0.65	0.70
Total nitrogen	0.74	0.59	0.80	0.81	0.65	0.65
Exchangeable potassium	0.79	0.72	0.82	0.80	0.77	0.71
Exchangeable sodium	0.72	0.84	0.72	0.74	0.88	0.74
Exchangeable calcium	0.42	0.14	0.49	0.49	0.18	0.29
Exchangeable magnesium	0.39	0.36	0.45	0.44	0.40	0.46
Cation exchange capacity	0.71	0.68	0.79	0.78	0.73	0.72

Correlations that exceed the absolute value of 0.32 are significant at the 1% confidence level.

is influenced by soil fertility. They observed that the height of the regenerating secondary forest on soil of high nutrient status (alfisols) during the first 15 years of secondary succession was 2 m higher than secondary forests on poor, nutrient-deficient soils (ultisols and oxisols). Also, the growth rate of different species in fallow vegetation varies with levels of nutrient availability in the soil.

The correlations between species diversity and number of tree species on the one hand, and soil chemical properties such as organic matter, total nitrogen, exchangeable potassium, sodium and cation exchange capacity (CEC) on the other are positive and generally exceed 0.64. As pointed out earlier, an increase in soil organic matter content results in increased nutrient levels and this not only makes it possible for the more demanding tree species to invade fallow vegetation and become more numerous, but also the soil becomes more capable of supporting a wider range of plant species. Hence, soil organic matter and nutrients are positively correlated with species diversity and the number of tree species in fallow vegetation. The high positive correlations between soil nutrients and fallow vegetation species diversity presumably suggest that a floristically diverse bush fallow vegetation may be more efficient than a monocultural community in restoring soil fertility under bush fallow. The findings of Ewel *et al.* (1991) tend to lend credence to this view, as their 5-year monitoring study of the fertility of volcanic ash-derived soil (inceptisol) in Costa Rica indicated that floristically diverse fallow vegetation generally had a higher nutrient status than monocultures, even of the successional tree species *Cordia alliodora*.

6.2 Fallow Soil–Vegetation Interrelationships: Multiple Regression Analysis

The relationships between pairs of fallow and vegetation variables were discussed in the preceding section. Using bivariate correlation analysis, it was not possible to examine the interactions between various soil or vegetation attributes that collectively or individually influence soil or vegetation dynamics, or to quantify how the different soil attributes affect vegetation characteristics during secondary succession and vice versa. The various soil attributes interact with one another and it is not very realistic to talk about the 'isolated' relationships between pairs of soil and vegetation characteristics. For instance, soil organic matter influences and interacts with soil nutrients such as ammonium nitrogen, exchangeable calcium, potassium and magnesium and available phosphorus – and even soil CEC and water-holding capacity – to influence vegetation parameters such as tree diameter or height. This section uses stepwise multiple regressions to evaluate the relative significance of vegetation factors that influence soil attributes during secondary succession and vice versa. Regression equations express the relationships between a dependent variable and a set of independent variables used for explaining the variation in the dependent variable. Regression models are used for explanation or prediction; that is, to elucidate the factors that account for the variation of the dependent variable, or to predict and provide estimates of the values of the dependent variable, given the values of a set of independent or explanatory variables. In this subsection, the primary objective is to use regression equations for explanatory purposes. Hence, emphasis will be on the signs of the regression coefficients and the proportion of the variance in the dependent variable that is explained by the independent variables, and the relative significance of the independent variables in terms of the variance of the dependent variable that they account for.

A number of assumptions underlie the application of the least square linear regression model and these have been discussed by Poole and O'Farrell (1971). One of these assumptions states that the independent variables should not be highly correlated with one another, and when this assumption is not satisfied a problem of multicollinearity is said to exist between the independent variables to be used for the regression. Hauser (1974) observed that multicollinearity is a serious problem in regression analysis when the correlations between pairs of independent variables exceed 0.80, and would result in errors in the estimates of regression parameters. Thirteen vegetation variables were to be used for explaining changes in soil characteristics during secondary succession, namely:

1. Megaphanerophyte proportion.
2. Mesophanerophyte proportion.
3. Microphanerophyte proportion.
4. Nanophanerophyte proportion.
5. Non-phanerophyte proportion.
6. Species diversity.
7. Ground cover.
8. Tree density.
9. Tree diameter.
10. Tree height.
11. Number of tree species.
12. Oven-dry weight of litter in quadrats of 30 cm by 30 cm.
13. Total number of tree species.

The pair-wise correlations between the variables listed above indicated that seven correlations were above 0.80, with the correlations between tree height on the one hand, and ground cover and tree diameter on the other, exceeding 0.94. This implied that multicollinearity was a serious problem in the set of independent variables to be used as explanatory variables in the regression. Hence, the entire set of 13 independent variables were subjected to principal components analysis in order to eliminate the problem of multi-collinearity by creating a new set of orthogonal variables (principal components) which were used as independent variables, instead of the original vegetation variables. Five rotated principal components were extracted as a result of principal component analysis. Soil properties were regressed on the vegetation components (using component scores) to yield reduced rank equations. The five principal components used in the regression analysis were: (i) mesophanerophytes and tree density; (ii) non-phanerophytes; (iii) tree size and vegetation cover; (iv) nanophanerophytes; and (v) total number of species. These five components account for 69.8, 9.9, 8.8, 4.0 and 2.6% of the total variance in the original set of vegetation data, respectively. The results of the regressions have been discussed by Aweto (1981d) and will be summarized below. The regression equations discussed below are for the topsoil (0–10 cm layer), as the effects of forest fallows in accumulating organic matter and nutrients in the soil is largely restricted to the top 10 cm of the soil profile.

The regression equation for soil organic matter (OM) was:

$$OM = 3.49 - 1.02\,T/GC + 0.41\,M/T \qquad (6.1)$$

where T/GC is tree size and vegetation cover and M/T is the proportion of mesophanerophytes and tree density. Only the two above components were significant in explaining soil organic matter out of the five vegetation components used for the regression, with tree size and vegetation cover accounting for 57%, and mesophanerophytes and tree density 10% of the variation in soil organic matter. Hence, both components explained a total of 67% of the variance of soil organic matter. The regression equation implies that soil organic matter is positively related to mesophanerophytes and tree density, as would be expected, but negatively related to tree size and vegetation cover. The results of the pair-wise correlation analysis discussed in Section 6.1.2 above indicated that soil organic matter had strong positive cor-relations with tree diameter, tree height and vegetation cover, implying that the three vegetation variables enhance organic matter build-up in fallow soil, contrary to the results obtained in the regression analysis. The negative relationship between soil organic matter on the one hand, and tree size and vegetation cover on the other, is due to the use of component scores rather than the actual data for the regression. Aweto (1981d) pointed out that component scores often have arbitrary signs because of standardization to mean zero and unit variance. He indicated that the use of principal components analysis, while solving the problem of multicollinearity, could result in errors in model specification, as the use of component scores instead of the actual data can result in regression coefficients with erroneous signs that intuitively run counter to theoretical expectation.

The regression equation for topsoil CEC was:

$$CEC = 12.76 - 2.56\,T/GC + 1.77\,M/T \qquad (6.2)$$

This regression equation is essentially similar to that for organic matter, and the same two vegetation components that significantly influence organic matter are those that influence soil CEC dynamics during the fallow period. This should be expected, as organic matter is the main determinant and contributor to the CEC of coarse-textured soils of the coastal plain of south-western Nigeria, which are dominated by kaolinitic clay minerals. Tree size/vegetation cover and mesophanerophytes/tree density accounted for 33.3 and 17.7% of the variation in CEC, respectively, implying that both variables explained a total of 51% of the variation in CEC. The high unexplained variance of 41% was largely due to the fact that the 40 fallow plots studied had varied cultivation histories, with short periods of cultivation alternating with varying periods of fallow.

The function of the bush fallow includes that of improving the water-holding capacity of the soil. This is particularly important in coarse-textured soils with low capacity to retain moisture, and in savanna and forest regions of the tropics that are characterized by a distinct seasonality in rainfall distribution. The regression equation for topsoil water-holding capacity (WHC) was:

$$\text{WHC} = 44.58 - 6.21\,\text{T/GC} + 4.10\,\text{M/T} \qquad (6.3)$$

The two vegetation components that emerged as the significant factors influencing topsoil organic matter, tree size/vegetation cover (T/GC) and mesophanerophytes/tree density (M/T) are again the most significant factors influencing soil water-holding capacity. This is not unexpected, as the capacity of sandy soils, such as occur in the coastal plain of south-western Nigeria in which the research on fallow soil–vegetation relationships was conducted, largely depends on the organic matter status. Both components accounted for 62% of the variation in soil water-holding capacity. As with the regressions for soil organic matter and CEC, the negative sign of the regression coefficient T/GC is an artefact, due to the use of component scores instead of the actual data for the regression, as explained above.

Because the use of component scores instead of the actual data resulted in errors in model specification, with the result that the regression coefficient of tree size/vegetation cover (T/GC) had an erroneous negative sign, Aweto (1981e) did a regression for soil total nitrogen using the actual data instead of the component scores. In this regression, only one of the two highly correlated variables whose correlation exceeded 0.8 was used, in order to eliminate the problem of multicollinearity. For instance, mesophanerophytes had a correlation

of 0.90 with tree density. Hence, the former was not used in the regression. Only five variables were regressed on soil total nitrogen:

1. Tree diameter.
2. Tree density.
3. Vegetation cover.
4. Number of legumes in fallow vegetation.
5. Length of the fallow period.

The regression equation for soil total nitrogen (N) was:

$$\text{N} = 0.73 + 0.06\,\text{TD} + 0.03\,\text{VC} \qquad (6.4)$$

where TD = tree diameter, and VC = vegetation cover.

Only two of the five variables used for the regression were significant in influencing soil total nitrogen, tree diameter and vegetation cover, and these accounted for 43.6 and 6.8% of the variation in soil nitrogen, respectively. It is instructive to note that when the actual data were used for the regression, the relationships between soil total nitrogen on the one hand, and tree diameter and vegetation cover on the other, were positive in the regression equation as would be expected. The use of the actual data instead of component scores has, therefore, helped to eliminate the problem of erroneous signs of regression coefficients. The regression equation indicated that an increase in tree diameter and vegetation cover would lead to an increase in soil organic matter for reasons adduced earlier, and hence in total nitrogen. The length of the fallow period, per se, was not a significant factor influencing total nitrogen accretion in fallow soil. What was more crucial was the vigour of vegetation re-establishment, evidenced in an increase in tree sizes and vegetation cover.

Finally in this sub-section, we shall consider the soil properties that influence the rate of regeneration of fallow vegetation. For the sake of parsimony, only one regression will be considered. A total of 27 soil variables consisting of 15 physical chemical properties of the 0–10 cm soil layer (topsoil), 11 characteristics of the 10–30 cm soil layer (layer II) and pH of the 30–60 cm layer will be used as explanatory variables. In order to achieve parsimony and eliminate the problem of multicollinearity, the 27 soil variables were subjected to principal components analysis. The 27 soil variables, as well as the eight principal components extracted from the original data set, are described in Aweto (1981d). The eight extracted components were:

1. Topsoil nutrient and organic matter status.
2. pH of all three layers.
3. Silt, clay and potassium of layer II.
4. Topsoil clay.
5. Topsoil silt.
6. Total nitrogen level of layer II.
7. Topsoil bulk density and porosity.
8. Layer II sand.

The eight soil components together explained 86.5% of the variance in the original soil data consisting of 27 variables. The reduced rank equation that was obtained by regressing tree size and vegetation cover (T/GC) on the eight soil components was:

$$T/GC = 0.02 + 0.57\ N/OS + 0.49\ CL + 0.26\ BD/P \qquad (6.5)$$

where N/OS is topsoil nutrient and organic matter status, CL is the percentage clay content of the topsoil and BD/P is topsoil bulk density and porosity. Only these three soil components were significant in influencing tree size and vegetation cover in fallow vegetation, and they accounted for 37, 24 and 7% of the variance, respectively. All three soil components together explain 68% of the variation in tree size and vegetation cover. The results of the regression indicated that the nutrient/organic matter status of the topsoil exerted the greatest effects on tree size and vegetation cover, followed by topsoil clay content and, lastly, by the physical status of the topsoil – bulk density and total porosity. The regression indicated that the clay content of the topsoil enhances tree growth in fallow vegetation, implying that in coarse-textured soils, tree regeneration in fallow vegetation would be faster and the trees become bigger on the more clayey sites. It is significant to observe that no property of the subsoil emerged in the regression analysis as a significant factor influencing tree size and vegetation cover in bush fallow vegetation. This does not imply that the physical and chemical properties of the subsoil do not exert any influence on tree growth or the development of fallow vegetation. In fact, trees often absorb nutrients and water from the subsoil and this may be crucial during the dry season when the topsoil layer dries up. They also obtain supplementary nutrient supplies from the subsoil and even from the weathering zone, as tree roots penetrate deep several metres into the subsoil. In the Amazon basin of South America, Lu *et al.* (2002) performed regression separately for each soil layer including the topsoil and several subsoil layers, and their results indicated that the properties of both the topsoil and the subsoil influenced the rate of tree growth in secondary regrowth forests. Generally, the subsoil variables exert less effect on tree growth in fallow than those of the topsoil. Hence, the subsoil variables did not emerge as significant variables in the above regression equation. This is not totally unexpected; as pointed out in Chapter 4, organic matter and nutrient build-up in fallow soil is largely restricted to the topsoil layer. It is also important to observe that in stepwise regression analysis, the variables are entered sequentially in the regression on the basis of their correlation with the dependent variable. Once the variables with the highest correlations have been entered, those with lower partial correlations are discriminated against and may not be entered into the regression equation.

Lu *et al.* (2002) used multiple regression analysis to relate the rate of forest regeneration in the Altamira and Bragantina areas of the Amazon basin to soil physical and chemical properties. Specifically, they regressed the vegetation biomass growth rate on soil texture and chemical properties. They did not, however, state the regression equations for each soil layer and for each area studied. For each of their two areas, they stated one correlation coefficient for each regression. They did not indicate the regression coefficients of the several soil variables entered into the regression equation for each layer. For the Altamira area that is characterized by alfisols, they did not indicate the regression coefficients of eight soil variables including organic matter, nitrogen, the pro-portion of clay and the levels of magnesium, potassium and aluminium in the 0–20 cm soil layer that were significant in explaining fallow biomass growth rate. Clearly, the effects of exchangeable aluminium on vegetation growth sharply contrast with those of magnesium and potassium. Since their objective was to develop regression models in which soil factors were used to explain vegetation growth, it would have been useful to state the regression equations and the regression coefficient of each significant soil variable, and also to indicate whether each regression coefficient was positive or negative, so that one could ascertain whether such soil variables enhance or reduce the rate of regeneration of fallow vegetation biomass. Given the limitations of their use of multiple regression analysis, their analyses only helped to identify significant soil variables which influence the rate of vegetation biomass growth. The results of Lu *et al.* (2002) could not be compared directly with those obtained by the author for forest fallows in south-western Nigeria, as their regression equations were not stated and they did not state the relative amount of the variance in tree growth explained by different soil characteristics.

6.3 Implications of Fallow Soil–Vegetation Interrelationships

In the concluding part of this chapter, some consideration will be given to the implications of the interrelationships between fallow soil and vegetation, both in terms of managing fallow vegetation or the soil under shifting cultivation and replacing natural fallow vegetation with planted fallows of selected species. The results of the correlation analysis showed clearly that while mesophanerophytes and megaphanerophytes are positively correlated with soil nutrients and organic matter, lower plant life forms such as non-phanerophytes that include grasses are negatively correlated with soil nutrients and organic matter. This finding suggests that the higher life forms – megaphanerophytes and mesophanerophytes – enhance the process of organic matter and nutrient accretion in the soil under bush fallow. The search for alternatives to replace the natural bush fallow with planted fallows should therefore focus on tree-based systems and not on grass-based systems, as grasses have a much lower efficacy in restoring soil physical and chemical status. Fortunately, much of the research-based advocacy to enhance the efficacy of the bush fallow has been in the direction of either integrating trees in farmlands to form agroforestry systems, replacing natural fallow vegetation with planted

fallows of tree legumes, or the introduction of alley cropping systems (which involves growing field crops such as maize and cassava between rows of planted trees) (Kang *et al.*, 1984; Floret and Pontanier, 2000; Brady and Weil, 2002; Garrity, 2010). It is instructive to observe that trees usually have several other uses to native farmers in the tropics, apart from helping to improve soil fertility or conserve soils. They yield fruits that can be sold for cash or provide food, and also provide other benefits to the farmer including medicine, fuelwood and poles for building houses or for constructing sheds and fences. Hence small-scale farmers in the tropics are more likely to embrace tree-based agroforestry systems designed to replace or intensify shifting cultivation, than alternative farming systems based on grasses.

The results of both the correlation and regression analyses indicated that two vegetation components, tree size/ground cover and mesophanerophytes/tree density are the most significant components of fallow vegetation that enhance organic matter and nutrient build-up, and increase soil water-holding capacity in fallow soil over time. The implications of these findings are twofold. First, during the course of secondary succession, fallow vegetation with an adequate ground cover is more efficient in restoring soil fertility than fallow with a low ground cover. Second, fallow vegetation in which trees regenerate quickly is more efficacious in restoring soil fertility than one in which trees regenerate slowly or which is not densely stocked with trees. Aweto (1981d) has pointed out that the strong relationship between soil nutrient accretion and vegetation cover should be taken into consideration when designing alternative farming systems in which planted fallows replace the natural bush fallow. Only fallows with an adequate ground cover are effective in protecting the soil against leaching and erosion, and hence in restoring soil fertility. It is gratifying to note that an innovative farming system developed by shifting cultivators in parts of South and Central America, the slash–mulch system (Section 2.4.5), is designed to ensure that the ground surface is adequately covered and protected by the cut slash of fallow vegetation, used as mulch during cultivation.

It is also instructive to note that soil physical properties such as textural composition and soil bulk density and total porosity, influence tree size and hence the rate of regeneration of fallow vegetation. Soil clay content enhances tree size and this implies that in coarse-textured soils, such as occur in the coastal plain of West Africa, tree growth in fallow vegetation would be enhanced on the more clayey sites which would naturally have a higher nutrient and water-retaining capacity. Fallow vegetation is sometimes a source of useful commercial timbers (Richards, 1996). Kennard (2002) estimated that about half of the commercial timber species in the Lomerio area in the eastern plains of Bolivia are long-lived pioneer fallow vegetation species, and she posited that it might be more expedient to use regenerating fallow vegetation as a source of such timber species instead of depending on mature forests, where the regeneration of such timber species might be slow and fortuitous. As with other trees in fallow vegetation, such commercial timber tree species are likely to grow faster and attain mercantile status for exploitation more quickly on the more clayey sites. Soil porosity and bulk density influence tree size in fallow vegetation. Soil bulk density is inversely related to tree growth, as an increase in bulk density leads to soil compaction, while an increase in total porosity enhances tree growth by improving soil physical status, especially by increasing soil permeability to air and water and by reducing surface runoff. When clearing land for arable farming or commercial timber plantations it is important to decide on appropriate methods that have minimal adverse effects on soil physical status. In most forms of shifting cultivation, land is cleared using simple implements such as axes and machetes which do not compact the soil or degrade its physical status. This enhances the resilience and sustainability of farming systems based on the bush fallow.

References

Aide, T.M., Zimmerman, J.K., Herrera, L., Rosario, M. and Serrano, M. (1995) Forest recovery in abandoned tropical pastures in Puerto Rico. *Forest Ecology and Management* 77, 77–86.

Aweto, A.O. (1981a) An ecological study of forest fallow communities in the Ijebu-Ode/Shagamu area, southwestern Nigeria. *Singapore Journal of Tropical Geography* 2, 1–8.

Aweto, A.O. (1981b) Secondary succession and soil fertility restoration in south-western Nigeria. I. Succession. *Journal of Ecology* 69, 601–607.

Aweto, A.O. (1981c) Secondary succession and soil fertility restoration in south-western Nigeria. II. Soil fertility restoration. *Journal of Ecology* 69, 609–614.

Aweto, A.O. (1981d) Secondary succession and soil fertility restoration in south-western Nigeria. III. Soil and vegetation interrelationships. *Journal of Ecology* 69, 957–963.

Aweto, A.O. (1981e) Total nitrogen status of soils under bush fallow in the forest zone South western Nigeria. *Journal of Soil Science* 32, 639–642.

Aweto, A.O. and Dikinya, O. (2003) The beneficial effects of two tree species on soil properties in semi-arid savanna rangeland in Botswana. *Land Contamination and Reclamation* 11, 339–344.

Brady, N.C. and Weil, R.R. (2002) *The Nature and Properties of Soils.* Prentice Hall, Upper Saddle River, New Jersey.

Congdon, R.A. and Herbohn, J.L. (1993) Ecosystem dynamics of disturbed and undisturbed sites in north Queensland wet tropical rain forest. I. Floristic composition, climate and soil chemistry. *Journal of Tropical Ecology* 9, 349–363.

Dean, W.R.J., Milton, S.J. and Jeltsch, F. (1999) Large trees, fertile islands, and birds in arid savanna. *Journal of Arid Environments* 41, 61–78.

Donfack, P., Floret, Ch. and Pontanier, R. (1995) Secondary succession in abandoned fields of dry tropical northern Cameroon. *Journal of Vegetation Science* 6, 499–508.

Elberling, B., Toure, A. and Rasmussen, K. (2003) Changes in soil organic matter following groundnut-millet cropping at three locations in semi-arid Senegal, West Africa. *Agriculture, Ecosystems and Environment* 96, 37–47.

Ewel, J.J., Mazzarino, M. and Berish, C.W. (1991) Tropical soil fertility changes under monocultures and successional communities of different structure. *Ecological Applications* 1, 289–302.

Floret, Ch. and Pontanier, R. (2000) *La jachere en Afrique tropicale.* John Libbey Eurotext, Paris.

Garrity, D. (2010) Growing crops under canopy. *Spore* 150, 12.

Golley, F.B. (1986) Chemical plant–soil relationships in tropical forests. *Journal of Tropical Ecology* 2, 219–229.

Hauser, D.P. (1974) Some problems in the use of stepwise regression techniques in geographical research. *Candian Geographer* XVIII, 148–158.

Johnston, M.H. (1992) Soil–vegetation relationships in a tabonuco forest community in the Luquillo mountains of Puerto Rico. *Journal of Tropical Ecology* 8, 253–263.

Kang, B.T., Wilson, G.F. and Lawson, T.L. (1984) *Alley Cropping: A Stable Alternative to Shifting Cultivation.* International Institute of Tropical Agriculture, Ibadan, Nigeria.

Kennard, D.K. (2002) Secondary forest succession in a tropical dry forest: patterns of development across a 50-year chronosequence in lowland Bolivia. *Journal of Tropical Ecology* 18, 53–66.

Lafon, C.W., Huston, M.A. and Horn, P. (2000) Effects of agricultural soil loss on forest succession rates and tree diversity in east Tennessee. *Oikos* 90, 431–441.

Lu, D., Moran, E. and Mausel, P. (2002) Linking Amazonian secondary succession forest growth to soil properties. *Land Degradation and Development* 13, 331–343.

Mesquita, R.C.G., Ickes, K., Ganade, G. and Williamson, G.B. (2001) Alternative successional pathways in the Amazon Basin. *Journal of Ecology* 89, 528–537.

Moran, E.F., Brondizio, E.S., Tucker, J.M., da Silva-Forsberg, M.C., McCracken, S. and Falesi, I. (2000) Effects of soil fertility and land-use on forest succession in Amazonia. *Forest Ecology and Management* 139, 93–108.

Moss, R.P. (1969) The ecological background to land use studies in tropical Africa, with special reference to the West. In: Thomas, M.F. and Whittington G.W. (eds) *Environment and Land Use in Africa.* Methuen & Co. Ltd, London, pp. 193– 238.

Poole, M.A. and O'Farrell, P.N. (1971) The assumptions of the linear regression model. *Transactions of the Institute of British Geographers* 52, 145–158.

Richards, P.W. (1996) *The Tropical Rain Forest.* Cambridge University Press, Cambridge, UK.

Roder, W., Phengchanh, S. and Maniphone, S. (1997) Dynamics of soil and vegetation during crop and fallow period in slash-and-burn fields of northern Laos. *Geoderma* 76, 131–144.

Schade, J.D. and Hobbie, S.E. (2005) Spatial and temporal variations in islands of fertility in the Sonoran desert. *Biogeochemistry* 73, 541–553.

Swaine, M.D. (1996) Rainfall and soil fertility as factors limiting forest species distribution in Ghana. *Journal of Ecology* 84, 419–428.

Tasser, E., Walde, J., Tappeiner, U., Teutsch, A. and Noggler, W. (2007) Land-use changes and natural reforestation in the Eastern Central Alps. *Agriculture, Ecosystems and Environment* 118, 115–129.

Uhl, C., Buschbacher, R. and Serrao, E.A.S. (1988) Abandoned pastures in eastern Amazonia. I. Patterns of plant succession. *Journal of Ecology* 76, 663–681.

Vasquez-Mendez, R., Ventura-Ramos, E., Oleschko, K. Hernandez-Sandoval, L., Parrot, J.-F. and Nearing, M.A. (2010) Soil erosion and runoff in different vegetation patches from semiarid Central Mexico. *Catena* 80, 162–169.

Wezel, A., Rajot, J.–L. and Herbrig, C. (2000) Influence of shrubs on soil characteristics and their function in Sahelian agro-ecosystems in semi-arid Niger. *Journal of Arid Environments* 44, 383–398.

Yimyam, N., Rerkasem, K. and Rerkasem, B. (2003) Fallow enrichment with pada (*Macaranga denticulata* (Bl.) Muell. Arg.) trees in rotational shifting cultivation in northern Thailand. *Agroforestry Systems* 57, 79–86.

7 Ecological Succession Theory and Models

Chapters 4–6 focused mainly on empirical studies on soil fertility restoration and secondary succession in farmlands under shifting cultivation, and the interrelationships between the soil and vegetation components of successional communities. This chapter considers the theory of succession and models that have been used to study the process, in order to gain insight into the very complex changes the plant community undergoes during the process of ecological succession. These considerations not only complement Chapters 4–6, but more importantly will serve as a theoretical and explanatory framework for unravelling the processes both fallow vegetation and the soil undergo during the fallow period. Such theoretical knowledge is important, not merely for the sake of acquiring knowledge per se, but more significantly to obtain a holistic knowledge of shifting cultivation or slash-and-burn agriculture, and especially the ecological and pedological aspects of this system. This will enable the system to be modified, intensified, or even replaced – when ecological and socio-economic conditions permit – with a more intensive and sustainable system of agriculture. Clearly, one can only make sound decisions to intensify or replace shifting cultivation with more sustainable systems of agriculture if such decisions are firmly anchored on a theory of the processes of succession and the mechanism of how soil fertility is restored under bush fallow vegetation. Similarly, the techniques of farming or farming systems designed to replace one of the several varieties of shifting agriculture must be based on knowledge of secondary succession and the process of soil fertility restoration under bush fallow. Otherwise, such novel techniques or adaptations of the current methods used by shifting cultivators are doomed to failure from the start. In explaining the results of empirical studies on secondary succession and soil restoration under bush fallow in Chapters 4–6, reference will have been made to some theories or models without stating them explicitly. In this chapter and in Chapter 8, such theories and models will be considered. We also consider some models or theories which may not have been explicitly referred to in the preceding chapters, if they provide useful insights into the processes of secondary succession or soil fertility restoration under natural or planted fallows, as the latter types of fallows are being proposed as alternatives to the natural ones.

This chapter will specifically consider the theories and models used for studying ecological succession while Chapter 8 will concentrate on the models of soil fertility restoration. Obviously, the two processes are intricately and intrinsically interrelated and some models or theories can be applied to both. However, for the sake of convenience, some distinction will be made between them. Chapter 5 considered the changes fallow vegetation undergoes during secondary succession, without discussing the basic concepts of ecological succession, and it is appropriate here to consider the basic features of ecological succession before treating ecological succession theory. This is important, as secondary succession, a major theme of this book, is an aspect of ecological succession.

7.1 Definition, Basic Features and Concepts of Ecological Succession

Ecosystems are dynamic ecological entities that develop in response to inputs of matter and energy. Ecological succession is an important process of ecosystem development over time, which ultimately results in the establishment of a relatively stable community in a state of dynamic equilibrium with the controlling environmental factors. Odum (1971) defined ecological succession as an orderly process of community development that involves changes in species populations and community structure, and modification of the environment over time, and ultimately terminates in a stabilized ecosystem. The above definition implies that several communities replace one another on the same site or area during the course of ecological succession or ecosystem development, until a terminal community, which is in a state of dynamic equilibrium with the environment, or (in some cases) some controlling biotic factor, is ultimately established. The first community to be established in an area, whether on land or in water, is called the pioneer community; the terminal, relatively stable community, that is in a state of dynamic equilibrium with the controlling environmental factor, is called the climax. The short-lived communities that replace one another on the same site before the climax stage is attained are collectively known as seral stages, or the sere. The various seral stages and the climax are characterized by different species populations as the floral and the faunal composition of the ecosystems changes markedly during the course of ecological succession, prior to the climax stage. Usually, the first community that initiates the process of

succession is simpler in structure and floral composition than the community that subsequently replaces it and, over time, the structure of the community becomes progressively more complex and the floristic and the faunal composition more diversified as succession progresses toward the climax stage. Ecological successions are usually directional and predictable (Margalef, 1968; Odum, 1975). This implies that successions usually lead ultimately to the terminal climax state, unless the process of ecosystem development is interrupted by some interposing biotic factor, such as burning or heavy grazing by livestock or wildlife, which may prevent or arrest successional development of the ecosystem. The basic features of ecological succession will be considered in Section 7.3.1 when considering the theoretical viewpoint of F.E. Clements on ecological succession.

7.2 Types of Ecological Succession

Ecological succession is initiated when plants invade and colonize an exposed site; that is, an area that is devoid of vegetal cover. Such an exposed surface may be dry land or an open water surface. When plants have become well established in such an area, animals can subsequently migrate and colonize it, as it is already inhabited by the plants that provide food for the first and later sets of animal migrants. Classification of types of ecological succession is usually based on whether the exposed surface on which the process of ecological succession takes place has been previously colonized by plants. In this regard, two main types of ecological succession can be recognized, namely: (i) primary succession; and (ii) secondary succession. The former is initiated on surfaces which have not previously been colonized by higher land plants. Such surfaces may be inselbergs or rock outcrops exposed as a result of erosion, or they may be produced as a result of lava outpouring or the deposition of a thick deposit of volcanic ash on land; from the silting and recession of lake shores and estuaries; or by deposition of sediments on the beds of rivers. Long-term climatic changes can also cause a drop in sea level, exposing land that was formerly beneath the sea and enabling it to be colonized by plants, and this would result in primary succession. Similarly, tectonic movements can result in the uplift of a part of the sea floor to form dry land which could be colonized by plants to initiate the process of primary succession.

In contrast, secondary succession is initiated on land that previously bore vegetal cover, which is subsequently cleared by humans or accidentally destroyed by natural catastrophic events such as tornadoes or thunderstorms. Secondary succession is central to the agricultural system of shifting cultivation or slash-and-burn agriculture, as native farmers in the tropics depend on the natural process of secondary succession to restore the fertility of soils which declined during cropping. It can be readily observed in most parts of the tropics where farmers practise shifting cultivation, as the farmers usually allow the fields to revert to natural bush fallow vegetation after harvesting their crops and allow the regenerating vegetation to develop and to help accumulate organic and nutrients in the topsoil. This book is primarily concerned with secondary succession, a process that usually proceeds at a much faster rate than primary succession. This is obviously because the latter occurs on relatively sterile surfaces, usually without a well-developed soil medium. Such surfaces may include bare rock surfaces or sand dunes on which soil has not formed. Consequently, primary succession usually takes several hundreds or thousands of years for succession to attain the climax stage. In contrast, secondary succession usually takes a couple of decades to attain the climax. In West Africa, Hopkins (1974) estimated that it takes about 100 years while in the Upper Rio Negro area of Colombia and Venezuela, Saldarriaga *et al.* (1988) estimated that it takes about 190 years for climax rain forest to be established, following the initiation of secondary succession on previously cultivated land.

Another distinction that can be made between types of ecological succession is based on the nature of the surface on which the process of ecosystem development is initiated. Successions that begin on surfaces such as dry land or rock surfaces are collectively referred to as xerarch successions, while those that begin on water surfaces such as lakes or ponds and estuaries are called hydrarch succession.

7.3 Theoretical Viewpoints on the Nature, Processes and Causal Factors of Succession

Various theoretical viewpoints and hypotheses have been proposed to explain the nature, mechanism and causal factors of succession. These range from the hypothesis of Clements (1916), which was the most comprehensive attempt to formulate a general theory of succession, to those of Egler (1954) or Tilman (1985). The latter two focused on specific aspects of succession, such as the nature of soil seed bank at or prior to the inception of succession, or competition between species in relation to the long-term supply of limiting resources. In the subsections that follow, the following theoretical viewpoints on succession will be considered:

1. The Clementsian hypothesis.
2. The individualistic theory.
3. The resource-ratio hypothesis.
4. The initial floristic composition.
5. Facilitation, inhibition and tolerance.
6. Initial soil-substrate conditions.
7. The nucleation model.
8. The spatio–temporal model.

7.3.1 The Clementsian holistic hypothesis

Clements (1916) was one of the earliest to propose a general theory of succession, which not only encompassed the nature and mechanism of succession initiated on different surfaces, but also the nature of the climax; that is, the terminal phase of succession. His theory applies to both primary and secondary succession and Clements can justifiably be regarded as one the leading pioneers of succession theory. His work is still relevant to an understanding of the complex ecological process of succession, although the validity of some of his theoretical postulations has been challenged in the light of empirical findings. His hypothesis is also known as the holistic viewpoint, because Clements compared the process of succession with the life cycle of an organism and maintained that the plant community, including the seral and climax stages, is similar to an organism. He observed that just as an organism is an entity, so the plant community is also an integrated ecological entity which functions as a whole, rather than a mere assemblage of plants that are independent of one another. Using the organism analogy framework, Clements maintained that the plant community is born, grows, matures, reproduces and finally dies. According to this analogy, the process of succession is developmental, linear, directional and terminates in the climax – the mature stage of ecosystem development. According to Clements, the climax community reproduces by repeating its successional development, following essentially the same stages as it did during its first process of colonization and development on the site. If a climax forest is cleared or part of it is destroyed by a violent storm and it is subsequently left to regenerate undisturbed, the forest repeats the successional development that ultimately leads to the re-establishment of a mature forest.

In his 1916 publication, Clements identified the following processes as the main components of ecological succession:

1. Nudation.
2. Migration.
3. Ecesis.
4. Reaction.
5. Competition.
6. Stabilization.

The first four processes operate during the beginning of ecological succession (referred to as the initiation stage), while the fifth and the sixth operate during the middle and terminal stages of succession (Fig. 7.1). Nudation is the process of creating a bare area, an essential precondition for the initiation of the process of succession, while migration involves the dissemination of seeds and propagules to the site to begin the process of plant recolonization. Ecesis refers to the germination of seeds and the growth and reproduction of plants on the site, while reaction is the modification of the environment (including the substrate and the

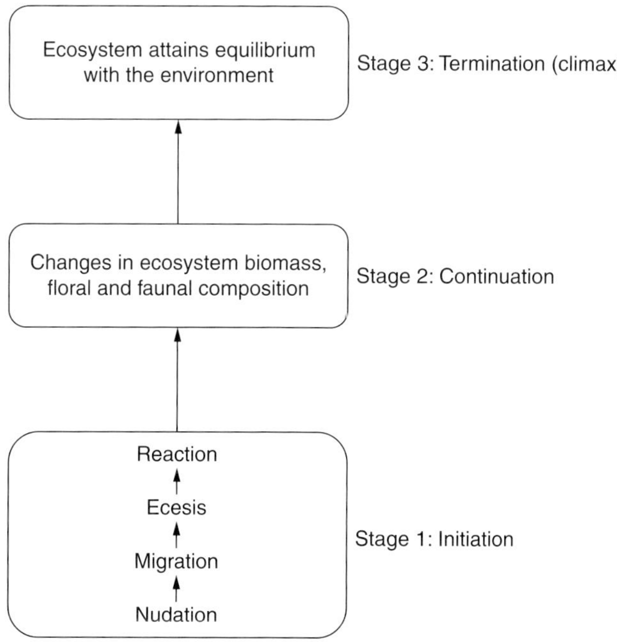

Fig. 7.1. The main stages of succession identified by F.E. Clements.

air) by the plants currently established in an area which was previously devoid of vegetal cover. Changes brought about in the soil, such as the addition of organic matter and nutrients and improvement in soil water-holding capacity, make possible the invasion by higher plant life forms that make more demands on environmental resources such as light, water and nutrients. This results in increased interspecific and intraspecific competition, and in turn this leads to the competitive replacement of some species populations by others. This process continues until the terminal stage, when the climax is established and the ecosystem becomes relatively stable, unlike the preceding transient stages. The stages that Clements identified as characterizing succession are still valid for both primary and secondary succession, although the details of the dynamics of species populations vary from one ecological zone to another, and within the same zone in response to factors including the nature of the substrate, the intensity of previous disturbance and the nature of the surrounding vegetation.

According to Clements (1916), the climax stage is attained when the reaction of the community currently occupying an area can prevent the invasion by a population of another species. This leads to the community perpetuating itself on the site indefinitely, at least until long-term climate change makes it gradually lose its identity and it is consequently replaced by another community. The successional communities that precede the climax usually modify the site, including the soil, such that it becomes more suitable for invaders that ultimately replace them. For instance, the chemical exudates secreted by the roots of certain plants may make the site more suitable for would-be invaders than for those currently occupying the site. In contrast, when the climax is attained, the plant species characterizing the community perpetuate themselves on the site and thereby prevent the invasion of other species populations. A fundamental hypothesis postulated by Clements regarding the nature of the climax is that all successions taking place in a given macro-climatic region, irrespective of the nature of the substrate on which they were initiated, ultimately terminate in the same climax that is determined by the regional climate. This is the monoclimax concept that

will be discussed in Section 7.5.1. This concept is based on the view that the regional climate is the primary determinant of the nature of the climax and that, over time, successions initiated in swamps or water surfaces, and sand dunes or mesic areas, terminate in the same climax stage. Clements observed that the overriding effect of the regional climate over a long period would eliminate differences in successional communities that reflect variations in topography or substrate conditions.

7.3.2 The individualistic theory

This was proposed by Gleason (1917, 1939), whose views on the nature of the plant community and succession represent a radical departure from those of F.E. Clements. Gleason considered the organism analogy framework of Clements for analysing ecosystems dynamics during succession to be misleading and inappropriate. In the view of Gleason, the plant community is individualistic, implying that it is unique and that no two plant communities are similar or identical. Unlike Clements, who stressed that all successions within a climatic region terminate in the same climax stage, Gleason held that ecological successions are neither directional nor predictable. Hence, he emphasized the view that succession is multidirectional, resulting in several end points and not necessarily the same climax community (which is determined by the climate, as Clements postulated). Also, Gleason questioned the validity of long-lasting climax communities because all plant communities (including the mature or climax stages) are in a constant state of flux, in response to environmental changes which may occur at varying time scales ranging from a few years or decades to several thousands of years. During the development of seral communities, several random or stochastic factors come into play. These include the arrival of seeds or propagules; availability of favourable sites for such seeds to germinate and establish themselves as new plant individuals; and site disturbances such as the creation of gaps as a result of treefall or destruction of parts of a forest by violent storms. These influence community dynamics, resulting in differences in the composition of seral or climax communities at the local level.

Presumably due to the towering status of F.E. Clements as one of the foremost ecologists during the first half of the 20th century, Gleason's views on succession were not appreciated by his contemporaries. Recent studies on primary succession, especially those by Whittaker *et al.* (1989) on the recolonization of the Krakatoa Islands in Indonesia, have tended to lend empirical support to Gleason's individualistic or disequilibrium theory of succession. The volcanic island of Krakatoa was totally destroyed as a result of a violent volcanic eruption in 1883 which reduced the original island to a submarine caldera. Subsequently, as a result of renewed magma outpouring, three volcanic islands formed on the site of the original island. Whittaker and colleagues studied these three islands and postulated that the pattern of forest vegetation establishment was patchy and not identical, suggesting that random factors such as treefall, patterns of seed arrival and subsequent establishment may have affected succession. The findings of Whittaker *et al.* (1989) suggested that the pattern of re-colonization of the three neighbouring islands of Krakatoa was multidirectional, rather than unidirectional as postulated by Clements.

7.3.3 The resource-ratio hypothesis

This recent and elegant theory on plant succession was proposed and developed by Tilman (1985), drawing mainly on results of his empirical research on primary succession on nutrient-deficient sites, especially sandy plains. It applies mainly to species that come into dominance during some stage of succession. The theory is based on two key elements, namely: (i) inter-specific competition for resources, especially for soil nutrients and for light; and (ii) the pattern of supply of limiting resources over a long period. The latter factor, relating to the relative availability of limiting resources over time, was referred to by Tilman (1985) as resource–supply trajectory. The central proposition of the resource-ratio hypothesis is that succession, and hence the replacement of one species population by another, is largely due to the temporal gradient in the relative availability of limiting resources. He further posited that succession would be directional or repeatable only if the resource–supply trajectory is directional or repeatable.

The hypothesis is based on the premise that each plant species is a superior competitor for a particular environmental resource, which may be limiting during a certain period of succession, and it predicts that the floristic composition of the community would change whenever the relative availability of two or more limiting resources changes. On well-drained land, with adequate soil depth and adequate annual rainfall – that is, on mesic terrestrial sites and habitats – the limiting resources are usually soil nutrients, particularly soil nitrogen and light. These resources are usually and naturally inversely related, as habitats with poor nutrient-deficient soils are usually characterized by high light availability, as the vegetation is usually low in respect to vertical development, while areas with fertile soils that are rich in nitrogen are characterized by tall and luxuriant vegetation, with low levels of light at ground level. A superior competitor for a particular resource is a species which tolerates low levels of the resource, or has the ability to meet its requirement of the resource, in spite of the fact that it is present in low levels. In addition, a superior competitor for a particular resource, such as a soil nutrient, is an inferior competitor for other environmental resources such as light. This is because of morphological or adaptive trade-offs in respect of resource utilization by plant species. Tilman (1990) observed that plant species that are superior competitors for soil nitrogen have large root biomass, which enables them to effectively exploit nitrogen in the soil to their advantage, although it is present in low levels. As a result of trade-offs, such species will of necessity be inferior competitors for light, as the proportion of their above-ground biomass will be small compared to their below-ground biomass. At the inception of succession on poor soils, especially those that are deficient in soil nitrogen, plant species which are superior competitors for soil nitrogen will become established in the site as the dominant species, as they can tolerate low levels of soil nitrogen. As succession progresses, soil nitrogen status will improve over time and light will become limiting in the habitat (as plant biomass increases), particularly at the ground level. The pioneer or early successional species to be established on the site are inferior competitors for light because of their comparatively small above-ground biomass; this is crucial for competing for light, just as a large root biomass is of pivotal significance for competing for soil nutrients. Hence, the earlier species that came into dominance in the site will be replaced by the later successional species which are more effective competitors for light. According to the predictions of the resource-ratio hypothesis, one species population will be replaced by another when their resource requirements are inversely related over time, but the equilibrium stage will be characterized by mixed populations if each species is limited by a different resource.

It would seem that one can use this model to explain the sequential changes that take place in plant species population during the course of secondary succession, particularly the replacement of pioneer species such as grasses, sedges and other herbs by later successional plants such as trees. However, Tilman (1985) has observed that the applicability of the model to terrestrial plant communities depends essentially on the identification of the resources that are limiting, and this has to be determined through controlled experimentation. Herein lies a major limitation of the resource-ratio hypothesis in terms of its applicability to bush fallow or slash-and-burn farming systems, which depend on natural succession on farmers' fields at the end of cultivation to rejuvenate soil fertility. Shifting cultivators often leave their land fallow as a result of the rapidity of weed regrowth which stifles the growth and productivity of planted crops. At the inception of the fallow period, soil fertility may still be high, and in some cases may be higher than that of pre-burn forest soil. This is usually the case where mature rain forest is cut, burnt and cultivated for a few years prior to fallowing, so soil nutrients may not be limiting at the inception of succession. Also, pioneer/ early successional and later successional species have contrasting light requirements. The former require a lot of light and become established in open spaces, while the latter require shade in order to be established in bush fallow during succession. Consequently the two groups of species are not competing for light at the inception of succession.

The two factors listed above imply that the sequence of change in plant species during secondary succession may not be due primarily to the relative ability of the species to compete for soil and other resources such as light, as proposed by the resource-ratio hypothesis.

7.3.4 The initial floristic composition hypothesis

This hypothesis, proposed by Egler (1954), focuses mainly on the dynamics of the replacement of one species population by another during secondary succession. According to this hypothesis, the seeds and vegetative propagules of plant species populations that come into dominance during secondary succession are usually present in the soil prior to the inception of the process of ecosystem development. Such seeds and propagules develop, germinate, sprout up and give rise to species which may later feature prominently at some stage during succession. It is important to stress that the seeds and propagules originally present in the soil before the initiation of succession, together with the seeds that may be later dispersed to the site, largely determine the sequence of replacement of species during secondary succession. Although the seeds of annual or perennial grasses, forbs, shrubs and trees are present in the soil before succession begins, these various plant life forms come into dominance at different periods during the course of succession. For instance, annual grasses and other annual herbs are usually the first to come into dominance because they have a much shorter life cycle and respond more quickly to the cessation of human disturbance than perennials. They are usually followed by perennial grasses which respond more quickly to cessation of human activities on a site than shrubs and trees. The seeds of trees and shrubs usually take a longer time to germinate than the seeds of annual or perennial grasses, and so shrub and tree species will be the last to germinate and establish themselves on a site devoid of vegetation. In some cases the seeds of trees remain dormant in the soil for a long time, as they require shade and certain micro-climatic conditions to germinate. As these do not obtain in the open, they may not germinate, as the seeds of annual and perennial grass species do.

The initial floristic composition is relevant in explaining the sequence of species dominance during secondary succession in shifting cultivation areas, not only because seeds of flowering plants are present in farmers' fields prior to abandonment, but also because farmers usually leave certain trees on the farm during cultivation to provide fruit, shade, medicine, or other valuable products. When the cultivated field is temporarily abandoned by the farmer, these trees become a source of seeds or vegetative propagules for the subsequent re-establishment of vegetation in the site during the fallow period. Besides, trees left on shifting cultivators' fields enhance nutrient and soil organic matter accumulation under their canopies, making such sites conducive for the establishment of their seedlings or the seedlings of other tree species. The study of Carrière *et al.* (2002) in southern Cameroon has shown that trees left by the farmer during cultivation subsequently influence the pattern of forest re-establishment after cropping. They observed that succession was more rapid under the canopies of remnant trees deliberately preserved by farmers on their fields during cropping. It is also important to note that shifting cultivators deliberately leave live stumps of trees on their farms during cultivation. Such stumps coppice during cropping and the coppices are pruned by the farmer and returned to the soil as mulch. At the end of cultivation, coppice regrowth from live stumps is unhindered, and such stumps later develop into trees, or in some cases suckers may regenerate from their roots, helping to revegetate the field. In addition, after crops such as cassava become mature, the farmer may not bother to weed the farm before harvesting them, a couple of months later. During this period, forbs and grasses usually invade the field. What is obvious from the foregoing is that the trees and live stumps left on the cultivated field by the farmer are not only an important element of the flora of the vegetation to colonize an abandoned field during the fallow period, but are also important sources of seeds and propagules for colonizing farmers' fields. Such trees retained by the farmer, and other plants that invade the fields before crops are harvested,

are perhaps more appropriately described as 'the initial floristic composition', rather than the seeds and vegetative propagules in the soil to which Egler (1954) applied the term. The 'initial floristic composition' hypothesis of Egler can be better referred to as the 'initial soil seed bank condition' hypothesis, as the theoretical postulate relates more to the latter than to the former.

7.3.5 The facilitation, inhibition and tolerance hypotheses

These three models were proposed by Connell and Slatyer (1977) to explain why species populations replace one another during the course of ecological succession, and also why some species populations may be able to prevent the invasion of the site they currently occupy by other species populations. The three alternative hypotheses explain the dynamics and the relative stability of species populations that colonize an exposed bare land surface.

Basically, the hypothesis of facilitation stipulates that the earlier successional communities established on a bare land surface modify the soil and substrate by improving its nutrient and water-holding status, and also improve the micro-climate over time. This makes it possible for higher and more demanding plant life forms to invade and colonize the site and ultimately replace the species that preceded them. This observation is clearly in line with empirical findings on secondary succession, which indicate the replacement of grasses and forbs that initiate succession by trees, when soil nutrient status is improved and the micro-climate ameliorated by the early colonizers.

In inhibition, the species currently on a site have competitive superiority over would-be invaders and are able to keep them at bay, sometimes by secreting chemical substances (e.g. root exudates) that inhibit the growth of other species – a phenomenon known as allelopathy. The leaf litter and barks of *Eucalyptus* are known to inhibit the growth of other species (Del Moral and Muller, 1970). The root exudates of eucalypts may also have allelopathic effects on other plants. The allelopathic effect of the leaf litter, and possibly of root exudates, largely explains why the floor of plantations of *Eucalyptus camaldulensis* (for which allelopathic effect has been reported) is usually bare, with minimal if any undergrowth.

In tolerance, different species and plant life forms may be able to coexist on the site for a long time, as they do not have adverse effects on one another, at least until the population or biomass of the species that are superior competitors for light or soil nutrients becomes very large. The tolerance hypothesis has also been supported by empirical findings on succession in natural bush fallows. In the forest zone of West Africa, for instance, perennial grasses such as *Panicum maximum* and *Andropogon tectorum*, and forbs such as *Chromolaena odorata* may coexist with trees such as *Baphia nitida*, *Anthonotha macrophylla* and *Antiaris toxicaria* for several years, until the canopy of the regenerating young secondary forest closes up, shading out of the grasses and forbs.

7.3.6 Initial soil–substrate conditions

As far back as the early decades of the 20th century, F.E. Clements recognized the importance of the substrate in ecological succession. He made a distinction between xerarch successions, which are initiated on dry land or bare rock surfaces characterized by soil moisture deficit, and hydrarch successions, which are initiated on surfaces with water, such as ponds, lakes and estuaries. Several empirical studies on primary or secondary succession have indicated that the nature of the soil influences the sequence of species that replace one another in a bare exposed site. Mesquita *et al.* (2001) have shown that the extent of soil nutrient depletion during the previous human disturbance influenced the floristic composition of the early successional stages. The study of Gehring *et al.* (1999) has shown that species in fallow vegetation respond differently to fertilizer application, and that nutrient availability at the inception of the fallow period may be a crucial factor determining the nature of the pioneer community, and hence which species come into dominance at the inception of succession. The nature of the pioneer community largely depends on the nature of the soil or the substrate. If a well-developed soil medium characterizes the bare land surface, as is usually the case with secondary succession, then higher land plants – flowering plants – initiate the process of succession. On the other hand, if the surface is characterized by a bare rock outcrop, without a soil layer, flowering plants cannot initiate the process of succession as there is no medium in which plant roots can be established. On such sites, lower plants such as mosses and lichens begin the process of succession. On the sand dunes that form on the shores of lakes or seas, grasses usually begin the process of succession, as the dunes usually lack organic matter and an adequate supply of essential plant nutrients. The foregoing suggests that soil and substrate conditions not only influence the nature of the pioneer community, but may also be important in ordering the sequence of replacement of one species population by another during the succession. Soil conditions include not only the physical, chemical and biological properties of the soil, but also the range and abundance of viable seeds and plant vegetative propagules, such as suckers in the soil.

7.3.7 The nucleation model

This was proposed by Yarranton and Morrison (1974) to describe primary succession on the sand dunes on the shores of Lake Huron, Ontario, Canada, which ultimately results in the establishment of mixed forests of pines and oaks. The process of primary succession, which takes over 3000 years to attain the climax forest stage, is initiated when grasslands with scattered juniper trees, *Juniperus virginiana*, are established on recent sand dunes on the shores of the Lake Huron. The juniper trees help to accumulate organic matter under their canopies, making it possible for seedlings of oaks to invade and become established under their canopies. Each juniper tree serves as a nucleus for the recruitment and establishment of oaks, which grow to become trees. As more oak trees are established, the juniper trees are ultimately eliminated and each nucleus of oak trees expands to form a clump of trees. This expands as more oaks and other trees become established in the wood clump due to the improvement of soil organic matter, and chemical, physical and biological status. Over time, adjacent wood clumps coalesce as they expand laterally until they merge to form a forest on the sand dunes. The process of nucleation which Yarranton and Morrison (1974) observed for succession on sand dunes is not unique to primary succession on nutrient-deficient substrates, neither is it restricted to the temperate regions. It has also been reported for secondary succession following shifting cultivation in Cameroon by Carrière *et al.* (2002), who reported that remnant trees on farmers' fields serve as nuclei for recruitment of trees during secondary succession. The study of Archer *et al.* (1988) has demonstrated that the process of nucleation is central to autogenic succession in subtropical savanna in Texas, USA, which results in the transformation of grassland vegetation into thorn woodland. In addition, as will be pointed out in the next subsection, remnant trees in farmland in shifting cultivation areas serve as foci for the regeneration of other trees – a process which is akin to, if not identical to, the process of nucleation.

7.3.8 The spatio–temporal model

This model is appropriate for analysing changes that take place in vegetation during succession and the changes that take place in the soil that result in the restoration of soil fertility. Essentially, it involves the integration of the concept of the single-tree influence circle (Zinke, 1962; Boettcher and Kalisz, 1990) and the nucleation model of Yarranton and Morrison (1974) described above. It also encompasses the basic ideas of the core–periphery model described by de Blij and Muller (2006). Essentially, the central idea of this model is that individual trees and shrubs serve as centres or foci for the recruitment of other trees and the restoration of soil fertility. Trees and shrubs accumulate organic matter and nutrients underneath their canopies and also reduce soil temperatures and improve soil water-holding capacity; this makes areas under tree canopies more suitable for tree seedling establishment than the open areas in farmers' fields that are outside their canopies. The spatio–temporal model, which encapsulates the basic propositions of the core–periphery model, will be described in the next chapter.

7.4 Changes that Occur in Ecosystems during Succession

The changes that take place in the vegetation and in the soil during secondary succession were considered in Chapters 4 and 5, using the results of several empirical studies drawn from different parts of the tropics. The changes that take place in soil and vegetation components are symptomatic of changes taking place in the entire ecosystem. The build-up of organic matter and nutrients in fallow topsoils over time, for instance, is indicative of the closing of organic matter and nutrient cycles which have become more efficient in conserving nutrients as succession progresses toward the climax. Similarly, the changes that take place in the vegetation and in the animal populations reflect changes that occurred in the substrate, especially in the soil. The process of succession has been aptly characterized by Odum (1971) as a process of ecosystem development. In Chapters 4 and 5, only changes that occur in soil and vegetation during secondary succession were discussed directly. In this subsection, the changes that take place in ecosystems during both primary and secondary succession will be briefly considered. These changes, which will be treated in outline form, are summarized in Table 7.1.

As pointed out earlier, one of the most noticeable changes that occurs during ecological succession is the one that takes place in the floristic composition of the plant community. This is associated with a change in the animal populations of the ecosystem that depend directly or indirectly on the plants for sustenance. The change in species populations is rapid during the early stages of succession but slows down considerably as the ecosystem matures and approaches the climax stage. Contemporaneous with the change in species populations over time is the change in plant life form composition. Pioneer communities are usually characterized by a preponderance of lower plant life forms such as annuals and forbs, but over time these are replaced by higher plant life forms, especially trees. Another major change in the ecosystem during the course of succession, that even the casual observer will not fail to notice, is an increase in the biomass of plants and of animals over time. The rate of increase in the biomass of autotrophs (green

Table 7.1. Changes in the ecosystem during ecological succession.

Ecosystem characteristic	Earlier successional stages	Later successional stages
Floral and faunal composition	Changes quickly	Changes slowly
Biomass	Total biomass is small, but increases rapidly over time	Total biomass is large and increases slowly over time
Species diversity	Low but increases fast over time	Stabilizes at a high level, well before the climax stage
Plant life form composition	Changes rapidly and is characterized by lower plant life forms	Relatively stable and is characterized by higher plant life forms
Soil organic matter	Usually increases rapidly, especially in the topsoil	Stabilizes at a level that is considerably higher than in earlier successional stages
Net primary productivity	High	Low or declining
Nutrient cycles	Inefficient and characterized by nutrient losses	Become more closed and efficient with minimal nutrient losses
Nutrient conservation	Low	High
Energy flow	Food chains are simple and linear	Food chains become interlinked to form webs
Ecosystem stability and resilience	Low	High

plants) is particularly dramatic during the early stages of succession, but declines as the climax stage is approached. Similarly, the biomass of heterotrophs, especially animals, will increase over time and tends to stabilize at a level (much higher than in early successional stages) that can be sustained by the ecosystem at the climax stage. As the floral and faunal composition of the ecosystem becomes more diverse over time, the simple and linear food chains that characterize early stages of succession become interconnected to form complex food webs. The changes in the floristic and the life form composition of the plant community are partly due to changes in the ecosystem, particularly an increase in soil organic matter and nutrient status, coupled with an increase in water-holding capacity. These make it possible for the higher and the more demanding plant life forms to invade the ecosystem, and subsequently replace the lower plant life forms and species that characterize the early successional communities. It is also important to note that changes induced by the community presently occupying a site, such as the addition of organic matter and nutrients to the soil referred to above, as well as the chemical substances secreted by the organisms currently in the ecosystem, can determine the sequence of

changes in species populations in the ecosystem over time.

A major determinant of the climax stage, as Odum (1975) pointed out, is the ratio of gross production to respiration. In early successional stages, this ratio usually exceeds 1, resulting in rapid increase in the standing biomass of living matter in the ecosystem over time. Both gross production and community respiration in the ecosystem increase over time, but stabilize at the climax when their ratio approaches 1, and this implies that increase in the biomass would be negligible over time. As biomass of living organisms increases over time, species diversity also increases. The diversity of plant species tends to attain the maximum level early during secondary succession in forest ecosystems, usually within the first 10–20 years, and diversity stabilizes thereafter. The diversity of animal species also increases and then stabilizes, usually before the attainment of the climax stage.

Nutrient cycles become more closed during succession and this results in greater efficiency in nutrient conservation. The build-up of organic matter and nutrients in the topsoils of bush fallow referred to in Chapter 4 suggests that nutrient cycles in the ecosystem are becoming more closed as succession progresses towards the climax. Young fallow vegetation with low ground cover is usually characterized by inefficient nutrient cycles, with considerable losses of organic matter and nutrients occurring from the ecosystem as a result of leaching and erosion. Over time, fallow vegetation becomes more densely wooded and vegetation cover increases appreciably, and this results in suppression or a considerable reduction in the leaching and erosion losses of nutrients. In addition, over time, substantial quantities of nutrients are absorbed from the soil in fallow vegetation and immobilized in trees. Such nutrients immobilized in the standing biomass of fallow vegetation are protected against losses from the ecosystem through leaching and erosion, and this indicates that nutrient conservation in the ecosystem increases over time as inferred by Odum (1969). During primary succession, nutrient conservation in the ecosystem also increases, especially in those ecosystems in which trees are an important component.

7.5 The Climax

As pointed out above, the climax is the terminal stage of succession which culminates in the establishment of a relatively stable ecosystem that is in a state of dynamic equilibrium with the controlling environmental factors. While the early successional communities in terrestrial ecosystems are characterized by a rapid change in their floral and faunal composition, with one species population replacing another in a matter of a few months or years, in the climax stage the same species usually persist in the ecosystem for several decades. The species populations are in a state of equilibrium with the controlling environmental factor, which may be the regional climate, substrate condition, topography or biotic influence, and they are in a state of balance with one another. The latter implies that although the populations of different species may fluctuate slightly over time, in response to fluctuations in the elements of the climate or other environmental factors, the ratios of one species population to other species populations remain fairly constant. The obvious implication of this is that the population of one species does not rapidly increase over time at the expense of the others so as to become overwhelmingly dominant. As Mueller-Dombois and Ellenberg (1974) aptly pointed out, this balance in the composition of the species can only be attained if the controlling environmental factors are also in a state of balance relative to one another. It is important to stress that stability in ecosystems does not mean constancy in the elements of the ecosystem, such as in species populations. Rather, resilience implies the ability of the ecosystem to return to a steady or equilibrium state after a disturbance. Mueller-Dombois and Ellenberg (1974) have pointed out that ecosystems are dynamic biological systems containing living organisms that undergo metabolic and seasonal changes such as leaf shedding, flowering and fruiting; and that in stable climax communities, micro-successions may be taking place in gaps created as a result of the falling of large trees. They further observed that two criteria that can be used for ascertaining the stability of terrestrial ecosystems are persistence in floristic/life form composition and structure over a period of several decades. Odum (1969) has established another criterion for assessing the stability of ecosystems. This is based on the ratio of gross production to community respiration. In early successional stages, it is usual for gross production to considerably exceed community respiration, resulting in a great increase in the biomass of plants and, by extension animals, over time. As the climax stage is approached, gross production roughly approximates community respiration and their ratio tends towards 1. This implies that the standing biomass of the ecosystem remains fairly constant over time in the climax, unlike in the earlier successional stages that are usually characterized by rapid increases in biomass over time.

Two alternative hypotheses have been proposed to explain the nature of the climax: (i) the monoclimax concept; and (ii) the polyclimax concept, and these will be considered in turn.

7.5.1 Monoclimax concept

F.E. Clements was one of the chief proponents of this concept. He believed that all successions taking place in a macroclimatic region terminate in the same climax stage, irrespective of the nature of the substrate on which they were initiated. He maintained that the regional climate is the ultimate and primary determinant of the nature of the climax community, and that whether successions are initiated on dry land, wet land, or even in open water surfaces such as lakes or estuaries, they will eventually terminate in the same climax stage – the climatic climax – due to the influence of the regional climate. He did not deny the existence of other stable communities that are primarily controlled by soil or edaphic factors (e.g. swamp forest or bog) or biotic factors such as burning or grazing, or topographic factors (e.g. montane forest), but he posited that they were seral communities that would ultimately be modified by the regional climate to assume the nature of the climatic climax vegetation.

Some ecologists who were contemporaries of Clements, such as A.G. Tansley, rejected the monoclimax concept. Today, in the face of abundant empirical observations that do not conform with Clements's basic theoretical postulate regarding the climax, the monoclimax concept has been criticized and is no longer widely accepted as a valid framework for assessing the status of climax communities. Some empirical cases that do not conform with the monoclimax concept will now be considered.

First is the observation that within a macroclimatic region, several stable communities which are in a state of equilibrium with factors other than regional climate have been observed; different climax communities have been observed on different soils or substrates in a climatic region. Mueller-Dombois and Ellenberg (1974) reported that in south-eastern Manitoba, Canada, jack pine communities occur on sandy upland soils, while mixed forest communities are characteristically associated with clayey soils. They went on to point out that the differences between the two communities reflect differences in soil parent materials, and that it is highly improbable that such textural differences between contrasting soil types would ultimately be eliminated by the regional climate. Hence, the two contrasting plant communities will continue to exist as climaxes or stable communities within the same climatic zone indefinitely. Singh *et al.* (2006) have also observed that in northern California, USA, two contrasting stable communities – a

stunted pigmy forest and the elegant giant redwood forest – exist side by side. The pigmy forest is associated with soils with iron-cemented hard pan that occurs near the ground surface which effectively reduces soil depth, while the giant redwood forest is characterized by deep soils. It is also improbable that a swamp forest would ultimately be modified to become identical with the climax community on dry terrestrial sites, as inferred by Clements, as palynological evidence suggests that wetland vegetation is less susceptible to change, even on a geologic time frame, than terrestrial vegetation on mesic sites (Singh *et al.*, 2006).

Another empirical observation which does not conform to the monoclimax concept includes the occurrence of a series of stable communities which exist side by side at different elevations on mountain slopes. These stable communities are in equilibrium with the micro-climates and the soil and substrate conditions at the various elevations, and it is very unlikely that they would be modified by the regional climate to assume the nature of the climatic climax that is typical of lowland areas.

7.5.2 Polyclimax concept

This concept was proposed by ecologists such as Tansley (1920) and Cain (1947) as an alternative to the monoclimax concept. The basic postulate of this concept is that within any climatic region, several self-perpetuating stable communities that are in a state of equilibrium with some local factor, such as soil or topography, can be observed in addition to the climatic climax. A given region with a uniform climate is therefore, characterized by a mosaic of several climax communities which are in equilibrium with either the regional climate or some local factor. In certain areas, a local factor may exert a more marked effect on the nature of some stable, self-perpetuating communities than the regional climate.

Apart from the climatic climax, the proponents of the polyclimax concept recognize edaphic climaxes: stable plant communities that are more closely controlled by local soil and substrate conditions than by the regional climate. They also recognize topographic climaxes, where a topographic factor such as high elevation is the primary factor determining the nature of the climax community and biotic climaxes, when the stable plant communities are in equilibrium with human or organism-related factors such as burning and grazing. Montane forests, which are topographic climaxes, are clearly structurally and floristically distinct from lowland forests in the same area, the latter being usually the climatic climax. Savanna vegetation in most parts of the tropics is usually burnt accidentally or deliberately, on an annual basis. As a result, species which are fire-tolerant have become established in the savanna vegetation. At the forest–savanna boundary, as a result of centuries of cultivation that involves slash burning, fire-tolerant savanna trees have invaded forest fallows and these have been transformed into savanna vegetation as described in Section 5.5. The resulting 'derived savanna' is stable, being prevented from reverting to forest by regular burning, and is an example of a fire climax: vegetation that is in equilibrium with a biotic factor – fire. In most parts of the world, grassland and savanna vegetation are grazed and the vegetation is prevented from developing further by grazing pressure. Hence, such vegetation is in equilibrium with grazing. The polyclimax concept provides a less rigid framework for studying succession and for assessing the status of climax communities.

References

Archer, S., Scifres, C. and Bassham, C.R. (1988) Autogenic succession in a subtropical savanna: conversion of grassland to thorn woodland. *Ecological Monographs* 58, 111–127.

Boettcher, S.E. and Kalisz, P.J. (1990) Single-tree influence on soil properties in the mountains of eastern Kentucky. *Ecology* 71, 1365–1372.

Cain, S.A. (1947) Characteristics of natural areas and factors in their development. *Ecological Monographs* 17, 185–200.

Carrière, S.M., Letourmy, P. and Mckey, D.B. (2002) Effects of remnant trees in fallows on diversity and structure of forest regrowth in a slash-and-burn agricultural system in southern Cameroon. *Journal of Tropical Ecology* 18, 375–396.

Clements, F.E. (1916) *Plant Succession: An Analysis of the Development of Vegetation.* Carnegie Institution of Washington, Washington, DC.

Connell, J.H. and Slatyer, R.O. (1977) Mechanism of succession in natural communities and their role in community stability and organization. *American Naturalist* 111, 1119–1144.

De Blij, H.J. and Muller, P.O. (2006) *Geography: Realms, Region and Concepts.* John Wiley & Sons, New York.

Del Moral, R. and Muller, C.H. (1970) The allelopathic effects of *Eucalyptus camaldulensis. American Midland Naturalist* 83, 254–282.

Egler, F.E. (1954) Vegetation science concepts: I. Initial floristic composition, a factor in old-field vegetation development. *Vegetatio* 4, 412–417.

Gehring, C., Denich, M., Kanashiro, M. and Vlek, P.L.G. (1999) Response of secondary vegetation in eastern Amazonia to relaxed nutrient availability constraints. *Biogeochemistry* 45, 223–241.

Gleason, H.A. (1917) The structure and development of the plant association. *Bulletin of the Torrey Botanical Club* 42, 463–481.

Gleason, H.A. (1939) The individualistic concept of the plant association. *American Midland Naturalist* 21, 92–110.

Hopkins, B. (1974) *Forest and Savanna.* Heinemann, Ibadan, Nigeria.

Margalef, R. (1968) *Perspectives in Ecological Theory.* University of Chicago Press, Chicago, Illinois.

Mesquita, R.C.G., Ickes, K., Ganade, G. and Williamson, G.B. (2001) Alternative successional pathways in the Amazon Basin. *Journal of Ecology* 89, 528–537.

Mueller-Dombois, D. and Ellenberg, H. (1974) *Aims and Methods of Vegetation Ecology.* John Wiley & Sons, New York.

Odum, E.P. (1969) The strategy of ecosystem development. *Science* 164, 262–270.

Odum, E.P. (1971) *Fundamentals of Ecology.* W.B. Saunders Co., Philadelphia, Pennsylvania.

Odum, E.P. (1975) *Ecology: The Link Between the Natural and the Social Sciences.* Holt Rinehart and Winston, London.

Saldarriaga, J.G., West, D.C., Tharp, M.L. and Uhl, C. (1988) Long-term chronosequence of forest succession in the Upper Rio Negro of Colombia and Venezuela. *Journal of Ecology* 76, 938–958.

Singh, J.S., Singh, S.P. and Gupta, S.R. (2006) *Ecology, Environment and Resource Conservation.* Anamaya Publishers, New Delhi.

Tansley, A.G. (1920) The classification of vegetation and the concept of development. *Journal of Ecology* 8, 118–144.

Tilman, D. (1985) The resource-ratio hypothesis of plant succession. *The American Naturalist* 125, 827–852.

Tilman, D. (1990) Constaints and tradeoffs: Towards a predictive theory of competition and succession. *Oikos* 58, 3–15.

Whittaker, R.J., Bush, M.B. and Richards, K. (1989) Plant recolonization and vegetation succession on the Krakatau islands, Indonesia. *Ecological Monographs* 59, 59–123.

Yarranton, G.A. and Morrison, R.G. (1974) Spatial dynamics of primary succession: Nucleation. *Journal of Ecology* 62, 417–428.

Zinke, P.J. (1962) The pattern of influence of individual forest trees on soil properties. *Ecology* 43, 130–133.

8 Theory and Models of Soil Fertility Restoration under Bush Fallow

In Chapter 7, the theories and models of secondary succession were considered. It will be useful to recall that the primary factors driving the process of ecological succession include: (i) changes that occur in the plant community itself, such as interspecific and intraspecific competition; and (ii) changes that occur in the environment that is external to the plant community, especially the soil and the micro-climate of the habitat. Changes that occur in the soil during the course of succession are of pivotal importance to the process of ecological succession and to the system of shifting cultivation or slash-and-burn agriculture, as they result in the restoration of soil fertility, hence obviating the need to apply chemical fertilizers. The current chapter considers theories and models which have been used to explain the process of soil fertility restoration under bush fallow. The models discussed include:

1. Guillemins's model.
2. The sigmoid model.
3. Trenbath's models.
4. The spatio–temporal model, which was referred to tangentially in Chapter 7.

8.1 Guillemin's Model

Guillemin (1956) was one of the earliest to formally propose a model of soil fertility restor-ation under bush fallow, based on the relation-ship between the length of cropping and that of cultivation, and also incorporating ideas relating to the organic matter equilibrium concept. His model is graphical and descriptive, and its main features are illustrated in Fig. 8.1. The central theme of the model is the decline of soil fertility during cropping and its rejuvenation during the subsequent fallow period. When the fallow period is sufficiently long relative to the period of cropping, soil productivity builds up to the maximum or equilibrium level for the given type of soil and vegetation. When this equilibrium or maximum level of soil productivity has been attained in the soil under bush fallow, leaving the land fallow for a longer period does not result in further increase in soil fertility. Hence, it is not be necessary to leave the land fallow much longer, if soil productivity has attained its maximal level. Guillemin (1956) observed that provided soil productivity is allowed to attain the maximum level during the fallow period (and this would normally require a minimum of 15 years of fallow for about 5 years of cultivation), soil fertility will build up to the maximum or equilibrium level prior to cropping, and the system of cropping to fallowing will be stable, without incipient decline in soil fertility occurring over time (Fig. 8.1b).

However, if the fallow period is considerably reduced relative to that of cropping, for example if a cropping period of 5 years is followed by a

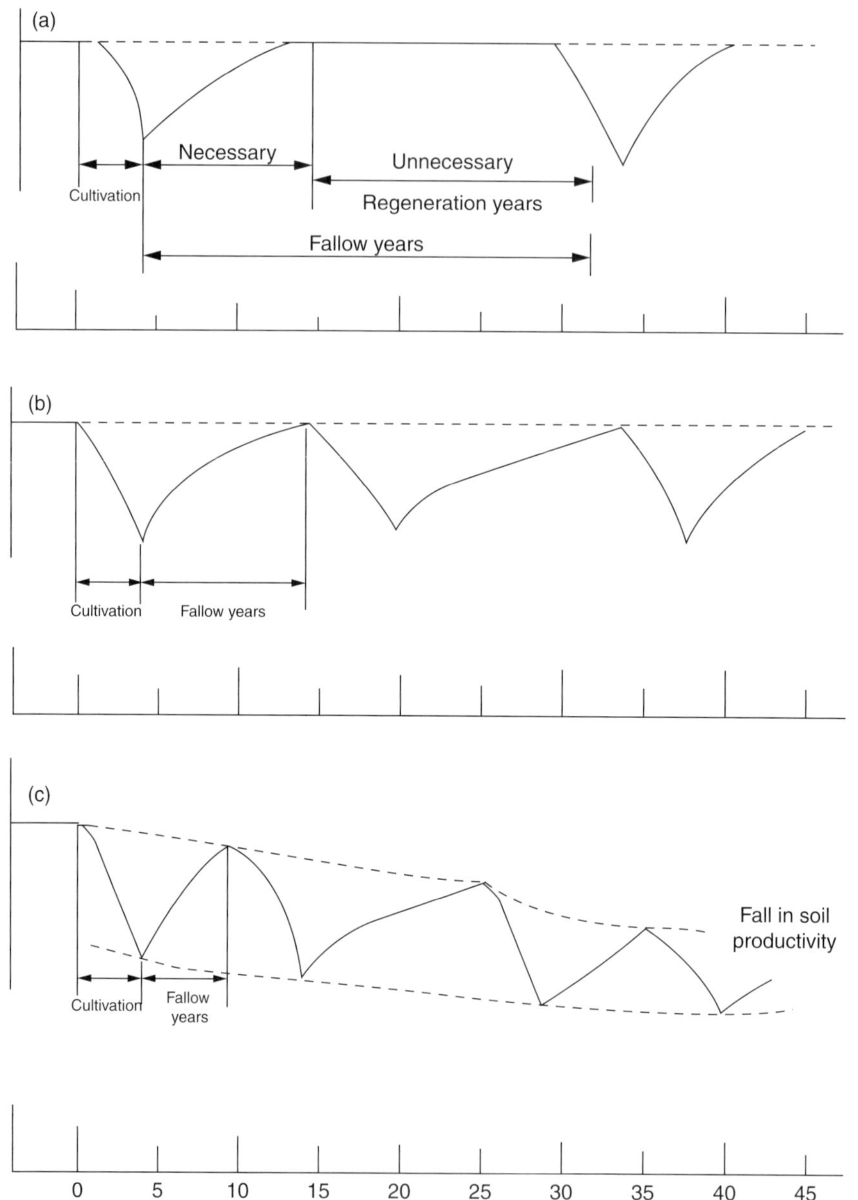

Fig. 8.1. Guillemin's (1956) model of the relationships between soil productivity and length of cropping and fallow.

fallow period of 5 years, there will be a progressive decline in soil productivity over time (Fig 8.1c). Guillemin's model clearly indicates that the system of shifting cultivation is sustainable when fallow periods are long relative to the period of cultivation, and that the system breaks down, resulting in progressive soil deterioration over time, when the fallow period is too short. His model indicated that when using natural bush fallow vegetation to restore soil fertility, a cropping period to fallow period ratio of 1:1 is clearly unsustainable, as experience has clearly shown in many areas where the fallow period is progressively shortened as a result of population pressure.

8.2 The Sigmoid Model

Guillemin's (1956) model hinges on the assumption that the improvement of soil fertility, and hence productivity, is contemporaneous with the inception of the fallow period. This explains why in the three graphs used to illustrate the model (Fig. 8.1), the year or point that marks the end of cropping is a 'change point' from soil productivity decline to productivity accretion. Several empirical studies have, however, shown that soil fertility improvement may not occur at fallow inception, and that soil organic matter and nutrient decline may last into the first 3–5 years of the fallow period, and sometimes for a longer period (Aweto, 1981; Ramakrishnan and Toky, 1981; Roder *et al.*, 1997; Brand and Pfund, 1998). Hence there is the need for a model that includes an element of organic matter and nutrient decline during the first few years following fallow inception. The sigmoid model depicted in

Fig. 8.2 addresses this issue of soil fertility decline at fallow inception prior to the latter part of the fallow, when there is a rapid build-up of soil fertility.

Three distinct phases can be observed in terms of fertility dynamics during the fallow period. The first phase, which immediately follows cultivation, is characterized by a gradual decline in soil organic matter and nutrient status. The decline in soil fertility during the initial part of the fallow period is largely attributed to the fact that an adequate vegetation ground cover may not have been established at fallow inception. This implies that the process of soil organic matter and nutrient decline that characterizes the cropping period lasts into the first few years of the fallow period, at least until fallow vegetation is developed enough to adequately protect the soil against accelerated organic matter and nutrient decline through leaching, erosion and thermal oxidation of soil humus. It is only when fallow vegetation adequately protects the ground, and especially when numerous trees have become established in it, that the process of soil fertility decline that characterizes the period of cultivation and the first few years of the fallow period is reversed.

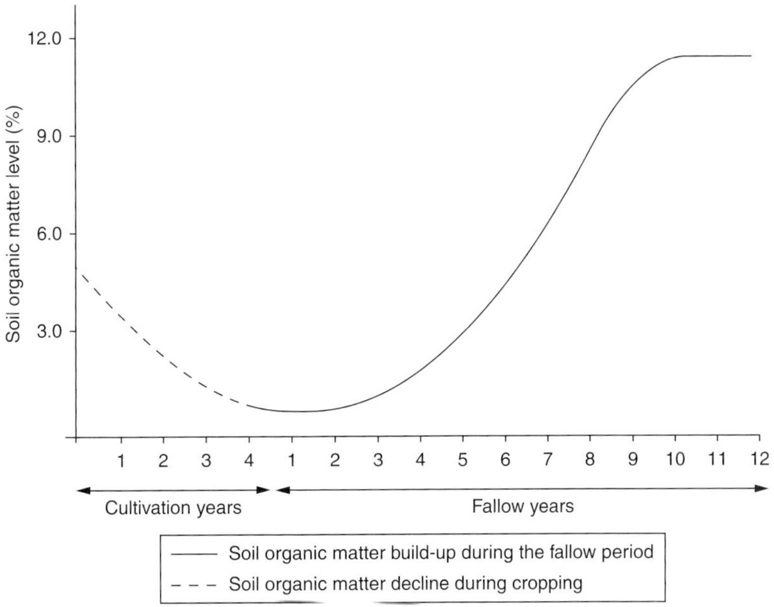

Fig. 8.2. A graphical illustration of the sigmoid model of organic matter dynamics during the shifting cultivation cycle.

The second phase of the model is marked by a rapid build-up of organic matter and nutrients in the soil. This phase usually occurs when trees and shrubs have become dominant in fallow vegetation and effectively suppress leaching and erosion of soil organic matter and nutrients, and when the rate of litter generation by fallow vegetation is much greater than the rate of humus decomposition in the soil. This results in accumulation of organic matter and nutrients in the soil, especially the topsoil, and an attendant increase in soil cation exchange capacity (CEC).

The last phase of the sigmoid model is characterized by a slow increase in soil organic matter, or stagnation in humus level over time, as humus in the soil has attained or approached the equilibrium level for the type of vegetation and soil occurring in the bush fallow ecosystem. As with Guillemin's (1956) model, the sigmoid model is based on the view that the soil fertility level builds up (subject to the proviso that fallow vegetation is not burnt, grazed or subjected to other biotic pressures such as intense fuelwood exploitation) and eventually attains the maximal level, when it stabilizes over time. This maximal level of soil fertility is largely determined by the equilibrium level of soil organic matter, which, as pointed out previously, is the main determinant of the nutrient-holding capacity and also of the water-holding capacity of soils, particularly sandy soils in the tropics. When soil fertility has attained the equilibrium level, leaving the land fallow for a much longer period does not improve fertility further, apart from increasing the nutrient capital stored in fallow vegetation, which can be subsequently added to the soil when it is burnt as part of land preparation for cultivation.

8.3 Trenbath's Models

Two models that Trenbath (1989) used to explain soil dynamics and the regeneration of fallow vegetation will be considered here.

The first is a graphical model that explains the restoration of soil fertility during the fallow period in terms of the relationship between the length of cropping and cultivation, and is based on the concept of soil organic matter equilibrium level under the climax (forest) vegetation. His

model clearly showed that when a cropping period of 1 year is accompanied by a fallow period of 10 years, soil organic matter and nutrient levels build up appreciably, and approach forest equilibrium level prior to the subsequent period of cultivation. Hence, the system of shifting cultivation would be sustainable as there would be no progressive deterioration in soil fertility over time. However, with intensification of the system to a cropping/fallow period ratio of 1:1 or 1:2 – that is, when the period of cultivation of 1 year is followed by 1 or 2 years of fallow – there would be substantial and progressive deterioration in soil fertility, leading to a marked decline in crop yields and an unsustainable system. This graphical model emphasizes the well-documented and well-known feature of shifting cultivation that long fallows are essential to the restoration of soil fertility to high levels that ensure the sustainability of the system of agriculture.

Trenbath's second model is an attempt to explain why forest fallow vegetation may fail to regenerate after a prolonged period of cropping and ultimately be replaced by a grass fallow. Essentially, this model uses graphs to relate the regeneration of forest fallows to the length of cropping preceding the fallow period. It is based on the observation that soil fertility declines progressively with the length of the period of cultivation. Consequently, as the length of cultivation increases, the vigour and rate of fallow regeneration slow down, resulting in delayed establishment of fallow vegetation after cultivation. The main prediction of the model is that five consecutive years of cultivation delay the regeneration of forest fallows substantially, but that 6 years of cropping effectively prevent forest fallow regeneration, and the cultivated field would be invaded by grasses, effectively preventing the re-establishment of forest fallow vegetation (Fig. 8.3).

Trenbath's observation that forest fallow may fail to regenerate and may be replaced by grass fallows is in line with several empirical findings. The regeneration of forest trees is usually very slow on sites that have been previously intensively disturbed by humans (Guariguata and Ostertag, 2001), and the vigour of forest recovery is inversely related to the intensity of the previous land use (Uhl *et al.*, 1988; Aide *et al.*, 1995). Furthermore, grass

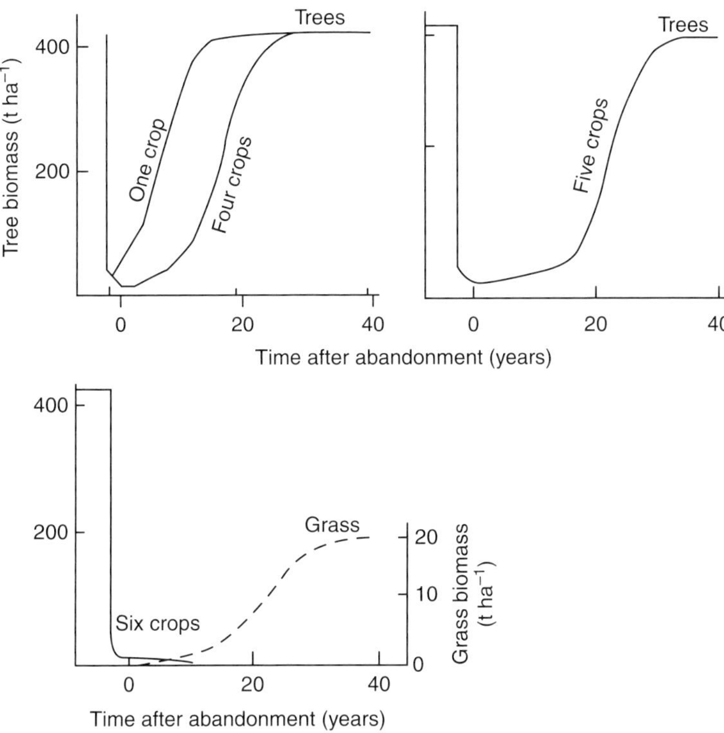

Fig. 8.3. Trenbath's (1989) graphical illustration of the effects of varying length of cropping on subsequent regeneration of fallow vegetation. (Data on biomass of secondary forests were removed from the two upper graphs.)

competition with tree seedlings usually delays or hinders tree regeneration, especially of tree species with small seeds (Nepstad *et al.*, 1990; Hooper *et al.*, 2002). Besides, the invasion of forest fallows by grasses usually enhances their proneness to fires, which hinders tree regeneration. In parts of Indonesia and the Philippines, forest fallows have been inadvertently transformed into grasslands dominated by *Imperata cylindrica* as a result of over-cultivation involving frequent burning of arable land (Beets, 1990). Usually, when forest fallows are invaded by grasses as a result of over-cultivation and the attendant depletion of the soil seed bank of tree seeds, the regeneration of woody fallow vegetation is hampered, and this tendency is reinforced by the progressive killing of tree stumps by burning as part of land preparation for cultivation. The study of Hooper *et al.* (2004) in Panama has shown that when an exotic grass (*Saccharum spontaneum*) invades forest clearings after cultivation, the re-

generation of woody vegetation is hampered and that burning reduces the number of tree recruits and tree re-sprouts in fallow vegetation.

However, one should not assume that whenever the land is cropped consecutively for more than 5 years, forest fallows are automatically transformed into grass fallows which hinder forest regeneration. In soils of moderate fertility, such as alfisols, the land can be cultivated for up to 6 years without immediate danger of forest fallow invasion by grasses. Burning appears to be a crucial factor in the transformation of forest fallows into grass fallows. If the land is not burnt every time the land is cultivated, or if enough trees are left on the farm during cropping, these serve as sources of seeds and propagules for tree establishment during the subsequent fallow period. Hence, a forest fallow may not be transformed into a grass fallow after 6 years or more of cultivation. In fact, in southern Benin Republic, the Adja people in Mono province cultivate their fields for up to

10 years after a palm fallow, and the forest fallows are not transformed into savanna or grassland vegetation.

Finally, Trenbath's model appears to be based on unrealistically high estimated rates of forest regeneration. His graphs (Fig. 8.3) indicate that following the harvest of one to four crops (presumably 1–4 years of cultivation), forest fallow vegetation attained the biomass of 400 t ha^{-1} after leaving the land fallow for only 10–26 years. The published data on the total biomass of secondary forests in different parts of the tropics, reviewed by Brown and Lugo (1990), clearly show that forest biomass was still less than 180 t ha^{-1} after 40–80 years of forest succession.

8.4 The Spatio–Temporal Model

The spatio–temporal model is an extension of the nucleation model described in Section 7.3.7, the core–periphery model and the concepts of facilitation and Egler's (1954) concept of initial floristic composition. Essentially, this is a descriptive model that hinges on tree invasion over time, with the trees serving as foci of soil fertility regeneration by accumulating organic matter and nutrients under their canopies. The model is both a spatial and a temporal model. It is spatial in that it involves changes in the spatial structure of the farm or abandoned fields over time as they are progressively invaded by trees. The spatial structure of the fallow community is altered over time as trees and other woody perennials gradually replace grasses and other herbaceous plants that initiated the process of vegetation recolonization. The changes in the structure of the fallow community take place sequentially over time. Hence, the model can also be described as a temporal model. The core–periphery analogy is also central to the spatio–temporal model and it will be briefly discussed here.

8.4.1 Core–periphery analogy

The core–periphery model is strongly rooted in the social sciences where it has been widely applied to analyse and depict the spatial disparities in the levels of economic development. Globally and regionally, the pattern of economic development is patchy, with a few areas that are highly developed being circumscribed by less developed areas or countries. The USA, Japan and the countries of Europe are the most developed countries and they are characterized by very high levels of economic and technological development, on account of which they dominate the rest of the world, particularly the developing countries in Africa, Asia and Central and South America. The economically and technologically developed countries of the USA, the EU and Japan that dominate world trade and manufacturing are core countries, while developing countries constitute the periphery (Knox and Marston, 2004). Within individual countries or regions, the pattern of economic development is very uneven, with highly industrialized areas (cores) that are surrounded by less developed agricultural areas that form the periphery. The core–periphery model is also applicable to a major industrial city which forms the core and the surrounding rural areas (periphery) that provide raw materials for the industries in the city. The growth and development of the core occur at the expense of the periphery (Fellman *et al.*, 2005). The latter exports semi-processed and unprocessed raw materials to the former and, in turn, receives high-value industrial goods from the core. The periphery, which is in dire need of highly skilled manpower, may lose a significant proportion of its skilled manpower to the core as a result of migration (brain drain) from developing countries to the core nations.

The main ideas of the core–periphery model are relevant to the process of secondary succession and soil fertility rejuvenation under bush fallow. As pointed out above, trees serve as foci of soil fertility restoration. Trees and shrubs exert a marked effect on nutrient cycling, especially beneath their canopies (Prescott, 2002). Here, they alter the nature of soils through the input of litter, capturing wet and dry deposition from the air and subsequently depositing it in the soil; they also lower soil temperatures and improve soil moisture under their crowns. Trees and shrubs also assist in recycling nutrients from the subsoil to the topsoil, and input organic matter into the soil through the death and decomposition of their roots. Trees, therefore, have a distinct circle of influence around them (Zinke, 1962; Boettcher and Kalisz, 1990).

Two distinct areas can be recognized in the zone of influence of a tree. At the ground surface, the zone directly below the canopy constitutes the core, while the zone beyond the outer limit of the crown or dripline as far as litter is dispersed, or as far as lateral roots extend from the tree bole, forms the periphery (Fig. 8.4). The core is the zone of litter and twig fall, and the tree's influence on the soil beneath is maximal in the core. With increasing distance from the tree and the dripline, the influence of the tree begins to wane as litter supply decreases due to the distance decay effect. The core – that is, the soil beneath the tree canopy – is characterized by an accumulation of organic matter, soil nutrients and elevated CEC levels relative to the soil outside the influence of the crown in the periphery (Belsky, 1994; Aweto and Dikinya, 2003).

The core and peripheral zones in the soil have been defined horizontally, that is, laterally on the ground surface. They can also be defined vertically, using the soil profile. The core corresponds with the immediate topsoil layer where there is a marked accumulation of organic matter. It is the humic layer to which nutrient accretion in the soil is largely restricted during the fallow period. This topsoil layer, in which there is a marked accumulation of nutrients and organic matter as a result of autogenic succession, is usually restricted to the top 20 cm of the soil profile in forest and savanna bush fallow (Aweto, 1981; Manlay et al., 2002). The underlying layers of the soil, including the zone of weathering deep below the ground surface, constitute the periphery (Fig. 8.5). Tree roots absorb nutrients from the subsoil and the weathering zone and subsequently concentrate them in the topsoil (core) through the fall and mineralization of litter (Nye and Greenland, 1960).

8.4.2 Stages of the process of soil fertility restoration

This section will describe the four main stages in the regeneration of soil fertility as a result of autogenic succession in bush fallow vegetation, which involves fallow invasion by trees. The processes that result in soil fertility restoration under bush fallow are closely related to those that take place in fallow vegetation. As pointed out in Chapter 6, the soil and vegetation components of bush fallow ecosystems are closely interrelated. The stages of the spatio–temporal model described below apply both to

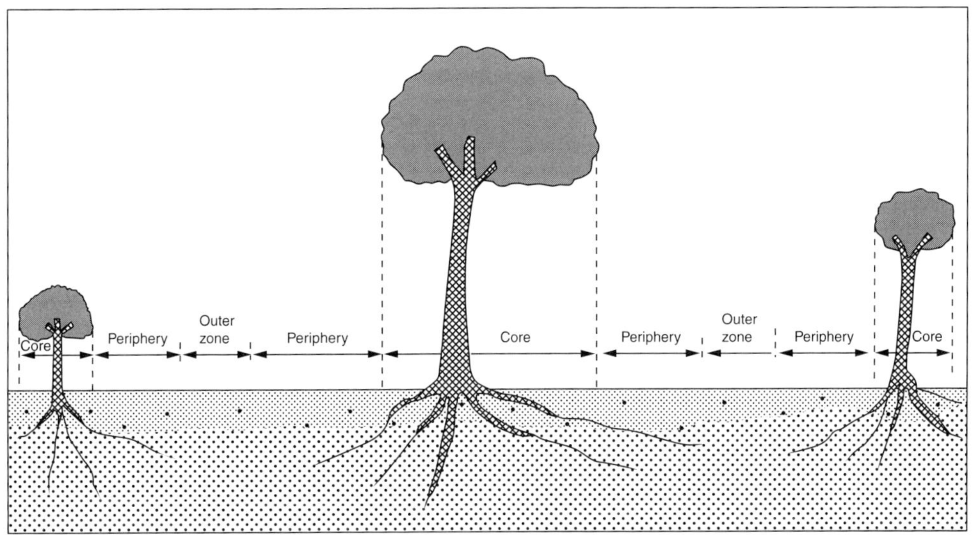

Fig. 8.4. The core and peripheral zones around isolated trees.

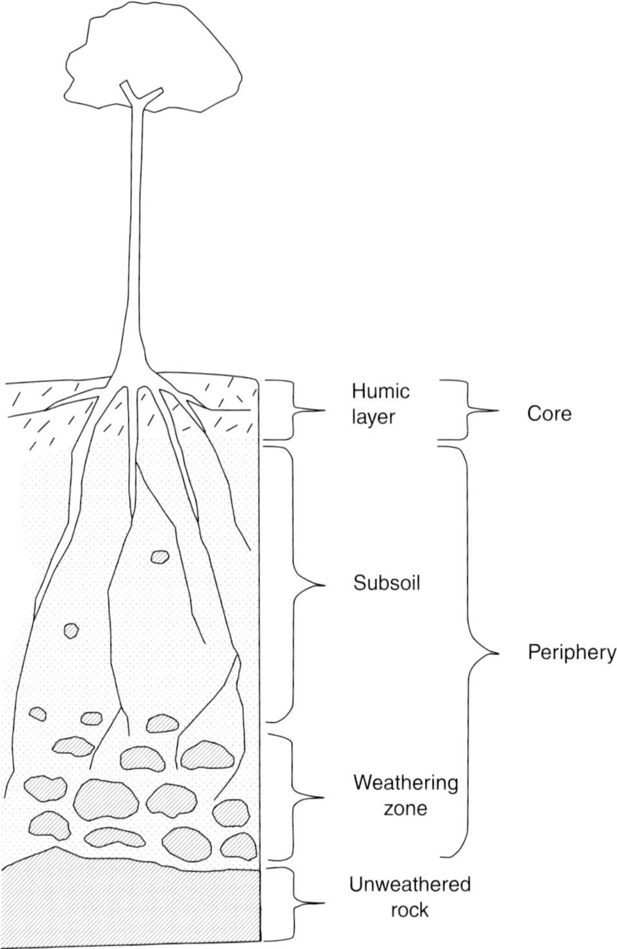

Fig. 8.5. The core and peripheral zones in the soil under a tree.

the processes of soil fertility restoration and to secondary succession. The stages of the spatio–temporal model are depicted in Fig. 8.6.

Stage 1

The first stage follows immediately from farm abandonment at the cessation of cropping. Only a few trees are present on the farmland and these were mainly left during cultivation to provide shade, fruits, or some other benefits for the farmer. At this stage there are only a few isolated stands of trees in the abandoned farm, which has become fallow vegetation because the process of natural re-vegetation has been initiated on the site. This is the pre-primary cores stage. As pointed out above, each tree serves as a focus of soil fertility regeneration as it concentrates organic matter and nutrients in the soil underneath its canopy. Hence, each tree can be regarded as a centre or focus for soil fertility regeneration. Since the trees were present in the cultivated field prior to its abandonment, they can appropriately be referred to as pre-primary foci or pre-primary cores of soil fertility restoration. By improving soil and microclimatic conditions beneath its canopy, each tree serves as a focus for the recruitment of forest or later successional species which may persist until the climax forest stage.

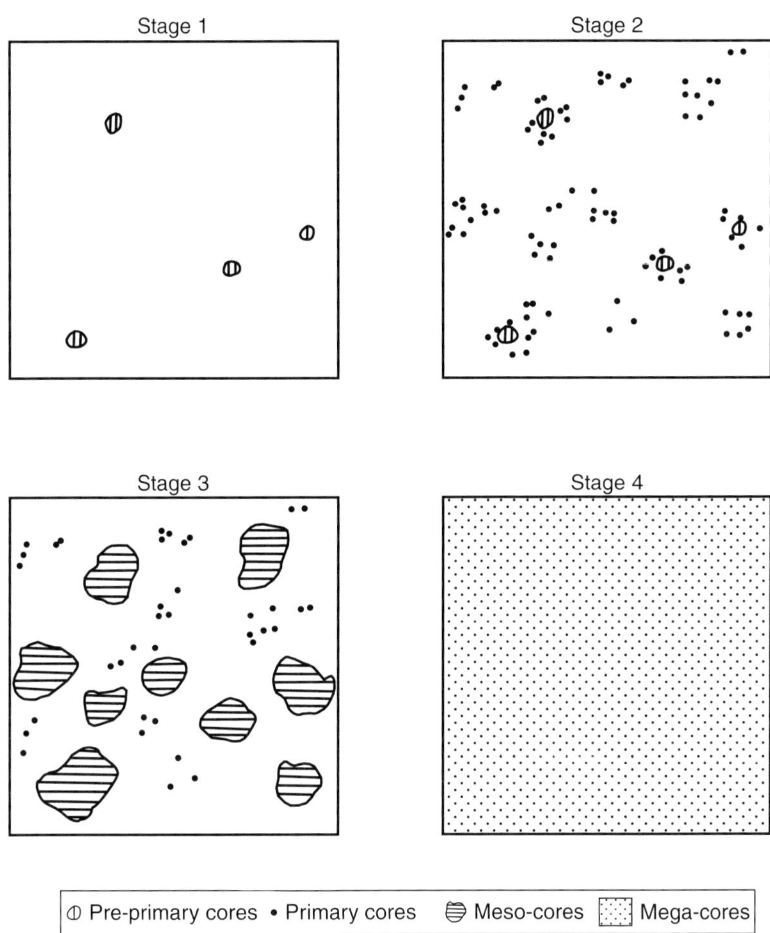

Fig. 8.6. The spatio–temporal model showing the stages of fallow soil fertility and vegetation regeneration based on the core–periphery analogy.

Stage 2

The second stage is that of primary foci or cores, which are trees established from seeds or root suckers present in the soil that were not killed during cultivation, or seeds that were dispersed to the abandoned field by wind or animals, such as birds from parent trees in adjoining or adjacent forest or savanna vegetation. Several primary foci or cores are usually established near or beneath the pre-primary foci, as the soil underneath the trees present in the field before the inception of the fallow period usually has more moisture and nutrients to facilitate seedling establishment. However, most of the trees

invading fallow vegetation at this stage are pioneer or early successional tree species which can establish themselves in the open (Richards, 1996). Such trees can establish themselves outside tree canopies but often take advantage of the more favourable soil conditions under tree canopies.

Once these pioneer trees have been established they serve as foci for recruitment of other pioneer trees or forest trees, as with the trees left on the farm prior to the inception of the fallow period. Vazquez-Yanes (1998) has observed that in Central and South America, the short-lived and fast-growing pioneer tree, *Trema micrantha*, improves soil nutrient and organic matter status under its canopy and also helps to ameliorate microclimatic conditions. This makes it possible for forest tree species, which cannot normally be established under the harsh microclimatic conditions that prevail in the open, to be established under its canopy. Such pioneer tree species are therefore of pivotal importance to the invasion and subsequent establishment of forest tree species in fallow vegetation. In abandoned pastures in Ecuador, Zahawi and Augspurger (1999) reported that species richness was not only higher under sites with *Psidium guajava* (guava) trees than in the open, but that tree sapling establishment was restricted to areas under the trees. In the subtropical savanna and shrubland vegetation of southern Texas, USA, which has become progressively more densely stocked with trees within living memory, Jurena and van Auken (1998) have reported that soil total nitrogen was 5–13 times higher under two species of acacia trees than in the open grassland, and that soil calcium, pH and sulfur were also higher in soil under the acacia trees. They further reported that they recorded more woody and succulent species under one of the tree species, *Acacia rigidula*, than in the open. This is evidently due to the amelioration of soil and microclimatic conditions by the trees, making it possible for the woody plants to establish themselves in sites under their canopies, in preference to the open grassland.

Carrière *et al.* (2002) also observed that remnant trees left on farmers' fields in southern Cameroon prior to the onset of the fallow period facilitated fallow vegetation regeneration, especially of woody perennials, by attracting seed dispersers such as birds and bats, and also by making sites beneath tree canopies more suitable for plant establishment. This clearly suggests that the initial trees to be established in fallow vegetation facilitate the establishment of invading trees by improving the nutrient and moisture status of the soil under their canopies, making such sites suitable for the establishment of the invading trees. Furthermore, the findings of Ganade and Brown (2002) have shown that plant litter that accumulated on the soil in both old fields (abandoned pasture) and forest plots in the central Amazon basin of South America enhanced the establishment of seedlings of mid- and late-successional tree species. Facelli and Facelli (1993) have also observed that availability of plant litter on the soil is an important factor that influences the establishment and subsequent development of late successional tree species during the course of vegetation recolonization of abandoned fields at the cessation of human interference. Since more litter usually accumulates under stands of trees present in pastures or fallow land at the inception of secondary succession than in open sites, such trees can also enhance seedling establishment of invading tree species that may later replace the pioneer tree and herbaceous species.

Stage 3

The third stage of the spatio–temporal model involves the lateral fusion of the crowns of clusters of both pre-primary and primary foci to form larger foci known as meso-foci or meso-cores. As clusters of adjoining tree crowns grow and expand over time, the rate of recruitment and establishment of more tree saplings increases, partly due to soil organic matter and nutrient accumulation in the soil. Contemporaneous with these changes, the microclimate of the expanding woodclump improves, with a further lowering of daytime temperatures and an increase in relative humidity, making the microclimate of the meso-foci (meso-cores) more similar to that of the forest. With this development, a much larger proportion of the abandoned field is now under the influence of tree canopies and experiences organic matter and nutrient accretion, as several adjacent tree canopies coalesce as a result of their lateral expansion. As with each pre-primary and primary focus, coalesced tree canopies forming meso-cores help to accumulate organic matter and nutrients in the soil below them, by trapping both dry and wet deposition from the atmosphere and by exploiting nutrients in the subsoil/weathering zone and the interspaces between the meso-cores, and subsequently recycling the nutrients to the top layer of the soil beneath them. During this stage of the process of soil fertility rejuvenation, fallow vegetation consists of patches of woodclumps, often with a closed canopy, and intervening areas of herbaceous vegetation or forbs with isolated trees or shrubs. Except when savanna woodlands (which represent the meso-core stage in the savanna environment) are protected from burning for a long period, the meso-core stage represents the last stage of soil fertility restoration and secondary succession in farmland. This is because, with repeated burning, savanna woodlands are prevented from being transformed into dry forests, which represent the last stage of soil fertility rejuvenation in the savanna environment. In fact, with repeated intense burning, savanna woodlands may be transformed into tree savanna, with isolated stands of trees that represent the stage before the meso-foci stage. In forest ecosystems – and especially in a rainforest environment – with prolonged fallowing, the last stage of the process of soil fertility restoration, described below, would be attained.

Stage 4

Ultimately, the meso-cores may fuse together as a result of further lateral expansion, to form a mega-core. At this stage, all the tree canopies in the community are overlapping and no part of the original fallow land is outside the influence of a tree canopy. Literally speaking, all the tree canopies have merged with one another to form a single, large overhead canopy – a mega-core. When this stage is attained, the lateral transfer of nutrients from the interspaces between the crowns to soil underneath the trees virtually ceases, as the interspaces between the individual tree crowns have been eliminated, having become part of the core of one tree or the other which fused together to form the mega-core. However, the transfer of nutrients from the subsoil and the weathering zone (periphery) still continues unabated, as it is now the major means of nutrient input into the ecosystem. The other major source of nutrient input is dry and wet deposition from the atmosphere. Atmospheric input is a major nutrient source for sustaining the ecosystem when it attains the mega-core stage, especially on soils developed on base-deficient parent materials such as sandstone, which usually contain low-weatherable mineral reserves. Such nutrient-deficient soils include oxisols and ultisols that often develop on deeply weathered coastal plain sediments.

Central to the core–periphery model is the idea that the relations between the core and the periphery are reciprocal and that the core grows at the expense of the periphery. The transfer of nutrients from the subsoil and the spaces between

tree crowns (periphery) to the topsoil underneath tree crowns (core) was referred to in the preceding paragraph. It is important to observe that during cultivation, nutrients are leached downwards from the topsoil – that is, from the core – to the subsoil that constitutes the periphery. Although the bulk of plant litter is deposited under the tree canopy, some litter is dispersed from the tree into the periphery. Hence the relationship in the soil between the core and periphery is also dynamic and reciprocal. Although the core exports nutrients to the periphery through leaching to the subsoil even during the fallow period, it is very likely that the bulk of the mineral nutrients that accumulate in the soil core zone is derived from the peripheral zones of the subsoil and the interspaces between tree canopies. Nye and Greenland (1960) observed that there is a net transfer of plant nutrients from the subsoil to the topsoil during the fallow period. In semi-arid savanna ecosystems, dust trapping by crowns of trees may be an important source of nutrient input to the soil underneath tree canopies (Wezel *et al.*, 2002). Even in arid and semi-arid savannas, Vetaas (1992) observed that trees and shrubs absorb nutrients horizontally and vertically in the soil (including the subsoil and interspaces between tree crowns) and ultimately concentrate them in the soil under their canopies through the processes of litterfall and decomposition. It would seem, therefore, that nutrient accretion in the soil core zone occurs at the expense of the periphery. Hence bush fallow cores can be compared, in a restricted sense, to industrial core nations or regions that grow at the expense of the periphery.

8.5 A Unified Theory of Succession and Soil Fertility Restoration

Secondary succession and the restoration of soil fertility in bush fallows are closely related and can be regarded as two sides of the same coin. They represent different dimensions of the same process of ecosystem dynamics and development over time. Both processes are mutually interdependent and are causally related to one another. They are also both driven by the input of solar radiation, water and gases such as oxygen, nitrogen and carbon dioxide from the atmosphere. Furthermore, both processes are interconnected by organic matter and nutrient cycles. The organic matter cycle begins with the process of biological production in fallow vegetation and terminates in litter input to fallow soil and the accompanying process of litter decomposition that results in organic matter accretion in fallow soil. Similarly, the cycling of nutrients such as nitrogen, phosphorus and potassium forges links between the vegetation and soil components of natural bush fallow ecosystems, and renders them functionally and developmentally inseparable from one another. The spatio–temporal model can be used to explain both the processes of secondary succession and the regeneration of soil fertility under bush fallow, both being dependent on the invasion of fallow vegetation by trees, which consequently become more numerous over time. It is obvious, therefore, that the two processes are closely related and are mutually reinforcing. Hence, it should be possible to develop a unified theory to explain both the dynamics of secondary succession and the process of soil fertility rejuvenation under bush fallow. This section presents a unified model of both processes.

Figure 8.7 depicts the essential features of the unified model of secondary succession and soil fertility restoration in bush fallow. As a result of the input of solar energy and precipitation (water) to the soil, seeds in the soil germinate, tree stumps coppice and, in some cases – especially in drier forests and savanna ecosystems – sucker regrowth begins from the roots of plants that survived the cropping period. The initiation of the process of fallow vegetation recolonization and development, which largely depends on the physical, chemical and biological status of the soil, also marks the beginning of the process of soil fertility restoration. The accumulation of organic matter in the vegetation biomass over time, resulting from net primary production, is closely related to the process of litterfall, which largely determines the build-up of organic matter in the topsoil of bush fallow during secondary succession. At fallow inception, soil organic matter and nutrient levels are usually low. The restoration of soil fertility under bush fallows depends mainly on the build-up of organic matter in the topsoil, and this, in turn, depends on the considerable input of plant litter from the standing biomass of fallow vegetation to the soil (Fig. 8.7).

The improvement in soil fertility and water-holding capacity over time enables higher and more demanding plant life forms, especially trees, to invade fallow vegetation. The early successional communities are transformed into a mid-successional vegetation in which trees are a dominant component. The trees grow and increase in abundance and density, and enhance vegetation ground cover; this reduces soil erosion, leaching and accelerated soil organic matter decomposition resulting from high soil temperatures in exposed sites. Improved litter supply to the soil, from an increase in the biomass and density of trees, further raises soil organic matter levels, leading to an increase in soil CEC and the possibility of nutrients building up in the soil.

It is important to observe at this juncture that most tropical soils used for shifting cultivation are intensely weathered and their clay minerals are dominated by low-activity minerals, such as kaolinite, with character-istically low capacity to adsorb plant nutrients. As a result, their capacity to retain plant nutrients, and subsequently make them avail-able for plant use, largely depends on their organic matter content. The increase in soil CEC makes it possible for the soil to retain plant nutrients transferred from the subsoil by the roots of trees, as well as those from dust trapped by tree canopies and ultimately washed into the soil. The improvement in soil fertility and water-holding capacity, coupled with modification of the microclimate by vegetation and competition between plant species and life forms, gradually transforms an early successional community into a mid-successional community, and the latter will be ultimately transformed into the climax stage, if the process of succession is allowed to proceed unhindered for several decades. This involves a substantial reduction in the rate of vegetation biomass increase, per-sistence in plant life forms, and minor fluctuations in vegetation floristic composition, when the climax state is attained. Although forest soil organic matter level may reach equilibrium level and attain about 80% of the forest fallow soil level after about 15 years of forest succession, fallow soil at this stage is far from attaining a 'nutrient equilibrium' with mature forest vegetation. This is mainly because fallow vegetation is growing vigorously and

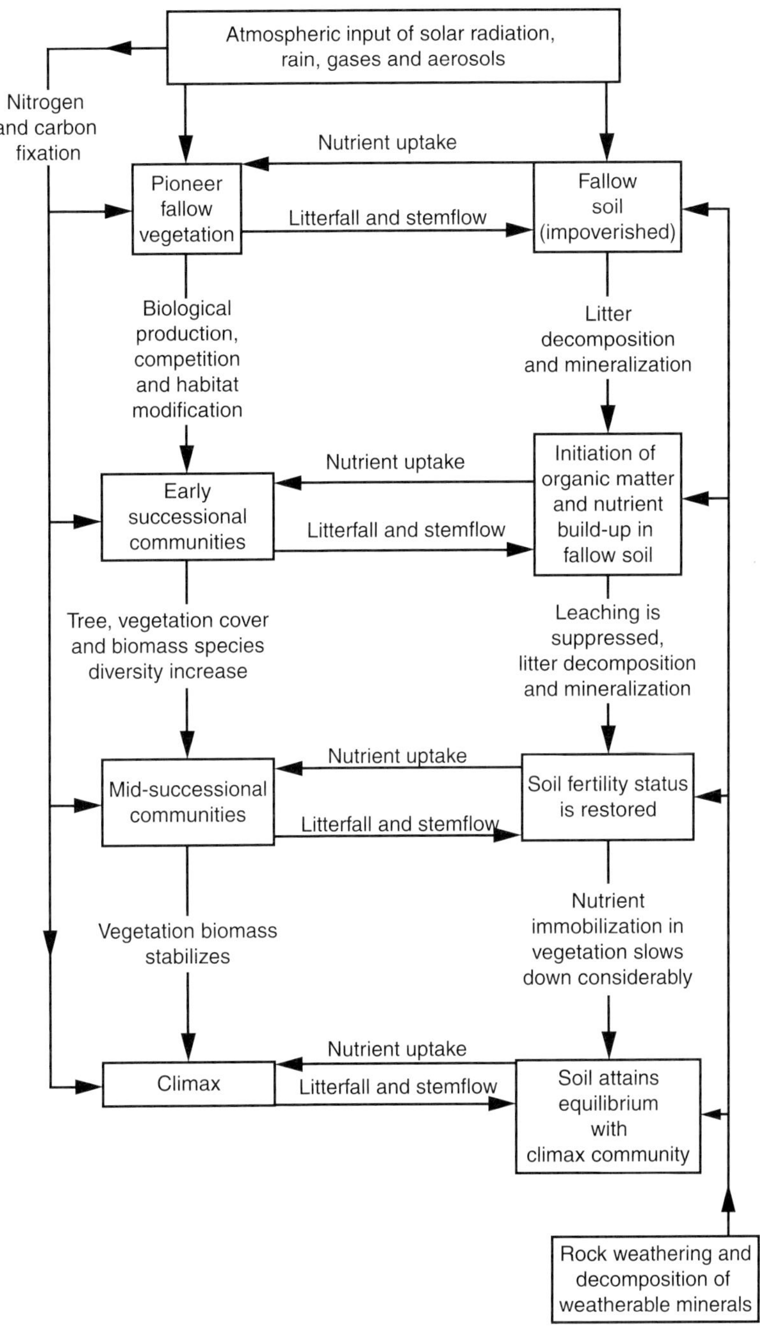

Fig. 8.7. A unified model of secondary succession and soil fertility restoration in the soil.

taking up nutrients from the soil and storing them in its biomass. Presumably it is only after the vegetation has attained the climax stage, when its biomass has stabilized, that soil nutrient status will attain a state of equilibrium under the forest vegetation.

It is important to briefly discuss the source of organic matter and nutrients that accumulate in soil during the fallow period. Undoubtedly, most of the humus that builds up in fallow topsoil is derived from plants, especially higher flowering plants that fix atmospheric carbon, synthesize other complex organic substances and subsequently incorporate them into their tissues. Organic compounds in plant tissues are returned to the soil to form organic matter through the processes of litterfall and decomposition. Trees and other plants also add nitrogen to the soil,

especially through the fixation of atmospheric nitrogen by legumes. The mechanism of nutrient accumulation in fallow topsoil during secondary succession is not well documented. Trees act as 'nutrient pumps' and they absorb nutrients from the subsoil and the weathering zone, immobilize them in their tissues and ultimately recycle them to the topsoil through litter and timberfall and through the decomposition and mineralization of litter. Tree roots also absorb plant nutrients from a large area and concentrate them in the soil beneath their canopies. Plants, especially trees with deep rooting systems or extensive root systems that usually extend well beyond the limits of their canopies, assist in redistributing plant nutrients in the soil and accumulate them in the topsoil, and especially in the soil underneath their canopies.

References

Aide, T.M., Zimmerman, J.K., Herrera, L., Rosario, M. and Serrano, M. (1995) Forest recovery in abandoned tropical pastures in Puerto Rico. *Forest Ecology and Management* 77, 77–86.

Aweto, A.O. (1981) Organic matter build-up in fallow soil in a part of south-western Nigeria and its effects on soil properties. *Journal of Biogeography* 8, 67–74.

Aweto, A.O. and Dikinya, O. (2003) The beneficial effects of two tree species on soil properties in semi-arid savanna rangeland in Botswana. *Land Contamination and Reclamation* 11, 339–344.

Beets, W.C. (1990) *Raising and Sustaining Productivity of Smallholder Farming Systems in the Tropics.* AgBe Publishing, Alkmaar, The Netherlands.

Belsky, A.J. (1994) Influence of trees on savanna productivity: tests of shade, nutrients and tree-grass competition. *Ecology* 75, 922–932.

Boettcher, S.E. and Kalisz, P.J. (1990) Single-tree influence on soil properties in mountains of eastern Kentucky. *Ecology* 71, 1365–1372.

Brand, J. and Pfund, J.L. (1998) Site- and watershed-level assessment of nutrient dynamics under shifting cultivation in Madagascar. *Agriculture, Ecosystems and Environment* 71, 169–183.

Brown, S. and Lugo, A.E. (1990) Tropical secondary forests. *Journal of Tropical Ecology* 6, 1–32.

Carrière, S.M., Letourmy, P. and McKey, D.B. (2002) Effects of remnant trees in fallows on diversity and structure of forest regrowth in a slash-and-burn agricultural system in southern Cameroon. *Journal of Tropical Ecology* 18, 375–396.

Egler, F.E. (1954) Vegetation science concepts: I. Initial floristic composition, a factor in old-field vegetation development. *Vegetatio* 4, 412–417.

Facelli, J.M. and Facelli, E. (1993) Interactions after death: plant litter controls priority effects in a successional plant community. *Oecologia* 95, 277–282.

Fellmann, J.D., Getis, A. and Getis, J. (2005) *Human Geography.* McGraw-Hill Higher Education, New York.

Ganade, G. and Brown, V.K. (2002) Succession in old pastures of Central Amazonia: role of soil fertility and plant litter. *Ecology* 83, 743–754.

Guariguata, M.R. and Ostertag, R. (2001) Neotropical secondary forest succession: changes in structural and functional characteristics. *Forest Ecology and Management* 148, 185–206.

Guillemin, R. (1956) Evolution de l'agriculture autochtone das les savannes de l'Oubangui. *L'Agronomie Tropicale* 11, 143–176.

Hooper, E., Condit, R. and Legendre, P. (2002) Responses of 20 native tree species to reforestation strategies for abandoned farmland in Panama. *Ecological Applications* 12, 1626–1641.

Hooper, E.R., Legendre, P. and Condit, R. (2004) Factors affecting community composition of forest regeneration in deforested, abandoned land in Panama. *Ecology* 85, 3313–3326.

Jurena, P.N. and van Auken, O.W. (1998) Woody plant recruitment under canopies of two acacias in a southwestern Texas shrubland. *Southwestern Naturalist* 43,195–203.

Knox, P.L. and Marston, S.A. (2004) *Places and Regions in Global Context: Human Geography.* Pearson Education, Upper Saddle River, New Jersey.

Manlay, R., Masse, D., Chotte, J.-L., Feller, C., Kaire, M., Fardoux, J. and Pontanier, R. (2002) Carbon, nitrogen and phosphorus allocation in agro-ecosystems of a West African savanna. II. The soil component under semi-permanent cultivation. *Agriculture, Ecosystems and Environment* 88, 233–248.

Nepstad, D.C., Uhl, C. and Serrao, E.A. (1990) Surmounting barriers to forest regeneration in abandoned, highly degraded pastures: a case study from Paragominas, Para, Brazil. In: Anderson, A.B. (ed.) *Alternatives to Deforestation: Steps Toward Sustainable Use of Amazonian Rain Forest.* Columbia University Press, New York, pp. 215–229.

Nye, P.H. and Greenland, D.J. (1960) *The Soil Under Shifting Cultivation.* Commowealth Bureau of Soils, Harpenden, UK.

Prescott, C.E. (2002) The influence of the forest canopy on nutrient cycling. *Tree Physiology* 22, 1193–1200.

Ramakrishnan, P.S. and Toky, O.P. (1981) Nutrient status of hill agro-ecosystems and recovery patterns after slash and burn agriculture (jhum) in north-eastern India. *Plant and Soil* 60, 41–63.

Richards, P.W. (1996) *The Tropical Rain Forest: An Ecological Study.* Cambridge University Press, Cambridge, UK.

Roder, W., Phengchanh, S. and Maniphone, S. (1997) Dynamics of soil and vegetation during crop and fallow periods in slash-and-burn fields of northern Laos. *Geoderma* 76, 131–144.

Trenbath, B.R. (1989) The use of mathematical models in the development of shifting cultivation systems. In: Proctor, J. (ed.) *Mineral Nutrients in Tropical Forest and Savanna Ecosystems.* Blackwell, Oxford, UK, pp. 353–369.

Uhl, C., Buschbacher, R. and Serrao, E.A.S. (1988) Abandoned pastures in eastern Amazonia. I. Patterns of plant succession. *Journal of Ecology* 76, 663–681.

Vasquez-Yanes, C. (1998) *Trema micrantha* (L.) Blume (*Ulmaceae*): A promising Neotropical tree for site amelioration of deforested land. *Agroforestry Systems* 40, 97–104.

Vetaas, O.R. (1992) Micro-site effects of trees and shrubs in dry savanna. *Journal of Vegetation Science* 3, 337–344.

Wezel, A., Rajot, J.-L. and Herbrig, C. (2002) Influence of shrubs on soil characteristics and their function in Sahelian agro-ecosystems in semi-arid Niger. *Journal of Arid Environment* 44, 383–398.

Zahawi, R.A. and Augspurger, C.K. (1999) Early plant succession in abandoned pastures in Ecuador. *Biotropica* 31, 540–552.

Zinke, P.J. (1962) The pattern of influence of individual forest trees on soil properties. *Ecology* 43, 130–133.

9 Intensification of Shifting Cultivation

Shifting cultivation is a sustainable system of agriculture where population density is low, enabling fallow periods to be sufficiently long to allow the soil to restore its fertility status adequately. Nye and Greenland (1960) observed that shifting cultivation is nicely adjusted to tropical conditions and is sustainable where population density is low. More recent studies on shifting cultivation (Fujisaka, 1991; Kleinman *et al.*, 1995) have also come to the conclusion that the system of arable farming is sound ecologically and economically in areas with low population densities, and hence is sustainable in the long term only in areas where population pressure on land is low. However, since about the turn of the 20th century, population has increased dramatically in most parts of the humid and sub-humid tropics, where the system of shifting cultivation is the dominant system of arable farming, making it virtually impossible to practise the extensive forms of slash-and-burn agriculture that require very long fallow periods.

In the forest zone of southern Nigeria, fallow periods that used to be up to 15–20 years have been reduced to 3–5 years due to population pressure, while in parts of south-eastern Nigeria, continuous cultivation has emerged and has largely replaced the system of shifting cultivation involving the bush fallow. Similarly, in northern Laos, consequent upon a 130% increase in population density between 1950 and 1990, fallow periods have decreased from 38 years to 5 years (Roder, 1997). Population increase over time is a major driver of agricultural change and intensification (Boserup, 1965). Due to the considerable increase in population density over the last five to six decades in areas where shifting cultivation is practised, the system of agriculture has undergone various forms of intensification in response to the challenge of increasing population pressure on land and the need to feed a rapidly growing population.

The aim of this chapter is not to provide a blanket or a uniform strategy for intensifying shifting cultivation. It is recognized that shifting cultivation, although usually characterized as a single agricultural system, consists of a multiplicity of agricultural practices, some of which reflect regional or local environmental conditions. Since environmental and economic conditions vary considerably in areas where shifting cultivation is practised in the tropics, the method adopted to intensify the system will vary and may be area-specific. Animal manure is readily available in the savanna regions of the tropics, where large numbers of livestock are kept. Hence, the application of animal manure is a viable option for intensifying shifting cultivation in the savanna regions, but not in the forest regions where livestock keeping is on a very low scale due to non-availability of natural grazing and, in the forest areas of tropical

Africa, the problem of tsetse fly infestation. In addition, *Elaeis guineensis* (the oil palm) has been integrated into farming systems in southern Benin and southern Nigeria to enhance the economic value of the fallow and the process of soil fertility restoration. One cannot recommend agroforestry systems involving the integration of the oil palm into cultivated land and bush fallow systems for Laos, Thailand and Vietnam, because the oil palm is not a native of those areas. It would be difficult, and perhaps un-economical, to obtain oil palm seedlings from West Africa or from Malaysia for distribution to small-scale farmers in these three Asiatic countries. Furthermore, there may be no ready market for oil palm products such as palm wine in Laos, Thailand and Vietnam.

This chapter discusses the various ways the system of shifting cultivation has been modified to make it a more intensive system of agriculture. Attempts to modify shifting cultivation include the integration of trees into farmlands to form agroforestry systems, enrichment of fallow vegetation with tree legumes or other species to enhance its soil restorative ability, and some practices that may be specific to certain areas, such as mound cultivation and composting. Alternatives which have been proposed as replacements for the natural bush fallow vegetation, such as alley farming, planted tree fallows and the intensive use of inorganic fertilizers to make continuous farming feasible, are considered in Chapter 10.

9.1 Palm Fallows

The planting of oil palm, as is usually done in southern Benin Republic and parts of south-eastern Nigeria, represents a bold attempt not only to enhance the efficacy of the natural fallow to improve soil fertility, but also to increase its economic value, in terms of yielding tangible and economic returns to the farmer. In southern Benin Republic, population density is high and fallow periods have been reduced to 3 years or less in some cases and, due to the short fallow periods, soil fertility is not adequately restored prior to cultivation (Edja, 2001). Farmers plant oil palm seedlings in fields which are cultivated for 3–6 years. The palms subsequently become established in fallow vegetation as trees, and

they enhance the restoration of soil fertility by the bush fallow. The density of palm trees may be up to 600 ha^{-1} (Muller-Samann and Kotschi, 1997) but is often considerably higher. In common with most trees and shrubs, palm trees help to accumulate organic matter and nutrients under their canopies, thereby helping to rejuvenate soil fertility in bush fallow vegetation in which they are an important component. The study of Kang (1977) in south-western Nigeria has shown that the levels of organic matter, exchangeable calcium, magnesium and potassium, available phosphorus and cation exchange capacity (CEC) in the 0–15 cm layer of the soil under oil palm canopies were 39–154% greater than their respective levels in adjoining soil outside the zone of influence of the tree canopy. The fibrous roots of the palms also help to reduce soil erosion and the attendant loss of organic matter and nutrients from the farmland, while the fronds and fruits help to add litter to the soil, thereby ensuring the long-term sustainability of the agroforestry system.

In most areas where shifting cultivation is practised, farmers deliberately retain a number of trees in farms to provide shade, fruits, medicine or some economic benefit. Such trees subsequently become an integral part of fallow vegetation at the end of cropping. It is important to recognize that they were not deliberately planted by the farmer in fallow vegetation. What is perhaps unique about the palm fallows of Benin Republic is that they were deliberately planted by farmers to perform the dual function of restoring soil fertility and yielding revenue. This is definitely an innovative approach to the intensification of shifting cultivation by small-scale traditional farmers.

As with most parts of the rainforest zone of West Africa, farmers in southern Benin also cultivate field crops such as maize and cassava in groves of oil palm, or palms that have been selectively retained in farmlands over the years. This type of traditional agroforestry ensures that field crops which are usually cultivated for food can be grown on the same piece of land as a tree crop, with the latter serving as a cash crop. In Mono Province of Benin Republic, the Adja people grow field crops such as maize and cassava in between oil palm seedlings established in their cultivated fields for 5–8 years (Kang *et al.*, 1991). They usually prune the branches of the palms when preparing land for re-cultivation after a few years of cropping. After about 6 years, the palms substantially reduce the yield of the field crops, due to shading, making further cultivation of field crops in the palm groves not worthwhile. At this stage, cultivation of field crops is discontinued and the palms are allowed to grow as a fallow crop for several years. Thereafter, the palms are felled and tapped for the production of local wine. They also yield other useful products such as palm oil and kernel and materials for making baskets and brooms and for roofing. The felled palm trees and their fronds are left to decompose on the farm to help replenish soil organic matter and nutrient status of the coarse-textured, nutrient-deficient soils. The nutrients accumulated in the soil during the palm fallow, as well as those released into the soil due to the decomposition of felled palms, make it possible for the farmers to cultivate the land again. The study of Quenum (1988) indicated that follow-ing the palm fallow, field crops can be cultivated for a period of 5 to about 12 years. After fallows of both low-density palms (500–700 palms ha^{-1}) and high-density palms (1500–2000 palms ha^{-1}), maize yields exceeded 1 t ha^{-1} during the first 5 years of cropping. Thereafter, yield after high-density fallow declined below that of low-density fallow, which stabilized at about 0.7 t ha^{-1} for up to the tenth year of cropping (Fig. 9.1). This type of agro-forestry has a major advantage in areas of high population density, such as the Gulf of Guinea coastal area of West Africa, particularly in the heartland of south-eastern Nigeria, where the land available for farming is very limited. Farmers therefore try to maximize the use of the available land by growing tree and field crops side by side on the same plot. It is important to observe that the oil palm, with its light canopy that allows light to penetrate to the ground, is suited to intercropping with field crops, if the density of palms is not too high.

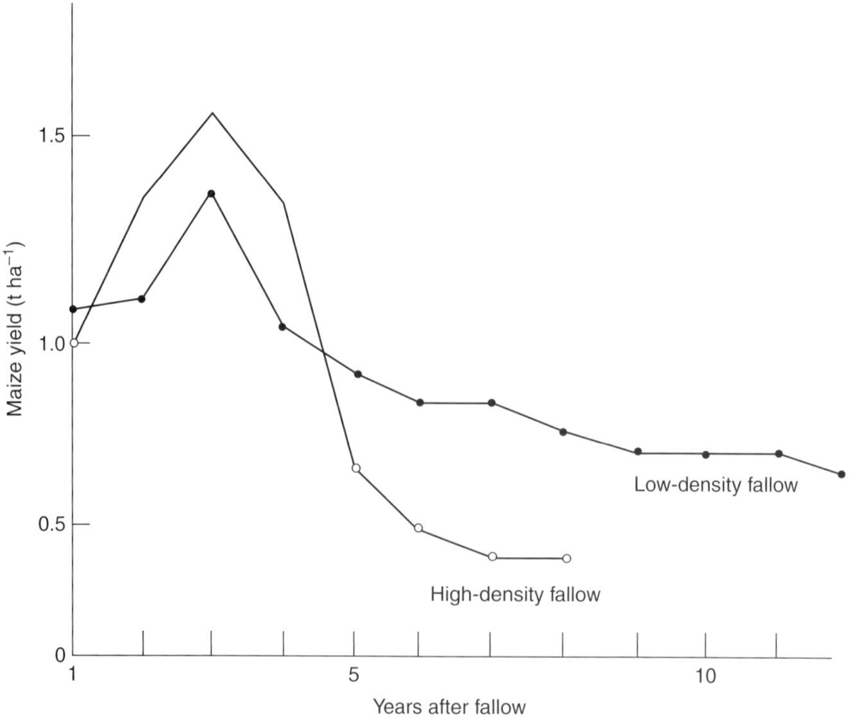

Fig. 9.1. Maize yield in Mono Province of Benin Republic following low-density (500–700 stands ha^{-1}) and high-density (1500–2000 stands ha^{-1}) palm fallow. (After Quenum, 1988.)

Agroforestry systems involving palms have also been developed in the Amazon basin of South America, particularly in north-eastern Brazil. Here, the babassu palm has been successfully and harmoniously integrated into farmlands and fallow land. Unlike in Benin Republic where farmers plant oil palms in their farms and leave them to yield fruits and wine during the fallow period, the babassu palm is not deliberately planted by farmers in north-eastern Brazil. Instead, farmers selectively retain babassu palms on their farms by protecting and encouraging them to regenerate during cultivation and they subsequently become an important element of the flora of fallow vegetation (May *et al.*, 1985). Owing to the selective preservation of the babassu palm in farmlands, its population increases over time, giving rise to what superficially looks like a palm 'forest'. The palms provide the farmers with oil and milk; some large rodents that inhabit the palm 'forests', feeding on the fruits of the palms, are hunted for food; while the larvae of beetles that are also associated with the palms are collected and eaten. The integration of palms into farmland in north-eastern Brazil will no doubt have some beneficial effect in improving soil fertility, thereby helping to make the traditional agroforestry system sustainable. Furthermore, the palms are a source of food and revenue for poor rural farmers (Albiero *et al.*, 2011) in addition to helping to conserve some species of wildlife.

Other palm-based fallow systems are also important in South and Central America. Such palms, as with the oil palm in West Africa, help in recycling nutrients from the subsoil. They most likely enhance nutrient accretion in the soil during the fallow period, and such nutrients subsequently become available to food crops planted in the site (Hecht *et al.*, 1988). The palms appear to play an important role in the regeneration of secondary vegetation after deforestation and cultivation, especially in nutrient-deficient forest and savanna soils which are common in the Amazon basin. Apart from the babassu palm, other palms which are an integral part of fallow vegetation in Latin America include *Attalea cohune* and *Attalea butyracea* in Central America and the northern part of South America, while *Mauritia flexuosa* and *Copernicia alba* are characteristic of flooded

or wetter areas (Kass and Somarriba, 1999). Mention should also be made of the American oil palm, *Elaeis oleifera*, which provides wax (Henderson *et al.*, 1995) and plays a role in the rural economy that is secondary to that of the babassu palm.

9.2 Traditional Agroforestry Systems Involving Other Trees

The agroforestry systems involving palm fallows discussed above were not established with the sole aim of improving soil fertility, but more importantly to provide additional revenue for the farmer, thereby transforming the bush fallow into a true source of income. Some traditional agroforestry systems have evolved mainly to conserve and to improve soil fertility. Cases in point include the enrichment planting of bush fallow vegetation with tree legumes, or with other trees which have proven abilities of improving soil fertility.

9.2.1 Tree legumes in farms in Java

A case of traditional agroforestry which deserves mention here is the system practised by the Baduy people who live in the relatively isolated area around the Kendeng Mountain, south of Jakarta, Indonesia. These people regard shifting cultivation as pivotal and central to their cultural identity, and their culture forbids the use of inorganic fertilizers and other modern chemical inputs (Iskandar and Ellen, 2000). In the face of increasing population and this taboo, the Baduy have embraced agroforestry as a means of intensifying their formerly extensive system of shifting cultivation and to prevent soil mining and progressive deterioration in soil quality over time.

They have introduced *Paraserianthes* (*Albizia*) *falcataria*, a commercially valuable tree legume, into their farmlands and fallow vegetation to help sustain or even improve soil fertility during cropping and to facilitate the process of soil fertility restoration during the fallow period, which has been consequently reduced. Rice is a chief crop of the Baduy, and they plant rows of *Paraserianthes* (*Albizia*) seedlings between the rows of rice. Sometimes

Paraserianthes (*Albizia*) seedlings are grown in nurseries and are transplanted to the rice field when they reach a height of about 20 cm. Up to 600 trees may ultimately be established per hectare of rice field (Iskandar and Ellen, 2000). Rice is usually cultivated for 2 years, and thereafter the land is fallowed for 3–4 years. During cultivation, *Paraserianthes* (*Albizia*) trees fix nitrogen and enhance soil fertility status. Farmers also prune small branches and leaves from the tree legume and return the prunings to the soil to help replenish soil organic matter and nutrient status. The branches and twigs of *Paraserianthes* (*Albizia*) trees are not pruned during the fallow period. After 3–4 years of fallow, when soil fertility is judged by the farmer to have been restored to a satisfactory level, the trees are felled for sale as timber, while the branches are used as fuelwood and the leaves and twigs recycled to the soil as mulch or burnt to fertilize the soil, prior to another phase of cultivation.

9.2.2 Trees in farms and fallows of Bora Indians, Peru

There is a large variety of traditional agroforestry systems but only a few more examples will be examined here. The Bora Indians of Peru, South America, have developed some interesting agroforestry systems. They cultivate several varieties of sweet and bitter cassava, which are intercropped with pineapples and fruit trees. *Theobroma bicolor* (Macabo tree) and peach palm are intercropped with cassava in farmers' fields to provide fruits. These trees are retained on the farm and become important elements of the regenerating fallow vegetation when farming is discontinued. At the inception of the fallow period, farmers also plant tropical cedar seedlings which are allowed to grow in fallow vegetation for about 30 years to yield valuable timber and income to the farmer or his children when they are felled for cultivation (Denevan *et al.*, 1984). The fallows are a source of fruits for the farmer and his family, but when the fallow vegetation attains the age of 20–30 years, collection of fruits from trees retained or planted on the farm during cultivation becomes difficult. Nevertheless, such old fallows serve as hunting grounds for the Bora

Indians as they provide habitat for a large amount of wildlife.

The Bora Indians' agroforestry systems are successional and consist of several layers of trees that are planted in their fields or fallows at different periods (Sanchez, 1994). The system produces field crops such as rice for food, edible fruits from which farmers earn cash income or which can be used to meet part of the family food requirements, firewood and, ultimately, lumber, which can be harvested two or three decades after the inception of the agroforestry system. Sequentially, it provides different products to the farmer at different stages of the cropping period–fallow period continuum. When land is cleared for cropping, the Bora Indians grow acid-tolerant trees such as peach palm with field crops such as rice or manioc. After a few years, when the trees cast a lot of shade on the ground on account of their increasing biomass, food crop production is discontinued and a ground cover of legumes such as *Centrosema macrocarpum* (kudzu) and, at the same time, the fast-growing tree legume, *Inga edulis*, is planted for fruit and charcoal production (Sanchez, 1994). The tree legume constitutes the second layer of trees in the agroforestry system. Finally, slow-growing timber-yielding trees are planted in the fallow, thus constituting the third tree storey, to produce lumber for the farmer or his children after about 20–30 years.

The 'successional' agroforestry system above is modelled on the *taungya* system in which trees and field crops are grown together for a few years, after which field crop production is discontinued and a tree plantation is established. The 'successional' agroforestry system of the Bora Indians, several variants of which exist in the Amazon basin, is discussed here to show the wide range of tree-based agricultural practices developed by shifting cultivators. Although such agroforestry systems mimic the natural rainforest in terms of structure and species diversity, they cannot be properly considered as viable alternatives to shifting cultivation, as food crop production is discontinued after a few years. In a somewhat restricted sense, they can be regarded as intensified 'fallows', which yield economic benefits to farmers after food crop production has ceased.

9.2.3 Casuarina trees in farms in Papua New Guinea

In Papua New Guinea, the casuarina tree (*Casuarina* spp.) is greatly treasured as a soil improver and a source of timber in agroforestry systems. The tree adds copious litter to the soil while nitrogen-fixing microorganisms associated with the tree roots fix atmospheric nitrogen and enhance nutrient accretion in the soil. The tree is planted by farmers in their fields, especially by the Wola people of central Papua New Guinea, to help accumulate nitrogen and – to a lesser extent – other nutrients, through the cycling of nutrients from the subsoil. It allows adequate light to penetrate to the ground surface on account of its light and fairly open canopy, so is ideally suited for integration into farmers' fields to support a field layer of cultivated crops (Sillitoe, 1995). The tree is often planted scattered in farmers' fields but also along field boundaries. After harvesting crops, the farmer may allow the trees to grow for another 10–20 years to yield valuable timber, which serves as a handsome source of revenue for the farmer, and other products such as firewood and building materials (Newton, 1960).

9.3 Enriched Fallows of Soil-improving Trees

The examples of traditional agroforestry systems considered in the preceding sections were those in which trees of economic importance to the farmer were integrated into fallow vegetation in order to transform the bush fallow into an important source of revenue for the farmer. The trees selectively retained or planted by the farmer in fallow vegetation have the additional advantage of helping to restore or improve soil fertility. In this section, consideration will be given to traditional agroforestry systems in which the primary objective of the farmer is to improve soil fertility. In essence, the farmers plant or retain some selected tree species which have been proven or perceived to have soil fertility-improving properties in their fields, either during cultivation, or prior to the inception of the fallow period. Two examples will be considered here.

9.3.1 Pada in rice fields and fallow vegetation in northern Thailand

First, we shall examine the case of the soil-improving tree known as pada (*Macaranga denticulata*) in northern Thailand, where the tree has been well integrated into fallow vegetation due to its soil-improving and rice yield-enhancing properties. Due to rapid increase in population over the last six decades or so, fallow periods have been shortened from 10–20 years to 7 years or less. The shifting cultivators in this hilly region believe that pada trees have been largely responsible for maintaining the productivity of rice under the current practice of a short fallow period, and have also prevented progressive soil degradation. Yimyam *et al.* (2003) carried out a study in farmers' fields in Mae Hong Son province of northern Thailand, in which they compared the soil properties of fallow vegetation with sparse stands of pada, with those of dense stands. Usually, pada seedlings emerge soon after the first rains following the planting of upland rice, and are protected by the farmer, being weeded out only when their density is too high. Sometimes farmers transplant pada to parts of the farm where the density of pada seedlings is low. The density of pada trees in fallow vegetation declines over time through competitive elimination of some individuals as the pada trees become bigger.

After 7 years of fallowing, the number of pada trees in fallow vegetation with a sparse stand of pada was 1000 trees ha^{-1}, compared to 4200 trees ha^{-1} in fallow vegetation with dense stands (Yimyam *et al.*, 2003). The biomass of pada in fallow with the dense stands was 43 Mg ha^{-1}, and this was 20% more than that of the fallow with sparse stands. The pada fallows did not consist exclusively of pada plants; they accounted for 45 and 67% of the trees in the fallow vegetation with sparse and dense stands of pada, respectively. Following 7 years of fallowing, the fields with both dense and sparse stands of pada were slashed and burnt, and planted to upland rice. The soil underneath the two types of fallow vegetation was also sampled and analysed. The results obtained by Yimyam *et al.* (2003) indicated that the dense stands of pada fallow vegetation accumulated about 50% more calcium and 141–185% more magnesium in the 0–30 cm and 30–60 cm layers of the soil (Fig. 9.2). Also, upland rice yield in the farms with dense pada was about three times that obtained in the plots with sparse pada stands. Obviously, pada trees accumulate nutrients in the soil, especially exchangeable calcium and magnesium, for the crop planted after the fallow. Similarly, the pada trees immobilize substantial amounts of nutrients in their standing biomass, which are released into the soil after the burning of pada-enriched fallow.

9.3.2 *Gliricidia*-enriched fallows of south-western Nigeria

Fallows in south-western Nigeria, especially in the Ibadan area, now contain a significant proportion of the tropical American tree, *Gliricidia sepium* (Fig. 9.3), originally introduced into western Nigeria to provide shade for cocoa seedlings (Gledhill, 1972). Local farmers later introduced it into their farms of field crops, primarily for the purpose of using it to train yam vines. The tree is resistant to burning, coppices profusely and has survived in cultivated fields near the city of Ibadan; yam, however, is no longer the main crop due to a decline in soil fertility caused by shortening the fallow period to about 3 years or eliminating it altogether. *Gliricidia sepium* is now the most abundant tree in cultivated fields in the Ibadan area, and also in fallow vegetation. Akpokodje and Aweto (2007) examined its effects on the soil beneath its canopy in farmers' fields in the Ibadan area. They observed that on the farms there was no significant build-up of organic matter, total nitrogen, exchangeable calcium, magnesium and potassium and CEC in the 0–10 cm layer of soil beneath the tree canopy compared to soil

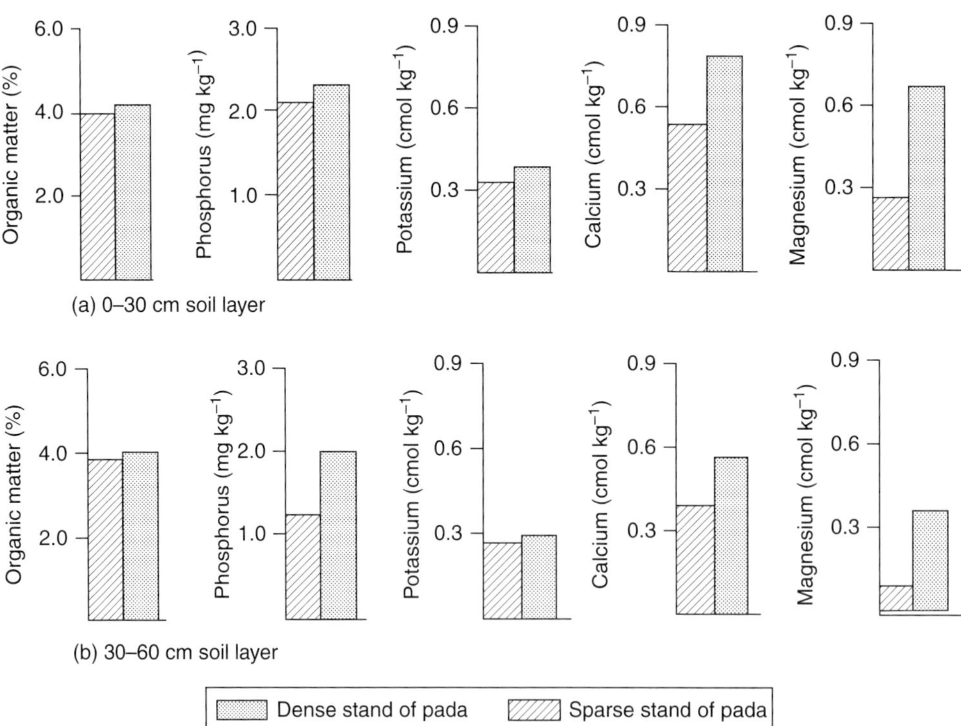

Fig. 9.2. Effects of sparse and dense stands of pada in a 7-year bush fallow vegetation on soil properties in Sob Moei district, northern Thailand. (Based on data of Yimyam *et al.*, 2003.)

Fig. 9.3. An enriched woody fallow vegetation consisting mainly of the tree *Gliricidia sepium*, with a field infested with *Imperata cylindrica* (spear grass) in the foreground.

outside the influence of the canopy. Similarly, they observed no significant accretion of the micronutrients iron, copper, manganese and zinc under *Gliricidia* canopies. The farms studied were farmed continuously, and Akpokodje and Aweto attributed the non-accumulation of plant nutrients in the soil under the tree canopy to site burning prior to cultivation, which destroys plant litter instead of allowing it to accumulate under the trees and subsequently decompose to form humus. However, when the land is left uncultivated during the fallow period, accretion of organic matter and plant nutrients occurs in the soil under *Gliricidia* fallows over time (Adejuwon and Adesina, 1990), implying that *Gliricidia* trees have a salutary effect on soil nutrient status.

9.4 Planted Fallows

Owing to the slow regeneration of natural bush fallow vegetation and the fact that it takes several years for it to adequately rejuvenate soil fertility, scientists and farmers have experimented with the idea of establishing fallow vegetation of fast-growing tree species, espe-

cially of legumes, to replace the natural bush fallow vegetation. Planted fallows are improved fallows; that is, fallow vegetation with enhanced capacity of restoring soil fertility within a relatively short period. There are two main types of planted or improved fallows, namely: (i) a seasonal fallow of herbaceous plants, usually legumes, which are established for a few months (usually for one growing season or part of this) prior to cultivation of field crops such as cereal crops (e.g. rice, maize, millet or other crops); and (ii) multi-seasonal fallows, which are usually of tree species, especially tree legumes that enhance nitrogen accretion in the soil, or deep-rooted trees that enhance recycling of nutrients to the topsoil for the subsequent use of field crops planted after clearing the planted fallow.

Such fallows may last from 2 to 5 years and, in several instances, provide additional benefits to the farmer such as fuelwood, fruits, or fodder for livestock. The use of planted fallows in place of the natural bush fallow vegetation is an attempt to intensify shifting cultivation rather than to replace it with a new system of agriculture. This paradigm of managing the soil under shifting cultivation attempts to manipulate the plant species in fallow vegetation, with a view to selectively utilizing those with proven ability in regenerating soil fertility within a reasonably short period, and replacing those in natural fallow vegetation. Such species may be indigenous to the area in question or may be exotics imported from other parts of the tropics where they have been demonstrated to have a beneficial, or even a dramatic effect, on the process of soil fertility rejuvenation.

9.4.1 Seasonal or short-duration fallows

These are short-term planted fallows of herbaceous plants which are established for a few months to improve soil fertility prior to the cultivation of field crops. Sometimes, a seasonal fallow crop may be planted a few months after a field crop such as rice or maize is sown. The seasonal fallow crop provides ground cover for the food crop and helps to protect the soil against erosion, as most are vines that creep on the ground surface. In addition the seasonal fallow crop, if it is a legume, may also help to increase soil nitrogen, a nutrient that is usually deficient in most soils used for shifting cultivation. The seasonal fallow crop is usually left on the farm after the field crops are harvested, to protect the soil and fertilize it during the subsequent fallow period, which usually lasts for a few months during the dry season.

Species planted as seasonal fallow crops in the tropics usually include *Mucuna pruriens* var. *utilis* (mucuna or velvet bean), *Pueraria phaseoloides* (kudzu) and other cover crops such as *Canavalia ensiformis* (canavalia). At the end of the fallow period, the planted fallow of cover crops is slashed and the cut plant material is usually incorporated into the soil as green manure (rather than being burnt, as in traditional shifting cultivation) to fertilize the soil on decomposition and benefit the field crops subsequently planted on the farm. Alternatively, the slashed plant material may be used as green mulch to protect the soil against erosion during the period of cultivation. Such mulch decomposes over time and helps to replenish soil nutrients and organic matter. Whether buried in the soil or used as green mulch, the slashed cover crop releases nutrients for planted crops over an extended period, with minimal loss of nutrients from the soil.

This is a major way in which the effects of seasonal fallows of cover crops differ from traditional shifting cultivation, which involves burning the cut slash of fallow vegetation and a substantial part of the ash derived from burning slashed vegetation is washed away by surface runoff or leached downward by percolating rain water. Steiner (1991) observed that planted seasonal fallows do not necessarily improve soil fertility as they are of very short duration, but rather they stabilize soil fertility, preventing it from further deterioration during the period of a couple of months when the planted seasonal fallow is allowed to remain on the field. More recent experimental work and farmers' experience with the use of seasonal fallows on their farmlands, however, suggest that they can improve soil fertility and substantially improve the yields of crops grown after the seasonal fallow of cover crops.

In northern Honduras, seasonal fallows of mucuna are used for maintaining and improving soil fertility and have been well synchronized with the bimodal rainfall regime that characterizes the area and the farming calendar. During the first part of the wet season, farmers establish mucuna fallows on their fields as seasonal fallows or cover crops. In the latter part of the wet season, the mucuna is slashed, incorporated into the soil and the land sown to maize. Over a period of 15 years of continuous use of mucuna seasonal fallow that alternated with maize production, soil pH and exchangeable calcium did not diminish, while soil organic matter, infiltration capacity and porosity improved on maize fields in which mucuna was relay cropped with maize, compared to fields without mucuna. In addition, maize yields in fields in which mucuna was continuously relay cropped with mucuna doubled during the 15-year period, compared to fields continuously cropped to maize without mucuna (Buckles and Triomphe, 1999). In the highlands of Guatemala, *Lathyrus nigrivalis* is widely used as seasonal fallow and cover crop. Seeds of the plant are sown in between maize rows 2–4 months after planting maize. The plant is allowed to remain on the farm after maize harvest and it grows on the field throughout the following dry season, as a seasonal fallow. At the inception of the wet season, the *L. nigrivalis* vines are slashed and buried in the soil, prior to planting maize (Kass and Somarriba, 1999). Through the use of seasonal fallows of *L. nigrivalis*, farmers have been able to increase maize yield from 2.6 to 5.1 Mg ha^{-1} (Flores, 1994). In Honduras and Guatemala, farmers also establish seasonal fallows of *Dolichos lablab* (lablab bean) which they alternate with maize crop. As with *L. nigrivalis*, the lablab bean is usually sown at the same time as a maize crop, and is allowed to grow on the field throughout the dry season after the maize is harvested. It is slashed and dug into the soil as green manure for a subsequent maize crop during the following wet season.

The sunflower-like *Tithonia diversifolia* (tithonia) is gaining increased acceptance as a seasonal fallow plant in tropical Africa and South-east Asia because of its remarkable ability to immobilize soil nutrients and accumulate them in its biomass within a few months. Many resource-poor farmers who cannot afford to buy inorganic fertilizers, especially in the thickly populated highlands of western Kenya, depend solely on tithonia short fallows for restoring and improving soil fertility, making it possible for them to cultivate the same land continuously. The leaves of tithonia are exceptionally rich in phosphorus (Gachengo, 1996). The plant also contains very high levels of nitrogen and therefore has the potential of supplying two of the major plant nutrients that are chronically deficient in most tropical soils used for shifting cultivation, especially those that are highly weathered. The plant grows readily as weeds on farms or roadsides, and can be readily established on farmers' fields or be gathered from roadsides and applied as green manure in farmlands. The study of Smestad *et al.* (2002) in the Vihiga District of the western Kenya highlands has shown that when 31-week seasonal fallows of *T. diversifolia* and of *Crotalaria grahamiana* were cut and dug into the soil (alfisol), soil fertility, as reflected in soil organic matter and total nitrogen content, was higher than in a natural weed fallow of 9–12 years. Maize yield following the fallow was also higher for the soil under the two types of seasonal fallows, mainly due to nitrogen accumulation in the 0–15 cm of the soil relative to soil under the natural weed fallow. In southern Cameroon, especially around Yaounde where traditional shifting cultivation is undergoing intensification, farmers plant pigeon pea on their farms of field crops and allow the pigeon pea to grow with the crops for about 11 months before it is cut (Franzel, 1999). The pigeon pea helps to fertilize field crops such as cassava and maize, and also provides some food for the farmer.

Seasonal fallows of mucuna and lablab have been reported to have beneficial effects on the soil in the Pacific islands, including Vanuatu and the Solomon Islands in Melanesia; and in the Polynesian islands of Samoa and Tonga and the Cook Islands. In these islands, the deployment of the seasonal fallows and the leguminous tree *Gliricidia sepium* to manage soil fertility has substantially increased soil fertility and improved cassava yield, thereby reducing the need to apply inorganic fertilizers (*Spore*, 2009). In the southern Guinea savanna zone in northern Nigeria, the study of Osunde *et al.* (2009) has shown that one season's fallow crop of mucuna has minimal effect on maize yield of an alfisol, compared to inorganic fertilizers. However, when mucuna and lablab fallows were established for 3 years in the northern Guinea savanna zone of northern Nigeria, the cover crops had beneficial effects on soil physical status. Odunze *et al.* (2009) reported that a 3-year fallow of lablab and mucuna had greater effects in conserving soil moisture and also had more water-stable aggregates in the 0–10 cm layer of an alfisol than a natural bush fallow of the same age. This implies that the planted fallow of mucuna and lablab improved soil aggregation and hence had greater potential for protecting soil against erosion than the natural bush fallow, when both natural and planted fallows are cleared for cultivation.

Seasonal fallows are established by some farmers in Mexico and Brazil. In most parts of the tropics where shifting cultivation is practised, farmers are generally unwilling to invest their time and energy in planting seasonal fallows, and applying the cut slash as green mulch or incorporating it into the soil, as this merely improves soil fertility, without yielding fruits or food and other economic benefits. Their use is, however, becoming popular in densely populated areas such as Nicaragua and the western highlands of Kenya, where shortage of land makes it virtually impossible to leave land under natural bush fallow for several years to ensure that soil fertility is adequately restored. In hilly areas with steep slopes such as in northern Honduras, farmers are also embracing the use of seasonal fallows to control soil erosion and to help improve soil fertility. *Canavalia ensiformis* (the jackbean) is gaining widespread acceptance as a seasonal fallow plant in semi-arid areas because it is drought tolerant and grows on poor soils, and has the ability to immobilize large quantities of nutrients in its biomass and to restore wasteland and degraded soil (Bunch, 2003).

9.4.2 Multi-seasonal or long-duration planted fallows

These are planted fallows that usually consist of trees, especially legumes, which are established for a few or several years with a view to accumulating organic matter and nutrients in the soil. Although they may have the additional benefit of providing the soil with an adequate cover, thereby preventing degradation in soil properties, they are established mainly to improve soil fertility at a rate that is faster than that of the natural bush fallow vegetation. Multi-seasonal fallows, unlike seasonal fallows, usually consist of fast-growing tree species such as *Gliricidia sepium*, *Inga edulis*, *Sesbania sesban*, *Tephrosia vogelii* and *Calliandra calothyrsus*, which are planted in monocultural stands. Non-leguminous tree species such as *Acioa barteri* and *Alchornea cordifolia* are also useful as multi-seasonal fallows. Such tree species are characterized by rapid growth and a deep rooting system which enables them to recycle nutrients leached deep into the subsoil back to the topsoil for the benefit of crops planted after the fallow.

These fallows are established for periods ranging from 2 to 5 years, and sometimes for a longer period. When they are cleared at the end of the fallow, they have accumulated enough nutrients in the soil so that burning vegetation biomass to release nutrients immobilized in the planted fallow may not be necessary. Also, there is no need for the farmer to expend so much energy incorporating the biomass of planted tree fallows into the soil as is the case with seasonal fallows such as mucuna or jackbean. The trees also provide the farmer with fuelwood, which is in critically short supply in densely populated areas that have been settled for a long period.

Some planted tree fallows, such as those of *Sesbania sesban*, have been proved to be superior to natural bush fallow vegetation in restoring soil nutrients, especially on nutrient-deficient soils (Kwesiga *et al.*, 1999); others, such as *Cassia siamea* and *Albizia lebbeck*, have about the same soil restorative capacity as natural bush fallow vegetation (Drechsel *et al.*, 1991). Only a few of the trees, or shrubs such as *Cajanus cajan* (pigeon pea), used as improved fallows produce fruits that can be used as food by farmers, and this has enhanced the acceptability of such plants as planted fallows. In some cases, planted tree fallows such as those of *S. sesban* improve soil properties remarkably in a matter of 2–3 years, while in others it may take up to 5 years for the improved tree fallow to rebuild soil fertility to a satisfactory level that guarantees adequate yields during the subsequent period of cropping.

A number of tree and shrub species have been tried as planted tree fallows on the poor soils of the Chipata area in eastern Zambia, and *S. sesban*, an indigenous tree that grows rapidly and produces large biomass and also fixes nitrogen (Evans and Rotar, 1987), was found to be one of the most promising trees in terms of its potential to improve soil fertility within a few years. The study of Kwesiga *et al.* (1999) has shown that a 2-year fallow of *S. sesban* established in the Chipata area of Zambia had a dramatic effect in improving soil fertility and increasing the yield of maize planted after the fallow, compared to a grass fallow and other planted fallows of pigeon pea (*Cajanus cajan*), *Tephrosia vogelii*, and even a fertilized plot continuously cropped to maize. The residual effects of soil improvement by *Sesbania* fallow were still marked in the second year of cultivation, as evident in the enhanced maize yields in the second year of cultivation after a fallow period of 2 years.

In the northern Guinea savanna zone of Ghana, Agyare *et al.* (2002) observed that following a 2-year planted fallow of pigeon pea, the maize yield was 3.02 t ha^{-1}, which was significantly greater than the maize yield from plots cleared from natural bush fallow (which amounted to 1.54 t ha^{-1}). In addition to the increased maize yield from the improved fallow plots of pigeon pea, they also produced fodder for livestock as well as grains. In the Yaounde area of southern Cameroon, fallows of *Calliandra calothyrsus* (calliandra) are established for 1–2 years to improve soil fertility, prior to cropping (Franzel, 1999). During cropping of field crops, calliandra is cut to ground level and the coppice shoots repeatedly cut to reduce competition with field crops. They are allowed to grow tall only during the fallow period. Hoang Fagerstrom *et al.* (2002) reported that planted fallows of *Tephrosia candida* have beneficial effects on soil fertility in Vietnam. They observed that a 2-year fallow of *T. candida* immobilized and stored 34% more nitrogen in its standing biomass and increased the nitrogen level of the top 5 cm of the soil by 20%, compared to a natural bush fallow vegetation of similar age.

It should be pointed out that it is not in all cases that improved fallows of planted trees have proved to be superior to natural bush fallow vegetation in restoring soil fertility status. A study by Drechsel *et al.* (1991) on the effects of 5-year planted fallows of *Azadirachta indica*, *Albizia lebbeck*, *Cassia siamea*, *Acacia auriculiformis* and natural bush fallow on the properties of soil (ultisols) in central Togo has shown that *C. siamea* and *Azadirachta indica* were superior to the other tree species in accumulating calcium and increasing the pH of the topsoil. The four improved fallows of the tree species were not significantly different from a natural bush fallow in terms of the ability to accumulate organic matter and nutrients in the soil (Drechsel *et al.*, 1991). Herein lies a major dilemma for native farmers who are advised by extension workers to adopt improved tree fallows as a means of intensifying shifting cultivation. In several cases, planted tree fallows are not necessarily superior to a natural bush fallow in terms of soil regenerative capacity. In others, with the exception of some trees such as *Sesbania sesban*, the planted tree fallow is marginally superior to a woody natural bush fallow in restoring soil fertility. Hence, shifting cultivators are generally reluctant to adopt improved tree fallow technology, unless the tree species to be established as improved fallows yields fruits or medicine or other tangible useful products to the farmer. Besides, as pointed out earlier, it is in hilly or mountainous areas with steep slopes, where soil erosion control is a major challenge for the farmer, or in densely populated areas with scarce arable land, that farmers are more likely to adopt improved fallow technology.

9.5 Agroforestry

Agroforestry is the integration of trees into fields of cultivated crops or into pastures so that the trees can help to maintain soil fertility and provide fruits, timber, or other additional economic benefits to the farmer. Young (1997) distinguished between two main types of agroforestry systems: (i) agricultural systems in which the culture of trees is alternated with crops; and (ii) systems in which trees and crops are grown simultaneously on the same piece of land. The term 'agroforestry' is used in this book to refer to the latter system, in which trees and field crops are grown together on farmers' fields, or where trees are integrated into pastures.

Although some authorities (e.g. Nair and Fernandes, 1984; Sanchez, 1994; Alegre and Cassel, 1996; Brady and Weil, 2002) have suggested that agroforestry is a stable alternative to shifting cultivation, agroforestry is not alien to the system of shifting cultivation. In fact, agroforestry is an integral part of shifting cultivation or slash-and-burn agriculture. Shifting cultivators usually leave trees on their cultivated fields for various purposes including provision of shade, fruits and medicine, and to help to conserve soil fertility (Fig. 9.4). The retaining of palms in farmlands in the Amazon basin and in southern Nigeria, the simultaneous cultivation of oil palm and maize in farmers' fields in southern Benin Republic, the integration of *Macaranga denticulata* (pada) into rice fields in northern Thailand, and of a tree legume, *Paraserianthes (Albizia) falcataria* into cultivated fields of rice by the Baduy people of West Java, Indonesia, are well-known examples of traditional agroforestry systems developed by innovative shifting cultivators. These and other forms of agroforestry are an integral part of shifting cultivation, which is an amalgam of various types of agroforestry, many of which are unique to certain areas. Several forms of traditional agroforestry have been discussed in this chapter, and other variations of agroforestry are common in the areas where shifting cultivation is practised, so it is not possible to

Fig. 9.4. Agroforestry involving the intercropping of cassava and some cocoyam with fruit trees: pawpaw; *Musa* sp. (plantain); a medicinal tree, *Newbouldia laevis*; and other trees.

discuss them all. In this subsection, the following types of agroforestry practices which are specific to certain areas will be discussed:

1. The *Faidherbia* (*Acacia*) *albida*-based system in the savannas of tropical Africa.
2. The leguminous/commercial trees-based system in the Chittagong Hill Tracts of Bangladesh.
3. Fodder trees in bush fallow in the Philippines.
4. The cassava–pigeon pea system in the Bas-Congo region of the Democratic Republic of Congo.
5. Maize cultivation intensification with soybean in Zimbabwe.
6. Hedgerow intercropping in south-eastern Nigeria.

9.5.1 *Faidherbia* (*Acacia*) *albida*-based agroforestry in tropical African savanna

Faidherbia albida, a tree legume, is common in the drier savanna lands of Africa, especially in West, East, Central and parts of Southern Africa, and it even occurs in the dry valleys in the Sahara desert. A peculiar feature of the tree, especially in West Africa, is that it sheds its leaves at the inception of the wet season and comes into foliage during the dry season (Wood, 1989), and this makes it ideally suited for integration into farms and pastures, as it does not shade cultivated crops during the growing season. Consequently, the tree has been widely established in farmers' fields, particularly in the Sudan and Sahel savanna zones of West Africa where maize, millet and other field crops are frequently grown under the trees. Being a legume, the tree enhances the build-up of nitrogen, which is usually deficient in soils in dry savannas on account of low levels of organic matter, under its canopy. The tree also accumulates organic matter and plant nutrients in the soil beneath it. The study of Depommier *et al.* (1992) in Burkina Faso has shown that the soil under the canopies of *F. albida* has enhanced levels of organic matter, total nitrogen, available phosphorus and exchangeable calcium and potassium compared to soil outside the canopies, and this effect is more marked in the 0–20 cm soil layer (Fig 9.5). Furthermore, the branches of the trees are lopped by farmers during the dry

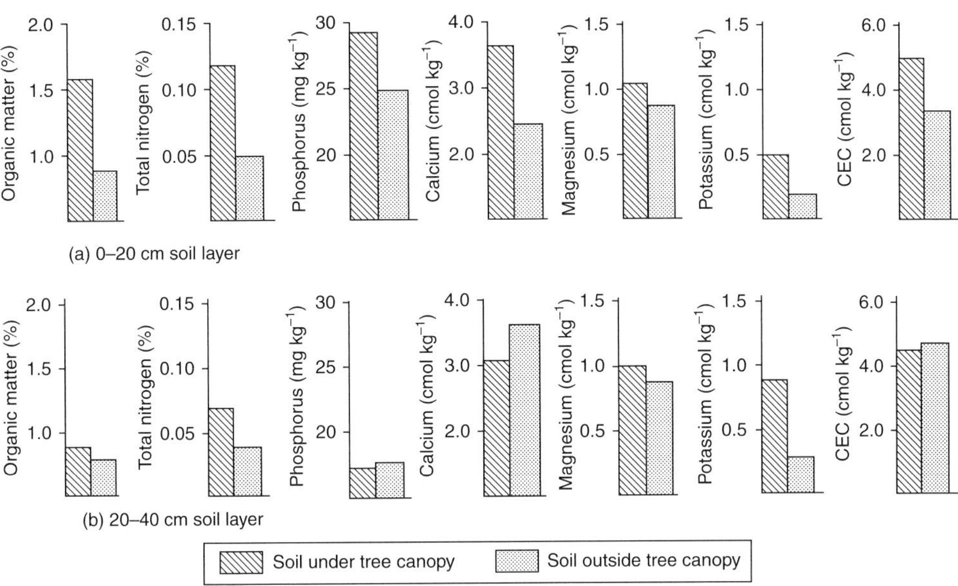

(a) 0–20 cm soil layer

(b) 20–40 cm soil layer

Fig. 9.5. Effects of *Faidherbia* (*Acacia*) *albida* on soil properties at Watinoma, Burkina Faso. (Based on data of Depommier *et al.*, 1992.)

season to provide fodder for livestock. *F. albida* has been integrated into farmlands in Malawi, Zambia, Tanzania and Ethiopia and in several countries, including those in West Africa, where it has helped to increase the yield of unfertilized maize (Garrity and Stapleton, 2011). Farmers in Niger Republic have embraced the integration of the tree legume into their farmlands and grazing land, and it is estimated that over 5 million ha of agricultural land in the county have sizeable stands of the tree.

9.5.2 Leguminous/commercial trees in farms in the Chittagong Hill Tracts of Bangladesh

The Chittagong Hill Tracts of Bangladesh is a mountainous region where the system of shifting cultivation has been practised for several centuries. Owing to the hilly and mountainous terrain, soil erosion is a major cause of soil nutrient decline during cropping, and long fallow periods are required in order to adequately restore soil fertility during the fallow period. Fallow periods, which used to be up to 10 years

and longer in the past, have been reduced to 4 or 5 years. Partly in response to the problem of declining productivity, the need to conserve the soils and increase income, farmers in the area have embraced agroforestry. Up to 40 types of crops are cultivated by farmers in the region but the major ones are rice, cucumber, chilli and ginger, and these are usually intercropped with trees such as *Ficus*, *Derris* and *Albizia*, a significant proportion of which are legumes (Mantel *et al.*, 2006).

The trees are retained in the farms when the farmers clear and prepare the land for cultivation, and they help to reduce soil erosion and add litter to the soil, which replenishes organic matter and nutrients on decomposition and mineralization. Some farmers even retain very large trees on their farms, instead of killing them by burning. They trim their branches to allow adequate light to reach the ground to sustain the production of various field crops. At the beginning of the fallow period, farmers plant commercial timber-yielding trees – especially teak and gmelina – on their farms where field crops are also grown, in spite of difficulties in accessing timber markets, partly in order to guarantee claim to ownership and use of land (Mantel *et al.*, 2006). With improved access to markets, more farmers are likely to embrace the planting of commercial timber trees in fallow vegetation, thereby enhancing the economic value of fallow vegetation.

9.5.3 Fodder trees in bush fallow in the Philippines

Trees have also been introduced into pastures and grazing land to enhance fodder supplies to livestock and to promote long-term sustainability of agropastoral systems. In Western Batangas, south-western Luzon, the Philippines, farmers practise shifting cultivation but also depend on sale of livestock to improve their economic earnings. They clear rainforest and cultivate maize, rice and beans in small plots that rarely exceed 1 ha for about 3 years, and thereafter leave the land fallow for 4–10 years. Not infrequently, weeds such as *Imperata cylindrica* are an important feature of young fallow vegetation, and this is an indication that the land is over-cultivated and is undergoing deterioration in quality. During the fallow period, livestock are usually allowed to graze on the regenerating fallow vegetation, causing land degradation due to overgrazing and trampling, and the bush fallow cannot adequately restore soil fertility (Calub, 2003). To improve the carrying capacity of the fallow vegetation dominated by grasses of low quality such as *Imperata cylindrica* and *Themeda trianda*, farmers plant fodder trees and shrubs such as *Leucaena leucocephala*, *Gliricidia sepium*, *Trema orientalis* and *Macaranga tanarius* in the fallows to provide supplementary feed for livestock. The trees are often planted along farm boundaries in the small farms, but also inside larger farms and inhospitable sites, especially those with shallow, stony soils. The branches of the trees and shrubs are cut at intervals of 60 or 90 days and the fodder used for feeding livestock (Calub, 2003). By integrating trees into bush fallow vegetation, farmers in south-western Luzon have not only helped to arrest or even reverse the trend of land degradation in their hilly farmlands, but have increased the number of livestock they can raise to earn cash.

9.5.4 Cassava–pigeon pea system in Bas-Congo, Democratic Republic of Congo

In the Bas-Congo area of the south-western part of the Democratic Republic of Congo, especially in the Songololo area that is characterized by hilly terrain and acidic soils with low-activity clays (ultisols), the indigenous farmers have developed an agroforestry system involving the integration of a cassava crop with pigeon pea. The former provides them with carbohydrates and the latter with proteins, especially after the long dry season when other sources of protein are difficult to come by. In addition the system provides some extra income, as harvested pigeon pea can be sold to enhance the farmers' income. Pigeon pea leaves are also applied to the soil as green manure for the cassava crop after the former have been harvested, while the pigeon pea plants themselves also help to accumulate nitrogen in the soil through symbiotic nitrogen fixation. This agroforestry system has been described by Kang *et al.* (1991) as one of the successful agroforestry practices in tropical Africa.

Farmers in the Bas-Congo region intercrop pigeon pea with cassava. Both crops are usually planted at the same time between October and January; that is, at the inception of the wet season. Farmers prefer a local variety of pigeon pea that matures slowly but branches profusely and also produces a lot of leaves. Pigeon pea is harvested after about 10 months and the leaves applied to the ground as green manure to help fertilize the cassava crop, which is left growing on the farm for another 8–14 months. The pigeon pea plants and the green manure derived from them help to sustain cassava production on a continuous basis, without application of inorganic fertilizers, and also increase cassava yield, as farmers who practise this system of agroforestry have observed (Kang *et al.*, 1991).

9.5.5 Maize cultivation intensification with soybean in Zimbabwe

Maize is the staple food crop in Zimbabwe, as in most parts of Southern Africa. The crop accounts for up to 80% of the area cultivated by small-scale farmers (Giller, 2008), who depend on following as a means of restoring soil fertility after a few years of cropping. Due to a rapid increase in population, particularly during the past two decades, coupled with the seizing of farms owned by white farmers by the government of President Robert Mugabe, Zimbabwe has been confronted with a worsening problem of food shortage. Having been a major food-exporting country, Zimbabwe became food-deficit and had insufficient foreign exchange to pay for food imports. In the face of growing food crises, attempts were made to intensify smallholder agriculture. One of the most successful agricultural policies aimed at boosting local food production was the introduction of *Glycine max* (soybean), which is relay-cropped with maize, the chief staple crop. The introduction of soybean into small-scale farms has the advantage of improving soil fertility status, in addition to boosting farmers' incomes as a cash crop. In this regard, its introduction into peasant farms in Zimbabwe has had a similar effect to the introduction of palms into fallows and farms cultivated to field crops in southern Benin Republic. Furthermore, soybean can be processed into food, which contributes to alleviating the problem of malnutrition in a country that is in the throes of severe political and economic crisis.

Small-scale agriculture, involving the use of bush fallow to regenerate soil fertility, can be intensified by shortening the fallow period, and so increasing the length of time land is cultivated; or by raising productivity per unit of land, by increasing inputs such as inorganic fertilizers or manure; or by introducing high-yielding varieties of crops. In Zimbabwe, the intensification of smallholder farming has been mainly through improving soil fertility by introducing soybean which is cultivated before the maize crop. A small quantity of fertilizers and lime is usually required for establishing the soybean crop on the sandy and infertile soils that characterize most areas where small-scale agriculture is practised, but cow manure can be substituted for inorganic fertilizers. Usually, when maize is cultivated for two consecutive seasons, the yield of the second maize crop averages 0.5 t ha^{-1}, but this increases substantially to 1.5 t ha^{-1} when soybean precedes the maize crop, to fertilize it (Giller, 2008). Due to the dramatic effect of soybean in improving

the yield of maize crop grown after the soybean crop, many farmers in Zimbabwe have enthusiastically embraced this system of farming. A major factor that has contributed to the widespread adoption of maize cultivation after harvesting soybean in Zimbabwe is the availability of a market for the soybean produced. It is used locally for producing vegetable oil in Harare, the capital city, and for baking soya-bread within the country. Hence, producers do not depend on exports for their produce, as there is a growing internal market for the crop.

9.5.6 Hedgerow intercropping in south-eastern Nigeria

In the heartland of Iboland in south-eastern Nigeria, rural population density is very high and often exceeds 500 persons per km^2, making it one of the most densely populated areas in West Africa. Land for arable farming is scarce and land holdings are very small and fragmented. Consequently, the people cannot fallow their land for a long period and have developed an ingenious traditional agroforestry which involves the cultivation of field crops of cassava, yams and maize in between hedgerows of the shrub *Acioa barteri*, which is an important element of the flora of bush fallow vegetation in the region. This shrub, rarely exceeding 4 m in height, has a very deep-rooting system which allows it to recycle nutrients leached beyond the rooting zone of crops back to the topsoil to benefit the crops. Although it is not a legume, the study of Okeke and Omaliko (1991) has shown that a 7-year stand of *A. barteri* produces a copious litter of up to 9.8 t ha^{-1} per year. This amount of litterfall is comparable with the annual litterfall of tropical rainforest in Malaya studied by Ogawa (1978) and higher than the annual rates recorded by Ewel (1976) for secondary forests aged 6–9 years in Guatemala. *Acioa barteri* appears to be very efficient in recycling mineral nutrients, especially phosphorus and magnesium, to the top soil. Apart from helping to restore and maintain soil fertility status through the addition of humus and nutrients to the soil, the shrub has other uses for training yam vines, weed suppression and as browse for livestock (Kang *et al.*, 1991).

Farmers in south-eastern Nigeria, particularly in the Mbaise area, retain the shrub or plant it in rows 2 or 3 m apart. Prior to cultivation, the shrubs are cut to the height of about 20 cm above the ground and the dried stems are burned to release ash to fertilize the soil. Cassava, yams and maize are then cultivated during the first year of cropping in the alleyways between the hedgerows, to be followed by cassava during the second year. After this the land is fallowed and *Acioa* shrubs are allowed to grow and regenerate from stumps and to cover the entire farm, together with other fallow trees, for about 4 years. As a result of *Acioa* hedgerow intercropping, farmers have effectively reduced the cultivation:fallow ratio to 1:2, which is usually considered unsustainable for shifting cultivation on the deeply weathered and highly leached, acidic soils that characterize this area. Nevertheless, the farmers have practised this system of hedgerow intercropping for several decades, and perhaps for several centuries, without a marked decline in soil productivity. It is important to observe that the hedgerow intercropping described above is a precursor of modern alley farming. The major difference between the two is that alley cropping does not involve burning of the hedgerow trees or shrubs. Instead, they are pruned regularly and their prunings applied to the soil as organic fertilizer to make continuous cultivation possible where conditions are suitable.

9.6 Compost

In the face of declining soil fertility due to the drastic reduction of fallow periods, shifting cultivators have devised ingenious ways of improving and sustaining soil fertility. In the preceding sections, various techniques used by shifting cultivators, including enriched fallows and various forms of agroforestry, have been discussed. Neither approach involves the direct application of manure or compost to the soil.

In this section, we shall consider composting as a means of enhancing or sustaining soil fertility. Compost is mainly organic residue derived from the decomposition products of plant materials buried in the soil. Although animal dung may be added to the mass of plant materials to facilitate its microbial decom-position, compost mainly consists of decomposed plant materials. Mineral fertilizers, such as phosphates, may be added to compost to improve its capacity to supply nutrients to the soil. Compost improves soil fertility over time, but it is bulky and often difficult to transport to the field.

The practice of composting is not widespread in shifting cultivation areas. The study of Nguyen and Shindo (2011) has shown that the long-term application of compost in a field cropped to paddy rice and barley in Yamaguchi, Japan, resulted in a marked increase in soil organic matter and macro-aggregates in the 0–15 cm soil layer. Two areas where composting has been adopted to intensify the system of shifting cultivation will be considered here, namely northern Zambia and central Papua New Guinea.

The Mambwe people of northern Zambia practise shifting cultivation, using the *chitemene* system. They clear fields and cut branches of trees from the surrounding vegetation, pile these in the middle of the cleared area, then burn them and subsequently cultivate crops where the burning released ash to fertilize the soil. The farmers then have to fallow the land for about 25 years for the vegetation to adequately regenerate soil fertility. Long fallows are no longer practicable because of the increasing population, and land is only fallowed for about 5–8 years, resulting in a decline in soil fertility, with an attendant decline in crop yield. In order to ensure that soil fertility is sustained, despite shortening of the fallow period, the Mambwe people have developed a system of mound cultivation, which is really a system of in-field composting. When land is cleared for cultivation following a few years of fallow, mounds of topsoil about 1 m high and 1 m wide are made towards the end of the planting season (Siame, 2006). Heaps of grasses and other plants are then made and the mounds of soil are overturned and used to cover the accumulated plant materials. The grasses and weeds buried below the mounds decompose to release nutrients into the soil during the dry season, which is an off-farm period. In the following wet season, the mounds with the decayed grasses and weeds are flattened and used for cultivation for about 4–5 years. Usually crops that make greater demands on soil nutrients, such as maize and *Eleusine coracana* (finger millet), are the first to be planted on the flattened mounds, before others that are less demanding (such as cassava) are planted later during the period of cultivation (Siame, 2006). In-field compost thus enables cultivation of the same field continuously for about 5 years without using inorganic fertilizers on the nutrient-deficient soils that characterize the northern plateau of Zambia that is covered with *miombo* woodlands. The Kaonde people of northern Zambia also practise mound cultivation (Oyama, 2007), which is similar to the in-field composting system practised by the Mambwe people described above.

Mound cultivation involving in-field composting is also an important feature of shifting cultivation in the central highlands of Papua New Guinea, especially among the Wola people. After short fallows they clear weeds and grasses and, on cultivated sites from which crops were recently harvested, crop residues such as potato vines are left on the farm to dry. Earth mounds of 2–3 m width are made by heaping up the soil in selected areas in the field, using digging sticks, small spades, or knives. Depressions are made in the centre of each mound, where grasses, crop residues and weeds have accumulated. Sometimes these materials are burnt, if they are sufficiently dry. Soil is scooped around the perimeter of the mound to cover the plant materials (Sillitoe, 1998). The plant materials buried under the mounds decompose over time and release nutrients into the soil. This in-field land management strategy concentrates plant nutrients in mounds and enables the the same fields to be cultivated on a semi-continuous basis; sweet potatoes can be grown on the same site for up to 10 years before the people have to relocate to another site or leave the land fallow (Sillitoe, 1996).

Composting is gaining widespread acceptance in the Sahelian region of West Africa on account of its benefits of improving soil organic matter, the levels of soil nutrients and soil physical status – in particular its soil water-retaining capacity – in an area that is prone to drought. In addition, compost improves crop yields and also reduces the amount spent by farmers on chemical fertilizers. In countries such as Burkina Faso, Chad, Mali, Niger and Senegal, composting has contributed to the successful reclamation of degraded soil, especially when used in conjunction with other soil management strategies such as mulching and bunding (*Spore*, 2011).

Farmers living near the Chitwan National Park in Nepal have commercialized compost production to a certain extent. They collect the dung of elephants from the park and use it for producing compost which they apply on their fields and also sell the remainder (Sharma, 2009). The production and utilization of compost by the farmers has reduced their dependence on inorganic fertilizers. Unlike the cases discussed above, compost production by these Nepalese farmers is not necessarily done in-field, at the site of utilization. This is because a significant part of the total output is intended for sale to generate additional income for the farmers.

9.7 Green Manure and Cover Crops

Cover crops and green manure were discussed above, in the section on seasonal fallows. Most seasonal or short-duration fallows involve the use of cover crops such as mucuna, jackbean and *Centrosema* to protect the soil against erosion and accelerated soil organic matter decomposition and mineralization due to high soil temperatures in exposed areas. Cover crops, therefore, afford ground protection against runoff, the impact of raindrops and elevated soil temperature during the off-farm season. They usually have beneficial effects on soil fertility status when they are slashed before cropping and incorporated into the soil as green manure. In addition, cover crops help farmers to control weeds, including obnoxious weeds such as *Imperata cylindrica* (spear grass) which grow with increasing rapidity and become a major challenge to the farmer as fallow periods are progressively shortened. The use of green manure is becoming increasingly popular in densely populated areas where farmers are resource-poor and cannot afford to buy commercial fertilizers. In the highlands of western Kenya, for instance, farmers apply the biomass of *Tithonia diversifolia* as green manure on their fields and this has enabled them to cultivate the same fields continuously, without depending on inorganic fertilizers or herbicides. In the hills of Nepal, farmers cut leafy shoots of trees such as *Adhatoda vasica* (malbar nut),

Albizia spp. and *Crateva unilocularis* (garlic pear), and of the herb *Artemisia vulgaris*, from secondary vegetation or forest. These are carried to their fields, usually over long distances, where they are incorporated into the soil as green manure or used as green mulch which will release nutrients into the soil on decomposition (Subedi, 1997). Some of these green manure species are so much in demand for soil manuring that they are now over-exploited and are gradually disappearing. In order to conserve these green fertilizer species, farmers need to plant them on their fields. This will also go a long way in reducing the time and energy they expend in sourcing the fertilizer plants from the wild.

9.8 Mulching

This is one of the techniques applied by shifting cultivators in managing cultivated land to reduce soil erosion by water and wind, conserve soil moisture and conserve or even improve the soil organic matter level, in order to enhance crop yields and prevent progressive deterioration in soil quality over time. Mulching was referred to in the sub-section on cover crops above. In fact, cover crops are live mulches.

It is appropriate to define mulch here, although the definition is implied in the opening sentence of this subsection. Mulch is anything, organic or inorganic, that is placed or spread on the ground surface to protect it from erosion, desiccation and adverse weather phenomena (such as the direct impact of raindrops which may lead to surface sealing and crusting). Cover or seasonal fallow plants are usually slashed and left on the ground surface as green mulch or incorporated into the soil as green manure. Many shifting cultivators depend on crop and weed residue which they apply on the ground surface as organic mulch. The practice of mulching assumes increased significance in arid or semi-arid areas, such as the Sudan and Sahel savanna zones of West Africa that are particularly prone to wind erosion and desertification. In marginal arable land, especially where slopes are steep, mulching is crucial in order to prevent land degradation. Over much of the densely populated central plateau of Burkina Faso, the land is continuously cultivated, especially in the vicinities of the

larger settlements. Farmers cut grasses from uncultivated areas and also collect the dry leaves of trees and transport them to their cultivated fields, sometimes using donkeys, where they are applied as mulch (Slingerland and Masdewel, 1996). Apart from protecting the soil, the mulch attracts termites that burrow into the soil and remove surface crusts and also help decompose mulch, thereby adding organic matter to the soil. Farmers in Niger, where shifting cultivation has undergone intensification within the past two decades or so, also apply mulch on their fields. They apply millet stalks from the previous harvest, and other crop residues, to their fields as a means of helping to conserve or improve the fertility of their cultivated fields. They also use the branches of trees pruned during site preparation for farming as mulch (Wezel and Haigis, 2002).

Black polyethylene film, an inorganic mulch, has been reported by Kim *et al.* (2008) to improve the growth and yield of mungbean. Although inorganic mulches such as polyethylene sheets are very effective in conserving soil moisture by drastically reducing evaporation from the soil (and hence potentially increase the length of the growing season), shifting cultivators are likely to find the application of such mulches technically and financially unfeasible.

9.9 Socio-economic and Technological Aspects of Intensification

Shifting cultivators are resource-poor farmers who do not have access to bank loans, and they farm very small holdings which rarely exceed a few hectares and are quite often considerably smaller. In order for an innovation to improve this system of agriculture or replace it entirely, the level of technology must be within the reach of these small-scale farmers. It is no use introducing mechanized farming based on the use of tractors and other farm machinery to replace hand cultivation. Mechanized clearing removes much of the topsoil, reducing soil productivity. In addition, shifting cultivators cannot afford to buy tractors. In part of the savanna region of south-western Nigeria, tractors are hired to farmers for a fee, but many farmers cannot afford to hire the tractor to clear their land. More importantly, tractors clear-fell all the trees in farmers' fields, while ploughing the land mechanically exposes the soil to erosion. The removal of trees from cultivated fields results in the collapse of nutrient cycling, as there are no trees to recycle nutrients leached into the subsoil back to the topsoil for crops. Chapter 10 will point out that the long-term application of several types of inorganic fertilizers simultaneously or sequentially on the same plot, as a strategy to achieve continuous cultivation, is not a viable alternative to shifting cultivation. The environmental problems resulting from the application of inorganic fertilizers are also discussed in Chapter 10. Furthermore, farmers who operate smallholdings do not have the resources to buy fertilizers.

However, they would find the adoption of technologies such as mulching, composting and planting of cover crops or soil-improving trees such as *Sesbania sesban* easy, as these technologies are not complicated nor do they require heavy financial outlay. It is important to point out that farmers will not usually grow soil-improving trees unless they have a dramatic effect on crop yield, and therefore have the potential of substantially increasing the revenue or amount of food produced by the farmer. Kang *et al.* (1991) have pointed out that farmers in tropical Africa are generally reluctant to adopt agroforestry technologies purely for the purpose of soil fertility restoration. This is true for most parts of the tropics, apart, of course, from densely populated areas where there is an acute shortage of land for farming. Even in areas of rugged topography that are prone to gullying, such as the watershed of the Uporoto mountains of south-west Tanzania, Mwanukuzi (2011) observed that small-scale farmers are reluctant to embrace effective methods of land conservation, except those that contribute towards meeting their immediate and basic needs. Roder (1997) also reported that alley farming has been promoted in northern Thailand but that the rate of adoption by farmers has been dismally low.

For agroforestry purposes, farmers prefer multi-purpose trees that yield additional benefits besides improving soil fertility. Trees that provide fruits, such as palm trees, or those whose seeds can be sold for cash, are readily retained or planted by shifting cultivators in their farms. In the Bas-Congo region of the Democratic Republic of Congo, farmers readily embraced the introduction of pigeon pea into their cassava fields to help fertilize the soil and lengthen the period of cultivation. A major factor influencing the adoption of the agroforestry system based on the intercropping of pigeon pea with cassava is the fact that pigeon pea serves as an important source of food for the people of the south-western Democratic Republic of Congo during a lean period before the harvest, when there is shortage of food from other sources (Kang *et al.*, 1991). In addition, there is a ready market for pigeon pea seeds, and farmers can easily sell the excess produce to enhance their income. In contrast, in northern Thailand, farmers have not readily adopted the agroforestry system based on the intercropping of rice with pigeon pea, in spite of the latter having a beneficial effect on rice yield. Roder (1997) observed that the reluctance of the Thai farmers to introduce pigeon pea into their rice fields was due to lack of market for pigeon pea seeds.

Finally, it is important to observe that whatever the method or technology that is introduced to intensify shifting cultivation, it should be socially and culturally acceptable to the native farmers. A new method of farming has to be economically and technological feasible as well as being culturally acceptable. An attempt to introduce an 'improved' farming technology (based on the application of inorganic fertilizers and other agrochemicals to replace traditional in-field composting between ridges) into Ileje in the southern part of Mbeya region in Tanzania, failed woefully. The farmers were unable to buy the agrochemicals required as inputs and many who had been given loans to buy fertilizers were unable to repay, and migrated out of the area for fear of litigation (Nwangosi, 2008). Consequently, food crop production in Ileje based on industrial agrochemical inputs failed to take root, and the people have returned to their traditional method of maize, millet and rice production based on ridges and in-field composting. When a new technology is introduced, especially when it runs contrary to the beliefs of small-scale farmers, an attempt should be made to demonstrate the usefulness or appropriateness of the new approach with the farmers as participant observers. Otherwise, such new technologies are doomed to failure from the start.

References

Adejuwon, J.O. and Adesina, F.A. (1990) Organic matter and nutrient status of soils under cultivated fallows: an example of *Gliricidia sepium* fallows from south western Nigeria. *Agroforestry Systems* 10, 23–32.

Agyare, W.A., Kombiok, J.M., Karbo, N. and Larbi, A. (2002) Management of pigeon pea in short fallows for crop-livestock production systems in the Guinea savanna zone of northern Ghana. *Agroforestry Systems* 54, 197–202.

Akpokodje U. E. and Aweto, A.O. (2007) Effects of an exotic legume on soil in farmland in the Ibadan area, southwestern Nigeria. *Land Contamination and Reclamation* 15, 319–326.

Albiero, D., Maciel, A. J. da S. and Gamero, C.A. (2011) Design and development of babacu (*Orbignya phalerata* Mart.) harvest for small farms in areas of forest transition of the Amazon. *Acta Amazonia* 41, 57–68.

Alegre, J.C. and Cassel, D.K. (1996) Dynamics of soil physical properties under alternatives to slash-and-burn. *Agriculture, Ecosystems and Environment* 58, 39–48.

Boserup, E. (1965) *The Conditions of Agricultural Growth.* Allen & Unwin, London.

Brady, N.C. and Weil, R.R. (2002) *The Nature and Properties of Soils.* Prentice-Hall, Upper Saddle River, New Jersey.

Buckles, D. and Triomphe, B. (1999) Adoption of mucuna in the farming systems of northern Honduras. *Agroforestry Systems* 47, 67–71.

Bunch, R. (2003) Adoption of green manure and cover crops. *LEISA* 19, 16–18.

Calub, B.M. (2003) Indigenous fodder trees for rehabilitation. *LEISA* 19, 22–23.

Denevan, W.M., Treacy, J.M., Alcorn, J.B., Padoch, C., Denslow, J. and Paitan, S.F. (1984) Indigenous agroforestry in the Peruvian Amazon: Bora Indian management of swidden fallows. *Interciencia* 9, 346–357.

Depommier, D., Janodet, E. and Oliver, R. (1992) *Faidherbia albida* parks and their influence on soils and crops at Watinoma, Burkina Faso. In: Van den Beldt, R.J. (ed.) *Faidherbia albida in the West African Semi-Arid Tropics.* Proceedings of a workshop, ICRISAT/ICRAF, Niamey, Niger, pp. 111–115.

Drechsel, P., Glaser, B. and Zech W. (1991) Effect of four multipurpose tree species on soil amelioration during tree fallow in Central Togo. *Agroforestry Systems* 16, 193–202.

Edja, H. (2001) *Land Rights Under Pressure: Access to Resources in Southern Benin.* International Institute for Environment and Development, London.

Evans, D.O. and Rotar, P.P. (1987) Productivity of sesbania species. *Tropical Agriculture (Trin.)* 64, 193–200.

Ewel, J.J. (1976) Litter fall and leaf decomposition in a tropical forest succession in eastern Guatemala. *Journal of Ecology* 64, 293–307.

Flores, M. (1994) The use of leguminous cover crops in traditional farming systems in Central America. In: Thurston, H.D., Smith, M., Abawi G. and Kearle S. (eds) *Tapado Slash/Mulch: How Much Farmers Use It and What Researchers Know About It.* Cornell International Institute for Food, Agriculture and Development, Ithaca, New York, pp. 149–155.

Franzel, S. (1999) Socio-economic factors affecting the adoption potential of improved tree fallows in Africa. *Agroforestry Systems* 47, 305–321.

Fujisaka, S. (1991) A diagnostic survey of shifting cultivation in northern Laos: targeting research to improve sustainability and productivity. *Agroforestry Systems* 13, 95–109.

Gachengo, C.N. (1996) Phosphorus release and availability on addition of organic materials to phosphorus fixing soils. MSc thesis, Moi University, Eldoret, Kenya.

Garrity, D. and Stapleton, P. (2011) More trees on farms. *Farming Matters* 6, 8–9.

Giller, K. (2008) The successful intensification of smallholder farming in Zimbabwe. *LEISA* 24, 30–31.

Gledhill, D. (1972) *West African Trees.* Longman, Harlow, UK.

Hecht, S.B., Anderson, A.B. and May, P. (1988) The subsidy from nature: shifting cultivation, successional palms forests, and rural development. *Human Organization* 47, 25–35.

Henderson, A, Galeano, G. and Bernal, R. (1995) *Field Guide to the Palms of the Americas.* Princeton University Press, Princeton, New Jersey.

Hoang Fagerstrom, M.H., Nilsson, S.I., van Noordwijk, M., Phien, T., Olsson, M., Hansson, A. and Svensson, C. (2002) Does *Tephrosia candida* as fallow species, hedgerow or mulch improve nutrient cycling and prevent nutrient losses by erosion on slopes in northern Viet Nam? *Agriculture, Ecosystems and Environment* 90, 291–304.

Iskandar, J. and Ellen, R.F. (2000) The contribution of *Paraserianthes (Albizia) falcataria* to sustainable swidden management practices among the Baduy of West Java. *Human Ecology* 28, 1–17.

Kang, B.T. (1977) Effect of some biological factors on soil variability in the tropics. II. Effect of oil palm tree (*Elaeis guineensis* Jacq.). *Plant and Soil* 47, 451–462.

Kang, B.T., Versteeg, M.N., Osiname, O. and Gichuru, M. (1991) Agroforestry in Africa's humid tropics: three success stories. *Agroforestry Today* 3, 4–6.

Kass, D.C.L. and Somarriba, E. (1999) Traditional fallows in Latin America. *Agroforestry Systems* 47, 13–36.

Kim, D.-K., Chon, S.-U., Lee, K.-D., Son, D.-M., Rim, Y.-S., and Kim K.-H. (2008) Effect of mulching and soil conditioners on yield and flavonoids content of mungbean. *Korean Journal of Crop Science* 53, 353–358.

Kleinman, P.J.A., Pimentel, D. and Bryant, R.B. (1995) The ecological sustainability of slash-and-burn agriculture. *Agriculture, Ecosystems and Environment* 52, 235–249.

Kwesiga, F.R., Franzel, S., Place, F., Phiri, D. and Simwanza, C.P. (1999) *Sesbania sesban* improved fallows in eastern Zambia: Their inception, development and farmer enthusiasm. *Agroforestry Systems* 47, 49–66.

Mantel, S., Mohiuddin, M., Alam, M.K., Olarieta, J.R., Alam, M. and Khan, F.M.A. (2006) Improving the jhum system in Bangladesh. *LEISA* 22, 20–21.

May, P.H., Anderson, A.B., Frazao, J.M.F. and Balick, M.J. (1985) Babassu palm in the agroforestry systems in Brazil mid-north region. *Agroforestry Systems* 3, 279–295.

Muller-Samann, K.M. and Kotschi, J. (1997) *Sustaining Growth: Soil Fertility Management in Tropical Smallholdings.* Magraf Verlag, Weikersheim, Germany.

Mwanukuzi, P.K. (2011) Impact of non-livelihood-based land management on land resources: the case of upland watersheds in Uporoto Mountains, south-west Tanzania. *The Geographical Journal* 177, 27–34.

Nair, P.K.R. and Fernandes, E. (1984) Agroforestry as an alternative to shifting cultivation. In: *Improved Production Systems as an Alternative to Shifting Cultivation.* FAO Soils Bulletin 53, FAO, Rome, pp. 169–182.

Newton, K. (1960) Shifting cultivation and crop rotation in the tropics. *Papua New Guinea Agricultural Journal* 13, 81–118.

Nguyen, T.H. and Shindo, H. (2011) Quantitative and qualitative changes of humus in whole soils and their particle size fractions as influenced different levels of compost application. *Agricultural Sciences* 2, 1–8.

Nwangosi, M.C. (2008) Traditional soil maintenance stands firm. *LEISA* 24, p. 32.

Nye, P. and Greenland, D.J. (1960) *The Soil under Shifting Cultivation.* Commonwealth Bureau of Soils, Harpenden, UK.

Odunze, A.C., Danjuma, D.J., Gauji, G.R., Abolarin, K. and Salawu, I.S. (2009) Effects of 3-year short fallows on physical properties of alfisols in Zaria, northern Guinea savanna of Nigeria. Paper presented at the *33rd Annual Conference of the Soil Science Society of Nigeria*, University of Ado-Ekiti, Ado-Ekiti, Nigeria, 9–13 March 2009.

Ogawa, H. (1978) Litter production and carbon cycling in Pasoh forest. *Malaysian Nature Journal* 30, 367–373.

Okeke, A.I. and Omaliko, C.P.E. (1991) Nutrient accretion to the soil via litterfall and throughfall in *Acioa barteri* stands at Ozala, Nigeria. *Agroforestry Systems* 16, 223–229.

Osunde, A.O., David, D., Adeboye, M.K.A. and Bala, A. (2009) Residual effects of a planted mucuna (*Mucuna pruriens*) fallow and inorganic fertilizers on the yield of maize (*Zea mays* L.) in a southern savanna alfisol of Nigeria. Paper presented at the *33rd Annual Conference of the Soil Science Society of Nigeria*, University of Ado-Ekiti, Ado-Ekiti, Nigeria, 9–13 March 2009.

Oyama, S. (2007) Ecological knowledge of site selection and cultivating methods of Kaonde shifting cultivators in northwestern Zambia. *Geographical Reports of Tokyo Metropolitan University* 42, 15–20.

Quenum, E.K. (1988) Role du palmier a huile dans l'economie des families paysannes du Plateau Adja. Ingenier agronomie thesis, Department of Rural Economics and Sociology, National University of Benin, Cotonou.

Roder, W. (1997) Slash-and-burn rice systems in transition: challenges for agricultural development in the hills of northern Laos. *Mountain Research and Development* 17, 1–10.

Sanchez, P.A. (1994) Alternatives to slash and burn: a pragmatic approach to mitigating tropical deforestation. In: Anderson, J.R. (ed) *Agricultural Technology: Policy Issues for the International Community.* CAB International, Wallingford, UK, pp. 451–479.

Sharma, S. (2009) Using the natural environment. *LEISA* 25, 27.

Siame, J.A. (2006) The Mambwe mound cultivation system. *LEISA* 22, 14–15.

Sillitoe, P. (1995) Fallow and soil fertility under subsistence cultivation in the Papua New Guinea highlands: I. Fallow successions. *Singapore Journal of Tropical Geography* 16, 82–100.

Sillitoe, P. (1996) *A Place against Time: Land and Environment in the Papua New Guinea Highlands.* Hardwood Academic Publishers, Amsterdam, The Netherlands.

Sillitoe, P. (1998) It's all in the mound: fertility management under stationary shifting cultivation in the Papua New Guinea highlands. *Mountain Research and Development* 18, 123–134.

Slingerland, M. and Masdewel M. (1996) Mulching in the central plateau of Burkina Faso. In: Reij, C., Scoones, I. and Toulmin, C. (eds) *Sustaining the Soil.* Earthscan, London, UK, pp. 85–89.

Smestad, B.T., Tiessen H. and Buresh, R.J. (2002) Short fallows of *Tithonia diversifolia* and *Crotalaria grahamiana* for soil fertility improvement in western Kenya. *Agroforestry Systems* 55, 181–194.

Spore (2009) Soil fertility: feeding the land. *Spore* 139, 8–10.

Spore (2011) Composting: From present to the future. *Spore* 153, p. 20.

Steiner, K.G. (1991) Overcoming soil fertility constraints to crop production in West Africa: Impact of traditional and improved cropping systems on soil fertility. In: Mokwunye, A.U. (ed.) *Alleviating Soil Fertility Constraints to Increased Food Production in West Africa.* Kluwer Academic Publishers, Dordrecht, The Netherlands, pp. 69–91.

Subedi, K.D. (1997) Farmers' local knowledge agrees with formal experimental results. *LEISA* 13, 16–17.

Wezel, A. and Haigis, J. (2002) Fallow cultivation and farmers' resource management in Niger, West Africa. *Land Degradation and Development* 13, 221–231.

Wood, P.J. (1989) *Faidherbia albida.* Technical Centre for Agricultural and Rural Cooperation, Wageningen, The Netherlands.

Yimyam, N., Rerkasem, K. and Rerkasem, B. (2003) Fallow enrichment with pada (*Macaranga denticulata* (Bl.) Muell. Arg.) trees in rotational shifting cultivation in northern Thailand. *Agroforestry Systems* 57, 79–86.

Young, A. (1997) *Agroforestry for Soil Management.* CAB International, Wallingford, UK.

10 Alternative Farming Systems and the Future of Shifting Cultivation

A number of alternatives have been proposed as more intensive systems of agriculture to replace shifting cultivation. As will be pointed out later, some of these systems are not necessarily alternatives, but intensified forms of one of the variants of shifting agriculture. The systems considered in Chapter 9 are regarded as attempts to intensify shifting agriculture insofar as they do not involve a total elimination of the fallow – natural, semi-natural or planted – which is central to the process of soil fertility restoration under the low-input system of organic farming that shifting cultivation encapsulates. Hence, a system of agriculture is not an alternative to shifting cultivation if it involves fallowing, but rather an intensified form of the primordial system. In this section, various alternatives to shifting cultivation including continuous farming based on manuring or the intensive use of chemical fertilizers, alley cropping and the *Quensungual* agroforestry system will be considered. Apart from continuous cropping based on application of manure or inorganic fertilizer, some of the systems considered here are not necessarily alternatives to shifting cultivation, but are considered as alternatives in this chapter because those who developed them regard them as alternatives.

10.1 Continuous Cultivation Based on Application of Inorganic Fertilizers

The application of inorganic fertilizers has traditionally been regarded by scientists and policy makers in the temperate regions as a panacea for the problem of food production and intensification of shifting cultivation in the tropics. Some tropical soil scientists sometimes share the illusion of soil scientists in the temperate regions that with adequate levels of fertilizer application, the soils in the tropics – including the intensively weathered soils of the humid tropics – can be used for continuous cropping on a sustainable basis. The Green Revolution advocated and promoted in sub-Saharan Africa, where per capita food production has lagged behind the rate of population increase during the past two decades, is based on substantially increasing the rate of fertilizer application. Quinones *et al.* (1997) suggested that the average rate of application of mineral fertilizers for food crops has to be stepped up from about 5 kg ha^{-1} to about 30–40 kg ha^{-1} in order to ensure adequate production of food. While it is advantageous to apply inorganic fertilizers to replenish soil nutrients removed by harvested crops or lost through leaching and erosion during cropping, the application of mineral fertilizers may not guarantee sustainability of food production in the long term. Social and economic constraints may make it difficult for small-scale farmers to access inorganic fertilizers, which are usually expensive and priced beyond their reach. In addition, tropical soils, especially in the humid tropics, differ strikingly from those in the temperate regions, particularly with respect to the nature of clay minerals and the rate of organic matter decomposition in the soil following site clearance. Furthermore, long-term application of fertilizers can result in environmental problems including soil degradation and environmental pollution, which would undermine the sustainability of a fertilizer-based farming system in the long term. These will be discussed later in this chapter.

Although there are numerous examples of successful long-term continuous cropping in swamps and lowland areas adjoining rivers with or without fertilizer application, especially those based on paddy rice, there are very few cases of long-term continuous cropping based on fertilizer application on well-drained upland areas in the tropics. In the case of the lowland areas adjacent to river valleys, alluvium is deposited on the floodplain annually, and this helps to replenish soil nutrients and sustain cropping for several hundreds of years with or without the application of inorganic fertilizers. In the case of shifting cultivation in tidal floodplains of the Amazon basin, Brazil, referred to in Section 3.7, there was a build-up of organic matter and nutrients in the soil under cultivation over time due to the yearly input of supplementary nutrients in the alluvium deposited on the farm. However, in upland sites where farming is based on shifting cultivation, there is rapid diminution of organic matter and nutrients in cultivated areas over time, for reasons covered in Chapter 3. Cases of successful long-term continuous cultivation based on fertilizer application on well-drained soils are rare in the tropics. This is mainly because of the problem of rapid organic matter diminution and the attendant rapid depletion of soil fertility, except on inherently fertile soils such as those derived from volcanic parent materials, as occurs in Java.

Sanchez *et al.* (1983) have expressed optimism about the possibility of practising continuous cropping on acidic, inherently infertile soils in the humid tropics. Their study indicated that the fertility status of acidic nutrient-deficient soil (ultisol) in Peru significantly improved after 8 years of continuous cropping of upland rice, maize, peanut and soybean on a rotational basis that involved liming and the application of eight different inorganic fertilizers, including micronutrients. They reported that after 20 harvests in 8 years, the levels of nutrients in the continuously cropped soil, including exchangeable calcium and magnesium, available phosphorus and the micronutrients zinc, copper and manganese, increased appreciably. On the basis of this finding, Sanchez *et al.* (1983) opined that continuous cropping on infertile soils of the tropics is possible when appropriate fertilizers are applied at adequate levels. It is significant to note that their results have shown that with heavy soil fertilization, the soil can be cropped for 8 years consecutively, and not indefinitely. The findings of Sillitoe (1998) have shown that the Wola people in Papua New Guinea cultivate their fields for more than 10 years without using inorganic fertilizers. They are able to do this through a system of in-field composting of grasses and crop residue in mounds, which helps to replenish soil nutrients.

It may be argued that the soil cultivated by the Wola people continuously for up to 11 years is characterized by a relatively high base status, having been derived from volcanic parent materials. In southern Benin Republic, farmers have intensified shifting cultivation by integrating palms into their fields and are able to cultivate the same field for up to 8 years or longer using soils that are similar to those studied by Sanchez *et al.* (1983) for their experiments of 8 years of continuous cropping, based on the application of several types of fertilizers. So if farmers can cultivate the same land continuously for 8 years or more using compost or by integrating soil-improving trees into their fields, it would not be absolutely necessary to apply an amalgam of a wide range of inorganic fertilizers in order to crop the land for 8 years or longer. The technology of intensive fertilizer application proposed by Sanchez *et al.* (1983) is, in fact, well beyond the technical and financial capability of the resource-poor farmers who practise shifting cultivation.

Malawi presents an interesting case of a tropical African country which was plagued by drought and famine 5 years ago, forcing the country to depend on international aid to meet its food requirements. From 2005, Malawi provided fertilizer subsidies for small-scale farmers who accounted for the bulk of food production in the country. In 2009, Malawi was not only self-sufficient in food production but had surplus maize, the main staple food, to export. The government of Malawi attributed the remarkable success in food crop production to its subsidy of fertilizers, which enabled small-scale farmers to buy fertilizer inputs to enhance crop yield. The World Bank, however, holds a contrary opinion, suggesting that good rains, rather than fertilizer application, accounted for the surplus maize production (*The Guardian*, 2009). While fertilizer application may have helped to increase maize yields in Malawi for the period 2005–2009, it is instructive to observe that without adequate rains, massive fertilizer application would have resulted in crop failure. Hence, it is more likely that good rains, rather than fertilizer application per se would have accounted for the surplus maize production in Malawi. Only about half of Malawi's small-scale farmers benefited from the government subsidy programme during the period (*The Guardian*, 2009), and this lends credence to the view of the World Bank.

10.1.1 Problems associated with the use of inorganic fertilizers

In the preceding section, the observation was made that the application of mineral fertilizers is not a universal and sustainable solution to the problems of increasing crop production. The environmental problems associated with the use of inorganic fertilizers will be discussed briefly in this subsection. In addition, the problems of adopting the mineral fertilizer-based paradigm as a strategy for increasing crop production in the humid and sub-humid tropics will be highlighted.

The application of moderate amounts of mineral fertilizers may enhance crop yield. However, long-term application, especially in large amounts, would adversely affect the soil. The soil is an ecosystem containing a vast array of living organisms including bacteria, fungi, nematodes, actinomycetes, arthropods and earthworms, which degrade dead plant and animal matter and transform it into humus, and also help in the mineralization of humus to release nutrients into the soil for plant use. Some bacteria, such as nitrogen-fixing bacteria, help to transform atmospheric nitrogen into nitrogen-containing compounds which can be absorbed and utilized by plants. As to be expected, long-term fertilizer use, or application of large doses of fertilizers, detrimentally affects the soil eco-system, and its beneficial organisms, making it less suitable for plant growth. Continuous use of fertilizers can also lead to soil acidification, micronutrient depletion, soil degradation and reduced crop yield and vigour (Reijntjes *et al.*, 1992). Ammonium sulfate, for instance, increases soil acidity (Ahn, 1970), exacerbating the problems of aluminium toxicity and nutrient availability in the soil. It is also a strong biocide that kills nematodes and earthworms and inhibits the activities of nitrogen-fixing organisms in the soil (Reijntjes *et al.*, 1992).

The application of mineral fertilizers can also lead to nutritional imbalance in the soil, as the levels of the nutrients supplied in the applied fertilizers become very high relative to the levels of other nutrients, resulting in low uptake of certain nutrients. Because of the deleterious effects of mineral fertilizers on soil organisms, especially those that mineralize humus or fix nitrogen and the problem of soil nutrient imbalance that may ultimately result, the long-term application of inorganic fertilizers may cause soil fertility to diminish. In addition, soil structure may deteriorate following a decrease in the population of worms and microbes that help to maintain soil structure. This effect would be particularly marked if compost or organic manure is not applied in conjunction with mineral fertilizers to help maintain soil organic matter status over time. Furthermore, mineral fertilizers applied in farmers' fields, in common with pesticides and herbicides, are washed down to the water table by percolating rain water, polluting groundwater; or they may be washed by runoff into rivers, lakes and other aquatic ecosystems. Phosphates and nitrates washed into water bodies or groundwater are a major cause of their pollution. In areas where mineral fertilizers are heavily used, nitrate levels are very high in groundwater, with fatalities resulting when water from such a source is used to mix infant formula for new-born children (Cunningham *et al.*, 2005).

Excessive use of inorganic fertilizers and pesticides by farmers in the Jaffna peninsula of Sri Lanka has resulted in soil and groundwater pollution. Vakeesan *et al.* (2008) described the soils in the area as 'sick' and also reported that over 65% of farm wells in the area have nitrate levels that are well above the WHO (World Health Organization) recommended level of 10 mg l^{-1}. The farmers in the Jaffna peninsula had to discontinue the application of inorganic fertilizers and pesticides and resorted to the use of green manure, especially *Crotalaria juncea* (sunn hemp), which is slashed and incorporated into the soil, in order to restore its fertility. Wardjito (2008) also reported that farmers who embraced the use of chemical fertilizers and pesticides in part of central Java observed, to their chagrin, that their soils were becoming harder and less fertile over time. They had to stop applying mineral fertilizers and pesticides and resorted to organic farming, especially the use of cover crops and organic mulches, in order to reverse the trend of soil fertility decline, and even to improve soil productivity. In the Philippines, Mendoza (2010) reported that soils are badly degraded due to the use of chemical fertilizers, and he suggested organic farming as an alternative to agrochemical-based agriculture.

The nature of soils in the humid tropics, where the system of shifting cultivation is prevalent, means that they are not particularly suited to continuous farming based on the application of inorganic fertilizers, compared with soils in the temperate regions. The soils of the humid tropics are dominated by low-activity clays, especially kaolinite, which have very low capacity to adsorb plant nutrients and subsequently make them available for plants. The cation exchange capacity (CEC) of kaolinite is about 8 cmol kg^{-1}, less than 10% of that of montmorillonite, which is the dominant clay mineral in soils of the temperate regions. The implication of the dominance of the clay fraction of soils of the humid tropics by kaolinite is that their CEC, and hence their capacity to retain plant nutrients, largely depends on their organic matter content.

As pointed out in Chapter 3, soil organic matter decomposes quickly after land clearing and following a few years of cultivation. The prevalent high temperatures usually cause soil organic matter to diminish at a much faster rate in the tropics than in temperate regions. In the Great Plains of the USA, organic matter levels in soils cultivated continuously for 30–40 years were reduced to 40–73% of the initial, or pre-cultivation levels (Haas *et al.*, 1957). In sharp contrast, after only 5 years of cultivation of a tropical soil in Sierra Leone, soil organic matter was dramatically reduced to 50% of the pre-cultivation level (Brams, 1971). Rapid decline in the levels of organic matter in the soil, even after a few years of cultivation, also causes soil cation capacity to decline quickly. Consequently, a substantial proportion of the nutrients in inorganic fertilizers applied to the soil would be leached away.

It seems that with the progressive decline in soil organic matter levels during the first few years of continuous cropping in the humid and sub-humid tropics, before soil organic matter levels stabilize at a low equilibrium level, the intensity of leaching also increases over time, making the application of inorganic fertilizers unsustainable and uneconomic. These major differences between tropical soils and those of temperate regions largely explain why it was possible to develop systems of sustainable continuous cropping in temperate regions, based mainly on the application of mineral fertilizers, while attempts to do so in the humid tropics have largely failed. It is important to point out that in the temperate regions, farmers are increasingly embracing organic farming owing to the deleterious environmental effects of mineral fertilizers and other inputs such as herbicides and pesticides.

10.2 Continuous Cultivation Based on Manure Application

This is a viable alternative to shifting cultivation where manure is readily available, especially in the savanna lands of the tropics where large herds of livestock are usually kept, and also in the vicinities of large cities where night soil (human faeces) is available for soil fertilization. Human faeces has to be properly treated to ensure its safe use as manure for crops. In the savanna regions of the tropics, systems of continuous cultivation based on the use of animal manure are well developed. Large cities, such as Kano and Maiduguri in northern Nigeria, are surrounded by extensive zones of continuous cultivation based on animal manuring. Around and near cities in the Sudano–Sahelian zone of West Africa, similar zones have evolved where continuous farming involves the application of animal manure. Continuous farming is also practised around smaller settlements and villages, but these continuously farmed zones are much smaller than the areas farmed around large cities because of the smaller population of livestock kept in villages.

Wezel and Haigis (2002) observed that in the southern and central part of Niger Republic, farmers have embraced continuous farming involving the application of animal manure. Their results indicated that about 74% of the farmers use animal manure to fertilize their fields and this has made continuous cultivation possible. In other cases, farmers have reduced their fallow period to 2 or 3 years, and as this short period is inadequate to restore soil fertility, manure is applied on cultivated fields. Manure is in short supply and farmers usually apply it to fields within 1 km of the compound where animals are kept (Wezel and Haigis, 2002). Distant fields are usually not fertilized, although sometimes farmers pay nomadic or sedentary herdsmen to corral livestock there and so help to manure them through the deposition of dung by the grazing livestock (Achard and Banoin, 2003).

In the Maiduguri area of Nigeria, the fertility status of alfisol soil under continuous cultivation based on manure application was examined by Aweto and Ayuba (1993), who compared the soil's physical and chemical characteristics with that of soil under savanna woodland nature reserve. Their findings indicated that the application of animal manure had beneficial effects on soil physical status, especially on the 0–10 cm layer whose bulk density and total porosity values of 1.10 g cm^{-3} and 58%, respectively, are similar to those of the soil under the savanna reserve. Although organic carbon is significantly higher in the soil under the savanna woodland reserve, soil total nitrogen is significantly higher in the manured, continuously cultivated soil. The total nitrogen level of 0.40% in the 0–10 cm layer of the cultivated soil is high and response to the application of nitrogen fertilizers would be unlikely. Young (1976) observed that when soil nitrogen content exceeds 0.20%, response to nitrogenous fertilizer application would be highly improbable. The mean levels of soil CEC and exchangeable potassium in the cultivated soil are low, implying that the application of higher levels of manure to the soil by the farmers would be beneficial to soil productivity.

In the forest region, animal manure is not readily available, especially in tropical Africa where the forest region is infested with tsetse flies, some species of which transmit sleeping sickness to humans and 'nagana' to cattle, except for a few hardy indigenous breeds of cattle that are native to the forest region. Ranches, including large commercial ones, have been established in the rainforests of the Amazon basin of South America, especially in Brazil, and this makes manure more readily available than in the forest regions of tropical Africa. This implies that the prospects of intensifying shifting cultivation through the application of animal manure are much brighter in the forest regions of South America than in the forest zones of tropical Africa.

Compost made from weeds and crop residue can be readily used as manure for crops, where animal manure is not readily available. Another possible source of manure is municipal solid waste. These wastes usually decompose in dump sites to form manure or compost. Adeoye *et al.* (2005) have evaluated decomposed organic wastes in municipal dump sites as manure for crop production. Their findings indicated that the decomposed organic wastes are rich in organic carbon, nitrogen, available phosphorus and potassium and, when used as a soil amendment, their efficacy in promoting crop growth is similar to compost.

10.3 Alley Farming

This is a type of hedgerow intercropping that has been suggested by scientists at the International Institute of Tropical Agriculture, Ibadan, Nigeria as a stable alternative to shifting cultivation. Essentially, alley cropping involves cultivation of food or field crops in between rows of trees or shrubs that are pruned during cropping to avoid shading the crops, and the prunings returned to the soil as mulch or green manure. The trees, which are usually legumes such as *Gliricidia sepium* and *Leucaena leucocephala*, but may include non-leguminous trees such as *Acioa barteri* and *Alchornia cordifolia*, are planted in rows that are 2 or 3 m apart, and food crops such as cassava, yams, cowpeas, rice and maize or sorghum are planted in the spaces between the rows of trees. The pruned shoots decompose over time and add organic matter and nutrients to the soil, and have the additional advantage of reducing soil temperatures, and hence accelerated soil organic matter decomposition. Furthermore, they reduce surface runoff and erosion, the rate of water loss from the soil through evaporation and the growth of weeds. Tree legumes have the additional advantage of harbouring nitrogen-fixing bacteria in their root nodules which help to accumulate nitrogen, a nutrient which is deficient in most tropical soils, in the soil for the planted crops. The trees are usually pruned at the height of about 1 m above the ground during cultivation but are allowed to grow unhindered during the off-farm season. They have to be pruned prior to a subsequent season of cropping to provide fuelwood for the farmer. Kang *et al.* (1984) observed that alley farming is a stable alternative to shifting cultivation and makes continuous cultivation possible with or without the use of fertilizers. Really, alley farming is an intensified form of hedgerow intercropping, which is one of the intensified varieties of shifting cultivation.

Hedgerow intercropping is an intensified form of shifting cultivation, practised by the Igbo people in south-eastern Nigeria. Kang *et al.* (1990) observed that they developed and have practised this form of traditional agroforestry for several generations. Benge (1987) also observed that the Nalaad people in the Philippines have developed a form of agroforestry which involves planting *Leucaena* hedgerows on steep slopes to control erosion, and that they have been pruning the trees regularly and returning the loppings to the soil as mulch since at least 1923. This form of agroforestry, which to all intents and purposes is alley cropping, was developed by these small-scale farmers well before the International Institute of Tropical Agriculture (IITA) was established in 1967. Hence, the system of alley farming was not, strictly speaking, initiated by scientists at IITA. Rather, they experimented on and improved upon the system of hedgerow intercropping developed by native farmers in the Philippines and south-eastern Nigeria, in order to improve our understanding of this peculiar system of traditional agroforestry. They also promoted and advocated its adoption as an alternative to shifting cultivation.

In a sense, alley farming integrates both the fallow and the cropping phases of the shifting cultivation cycle. The hedgerows of trees in the system represent the fallow vegetation, which helps to restore and maintain soil fertility, while the cultivated corridors represent the cropping phase. The prunings of the planted trees in alley systems add organic matter and nutrients to the soil on decomposition; similarly, trees in bush fallow vegetation accumulate organic matter and nutrients in fallow soil over time.

The data published by Kang *et al.* (1990) indicated that the yield of maize was maintained at an average of 2 t ha^{-1} over a period of 7 years in continuously cultivated alley plots established on alfisols where *L. leucocephala* prunings were applied to the soil, compared with an average yield of 0.5 t ha^{-1} in plots from which the prunings were removed (Fig. 10.1).They proposed alley cropping as a stable alternative to shifting cultivation mainly on the basis of these results. It is important to observe that the main advantage of alley farming lies not in cultivating field crops between rows of trees or shrubs, but in returning tree prunings to the soil as green manure or mulch, as Fig 10.1 clearly shows. This presumably suggests that when adequate levels of organic residues are returned to the soil, cultivation can be prolonged well beyond a few years. In northern Mindanao, the Philippines, MacLean *et al.* (2003) reported the beneficial effect of alley cropping using *Gliricidia sepium* and *Cassia spectabilis* in improving rice yield on degraded and eroded soils (oxisols and inceptisols) when the prunings of these two species were applied to the soil as mulch and green manure. Their results suggest that alley

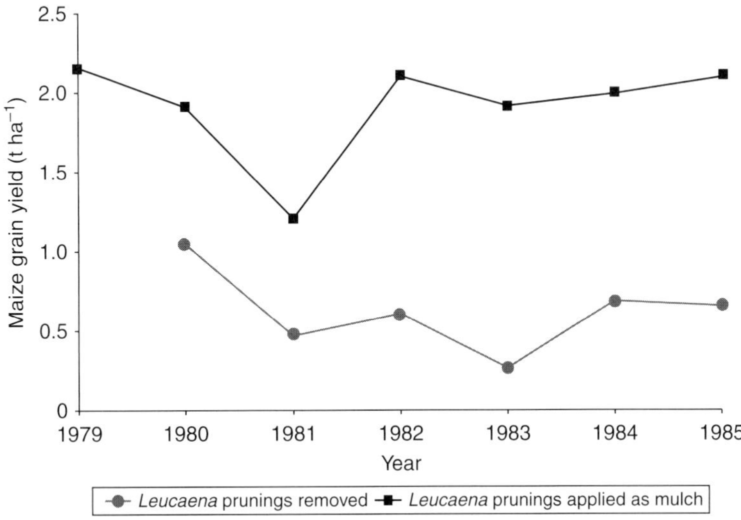

Fig. 10.1. Maize grain yield in alley farming. (Based on data of Kang *et al.*, 1990. The plots were fallowed in 1985.)

cropping has the potential of restoring and improving the fertility status of degraded soils in the humid tropics. The results obtained by Evensen *et al.* (1995) for a highly weathered oxisol in West Sumatra, Indonesia, seem to indicate that alley farming may decrease yields of crops, instead of improving them, unless inorganic fertilizers are applied to augment soil nutrients. On the highly acid soil in the equatorial region of West Sumatra, they observed that yields of cowpea and rice, as well as the soil exchangeable cations – calcium, magnesium and potassium – declined in alley cropping systems during the first 3 years of cultivation. This was in spite of the application of the prunings of *Paraserianthes falcataria*, *Gliricidia sepium* and *Calliandra calothyrsus* to the soil four to six times a year, coupled with lime application. The decline in crop yield was reversed during the fourth year by fertilizer application, after which crop yields and soil nutrients returned to their original level. In fact, alley cropping does not appear to be a stable alternative to shifting cultivation everywhere in the tropics. In semi-arid regions, it is difficult to establish the hedgerows of trees and, when established, they may compete with cultivated crops for nutrients (*Spore*, 2009a) and this would result in reduced crop yields.

10.3.1 Adoption of alley farming by small-scale farmers

Farmers have been generally reluctant to embrace alley farming, except perhaps in densely populated areas where land is in critically short supply. A study by the author of over 100 farms on the northern fringe of Ibadan revealed that no farmer has embraced alley farming, proximity to the International Institute of Tropical Agriculture notwithstanding. It is important to observe that alley farming is labour intensive, particularly during the first year when farmers have to establish the hedgerows of trees or shrubs in their fields. Small-scale farmers are generally wary of adopting new farming technology intended to intensify traditional agriculture, if it does not yield immediate financial benefits to them. David (1995), who evaluated the adoption of alley farming in the densely populated Machakos district of Kenya, observed that although the farmers were convinced that hedgerow intercropping has benefits in reducing soil erosion and improving crop yields, they were of the opinion that the long-term benefits to be derived from the system were not worth the extra labour required for establishing the rows of planted trees on farms.

10.3.2 Problems associated with alley farming

Apart from the issue of low adoption by small-scale farmers, alley farming is bedevilled by a number of problems. First, the hedgerows of trees do not adequately replicate the bush fallow vegetation, particularly with respect to floristic diversity. In several instances, only one tree species is established in hedgerows in alley cropping systems. This makes the system vulnerable to attack by pests and plant diseases. Stands of *L. leucocephala*, a tree legume that is widely promoted in alley cropping, have been reported by Reijntjes *et al.* (1992) to have been devastated in Fiji, Indonesia and the Philippines as a result of insect attack. If hedgerows are made of several rows consisting of various species, they will be less prone to insect attack and the alley system would become more stable. Furthermore, establishing a mixture of species in the hedgerows would ensure that the alley farming system more closely mimics the natural bush fallow, and the greater diversity of tree species would guarantee greater stability of the agricultural ecosystem, in addition to ensuring a more efficient cycling of nutrients from the subsoil layer to the alley tree crops and ultimately to the cultivated crops. Finally, alley cropping has been reported to encourage insect pests and crop diseases. MacLean *et al.* (2003) observed that the hedgerows in an alley cropping system in northern Mindanao, Philippines, encouraged the incidence of rice blast disease by reducing air movement and also enhanced the population of insect pests of rice.

10.4 *Quesungual* Slash-and-Mulch Agroforestry System

The *Quesungual* slash-and-mulch agroforestry system was developed in the Lempira district of Honduras by small-scale farmers who previously practised slash-and-burn agriculture. An important feature of the area is that slopes are steep, sometimes exceeding 45°, with the result that when farmers burn the cut slash of fallow vegetation before cultivation, considerable soil erosion results and fields usually undergo rapid degradation, with an attendant decline in soil fertility. The *Quesungual* agroforestry system was developed mainly to combat and even reverse this trend.

Farmers in the small village of Quezungual decided to use the slash of fallow vegetation as mulch on their cultivated fields instead of burning it. At the same time, they retained trees on their fields to further check erosion and enhance the cycling of nutrients from the subsoil to the topsoil for cultivated field crops. The resulting agroforestry system, in which the slash of cleared fallow vegetation is used as mulch to protect the ground surface, and field crops such as maize, beans and millet are sown in the soil through the ground layer of mulch is called *Quesungual* after the village where the system was first practised in western Honduras (Ayarza and Welchez, 2004). The system consists of three layers:

1. A ground layer of mulch derived from crop and weed residue and the cut slash of bush fallow vegetation.
2. A layer of field crops such as maize and millet, interspersed with pollarded trees, over-lying the mulch layer.
3. A layer or two of shrubs and trees retained on the site during site clearance.

Many trees, except those that produce fruits or timber, are pollarded by farmers to prevent the shading of cultivated crops, and the pruned branches are used as mulch or piled together on the slope to help check erosion (Hellin *et al.*, 1999). It is pertinent to note that Aweto (1981) observed that the maintenance of soil organic matter during cropping is pivotal to maintaining, improving and sustaining soil fertility status, and he suggested that the cut slash of fallow vegetation should be used as mulch, instead of being burnt. The *Quesungual* agroforestry system takes cognizance of this suggestion, and the productivity of the system depends primarily on the use of the slash of fallow vegetation as mulch, which not only protects the soil against erosion, but also adds organic matter and nutrients to the soil on decomposition.

The system involves zero tillage of the soil and this helps to conserve soil physical status during cultivation. The surface layer of mulch also helps to prevent degradation of soil structure and may even contribute towards improving soil physical status through the addition of organic matter to the soil over time.

The use of the cut slash of fallow vegetation as mulch helps not only to reduce erosion, but also to conserve soil moisture. Furthermore, because covering the soil with mulch leads to organic matter conservation and accretion when mulch decays and is incorporated into the soil, the water-holding capacity of the relatively shallow and stony soils that characterize western Honduras – entisols – has increased since the farmers in the area began to practise this system of agroforestry. Soil organic matter has been reported to have increased from 1 to 3% and, on account of this, soil water-holding capacity has also increased from 8 to 29% and soil erosion losses reduced drastically (Ayarza and Welchez, 2004). The crops are less prone to drought, mainly because the system is conservative of soil moisture, and the population of soil organisms is also enhanced by the system. Fonte *et al.* (2008) reported that the population of earthworms in the soil under *Quesungual* agroforestry is significantly higher than that under the slash-and-burn system. This obviously implies that the population of soil organisms, including microorganisms, is likely to be higher under the *Quesungual* agroforestry system. Hence, the mineralization of soil organic matter, which releases mineral nutrients for crops, will be more efficient under the *Quesungual* system.

Owing to the improved soil quality, crop yields have improved considerably and it has been reported that the yields of maize and beans have doubled since farmers adopted the *Quesungual* agroforestry system in western Honduras (Ayarza and Welchez, 2004). In addition, the system allows farmers to produce fruits and timber simultaneously with food crops. Timber-yielding trees which the farmers retain on their cultivated fields include *Cordia alliodora* (laurel), *Swietenia* spp. and *Cedrela odorata*, while the fruit trees usually planted by the farmers include *Persea americana* (avocado), *Mangifera indica* (mango), *Carica papaya* (pawpaw) and *Anacardium occidentale* (cashew) (Hellin *et al.*, 1999). The system also yields firewood, and allows farmers to cultivate their fields for a longer period before the land is left fallow. Some of the branches of pollarded trees are taken home for use as firewood, and this relieves pressure on natural forests for fuelwood, thereby helping to reduce the rate of deforestation in the area.

The *Quesungual* agroforestry system is a relatively simple and yet effective way of intensifying shifting cultivation. Strictly, it is not an alternative to, but an intensified form of, shifting cultivation, as farmers still have to fallow their plots to regenerate lost fertility, and at that time they have to relocate to cultivate other plots. Burning the slash of fallow vegetation or secondary forest is a quick though inefficient means of releasing nutrients in plant biomass into the soil, for a significant proportion of the nutrients in the ash produced by the burnt vegetation slash is leached away by percolating rainwater or blown away by the wind. Using the slash of cleared vegetation as mulch is an effective way of overcoming this limitation of slash-and-burn agriculture. The vegetation slash used as mulch decomposes slowly on the ground to form organic matter, which is subsequently mineralized by microorganisms to release mineral nutrients for crop use. The main limitation of the *Quesungual* system is that the release of plant nutrients through the decomposition and mineralization of organic matter may not be synchronized with the peak period of nutrient demand of cultivated crops. Another limitation of the system is that in acidic soils, such as those that characterize most parts of the humid tropics, soil pH is low at the inception of cropping, implying that the soil base saturation would also be low. Hence many plant nutrients will not be readily available for plant use, as would be the case when fallow vegetation is burnt to release ash into the soil. However, the doubling of bean and maize yields in western Honduras since the inception of the *Quensungual* system (Ayarza and Welchez, 2004) clearly shows that the two factors noted above do not constitute major constraints to soil productivity under the slash-and-mulch system.

Considerable amounts of carbon dioxide are released into the atmosphere when the slash of fallow or forest vegetation is burned before cultivation (Kotto-Same *et al.*, 1997; Murdiyarso *et al.*, 2005). Hence, shifting cultivation has been blamed as one of the major contributors to global warming. The *Quesungual* system does not involve burning cleared vegetation, and also helps to increase carbon storage in the soil, so helping to mitigate the threat of global warming. It is pertinent to note that the FAO (Food and Agriculture Organization) is promoting the adoption of this low-input alternative to slash-and-burn in hilly or mountainous areas where shifting cultivation is practised.

Under the Kyoto Protocol, which aims to limit greenhouse gas emission into the atmosphere, reforestation projects (which help to sequester atmospheric carbon dioxide) can earn carbon credits that can be sold for cash in the carbon market. Currently, only reforestation and industrial projects which limit carbon emission are eligible for carbon credits under the first phase of the Kyoto Protocol (*Spore*, 2007). Hence, small-scale land holders who have developed ingenious agroforestry systems such as the *Quesungual* system – in which the soil is kept covered with mulch during cropping and so becomes a carbon sink – are not entitled to carbon credits that can be sold to European or other buyers. In order to popularize agroforestry schemes such as the *Quesungual* system in the tropics, it is necessary to regard it and similar schemes as 'clean development mechanisms' that are eligible for carbon credits which can be sold on the carbon market. It is to be hoped that the additional revenue derived from the sale of carbon credits would make resource-poor farmers less dependent on biomass energy such as fuelwood, and so additionally contribute towards mitigation of global warming.

10.5 Shifting Cultivation in Retrospect

It is appropriate to re-examine shifting culti-vation, highlighting its status and current significance, before attempting to assess the future of the system of agriculture.

Essentially, shifting cultivation is synony-mous with small-scale agriculture that is practised by farmers in the tropics who produce field crops on a subsistence basis, using fallow vegetation to rejuvenate soil fertility. Shifting cultivation may seem insignificant globally, in terms of the total contribution to world food output, but not with respect to the total work force engaged or food produced in the tropics. Francis (1986) estimated that traditional multiple cropping systems, which are char-acteristically associated with shifting cultivation, account for 20% of the world food supply. This relatively low proportion tends to mask its importance in the tropics, where it is the major source of food supply. In Latin America, small-scale farmers, the vast majority of whom are shifting cultivators, account for 51–77% of the maize, beans and potatoes produced for consumption (Altieri, 2008). In West Africa, as with most parts of tropical Africa, shifting cultivators account for more than 80% of the food produced for consumption. About three billion people live in rural areas of the tropics, and 80% of them live in households directly engaged in small-scale agriculture, especially shifting cultivation.

10.5.1 Intercropping

An important feature of shifting cultivation is intercropping or polyculture. Several species of crops, sometimes up to 20 or more, are grown together on the same field. Hence, the cultivated field is characterized by a high diversity of cultivated crops rather than a monoculture. This diversity is further enhanced by the practice of retaining or planting trees in the fields cultivated by farmers who practise slash-and-burn agriculture. Intercropping confers a higher degree of resilience on the agroecosystem than do systems based on monoculture, as high species diversity ensures ecosystem stability. Holt-Gimenez (2001) has shown that farms characterized by a high diversity of cultivated crops are resilient to climate-related hazards such as hurricanes, floods and droughts, which are becoming more frequent in the wake of global warming. Another major advantage of intercropping is that the overall yield of all crops is higher than that of a single crop grown in monoculture (Francis, 1986; Altieri, 2008). Furthermore, high biodiversity provides a sound base for key ecosystem processes to function effectively, including natural control of pests; also, the more diverse the population of plants, animals and soil organisms in an agricultural ecosystem, the more diverse and efficient is the community of beneficial, pest-fighting organ-isms (Altieri *et al.*, 2006).

10.5.2 Reduced tillage

Shifting cultivators do not usually plough their fields. They sow their seeds using simple implements such as digging sticks or machetes which cause minimal disturbance to the soil. They also leave stumps of trees and stands of live trees on the farm to help check erosion during cultivation. Therefore they practise minimum or reduced tillage, which not only checks erosion but reduces soil organic matter decline during cultivation. In contrast, ploughing, which is associated with mechanized agriculture in the temperate regions of the North, enhances compaction of the subsoil, the rate of erosion and organic matter decline, reduction in the population of soil organisms and nutrient loss in cultivated soil (Brady and Weil, 2002; Uphoff, 2006).

Tillage also results in deterioration of soil structure. Due its deleterious effects on the soil and crop yield, zero till or no tillage, which involves the direct sowing of a new crop on the residue of a previous crop without tillage, is gaining acceptance by farmers in different parts of the world. Zero till increases crop yields and soil quality (Zoebisch *et al.*, 2008). More than 30% of the cultivated land in the USA, where conventional agriculture based on large-scale mechanized tillage is the main feature of arable farming, is now under zero till or reduced tillage (Uphoff, 2006). The practice of minimum tillage by shifting cultivators was initially criticized by colonial agricultural officers and experts from temperate regions, especially Europe, who did not appreciate that minimum or zero tillage conserves soil fertility and enhances crop yield. Dudgeon (1911), an agricultural adviser for British West African territories, criticized the farming practice of the Bini people and their neighbours in the forest zone of southern Nigeria, for growing grain crops in partially cleared land, and encouraged the practice of 'deep cultivation'. What he failed to realize was that the practice of cultivating partially cleared land, a type of minimum tillage that involves retaining live trees in the farm and planting crops in shallow holes, instead of hoeing or ploughing the entire field, helps to reduce soil erosion and conserve organic matter and nutrients during cultivation. Agricultural experts from Europe attempted to introduce large-scale mechanized agriculture involving deep ploughing into tropical Africa during the colonial period. Many of these projects, including the Mokwa agricultural mechaniz-

ation project in Nigeria, which was intended to produce groundnuts for export and local staples such as guinea corn to help boost local food production, failed and were subsequently abandoned. It is pertinent to note that scientists in the temperate regions now promote zero or minimum tillage – a practice they previously criticized in small-scale tropical farmers. Shifting cultivators, therefore, helped to draw the attention of agronomists from the temperate regions to the conservation value of zero or minimum tillage. The practice of reduced till by shifting cultivators also helps to reduce soil organic matter decline in the soil during cultivation and the attendant emission of carbon dioxide into the atmosphere.

10.5.3 Organic farming

Shifting cultivation does not involve the use of agrochemical inputs such as inorganic fertilizers, pesticides and herbicides; it is organic farming, which does not pollute the soil or the other components of the physical environment. Admittedly, carbon dioxide and other gases are emitted into the atmosphere when the cut slash of fallow vegetation is burnt and as a result of organic matter decomposition in the soil during cultivation. However, a substantial part of the carbon dioxide released as a result of shifting cultivation is sequestered by regenerating fallow vegetation in various stages of succession. Hence, the contribution of shifting cultivation to global warming is minimal compared to commercial agriculture in the temperate regions, which is heavily dependent on the input of fossil fuels. The health and fertility of the soil, and hence, the health of humans, animals, plants and the ecosystem in general are a major concern and an overriding objective of organic agriculture; furthermore, the health of humans is regarded as inseparable from soil health and fertility (Wright, 2008).

A major challenge of shifting agriculture as organic agriculture is to produce rich and nutritious foods, free from chemical contaminants such as pesticide residues, with balanced essential nutrients and vitamins that will promote human health and well-being. As pointed out in Chapter 1, the continent of Africa, especially sub-Saharan Africa, has been devastated by the AIDS pandemic, the prevalence of HIV/AIDS in the continent being the highest in the world. The efficacy of antiretroviral drugs is enhanced by good and balanced nutrition, and the contraction of AIDS by people who are HIV-positive is delayed by a good diet of rich, nutritious food (Wright, 2008). It is therefore expected that shifting cultivators, the main producers of food in Africa and even in Central and South America, should produce crops without using pesticides and inorganic fertilizers, in order to produce nutritious food that will sustain the health of the people and those who are HIV-positive. They should use compost with livestock manure to help correct nutrient deficiency in the soils they cultivate instead of resorting to the use of inorganic fertilizers and pesticides, in their bid to increase food production. It is also important to note that many shifting cultivators are HIV-positive and do not have ready access to antiretroviral drugs, as these are expensive. The onus is therefore on shifting cultivators to produce nutritious crops and fruits by integrating fruit trees into their farmlands. The consumption of nutritious foods by shifting cultivators will help to delay the contraction of AIDS by farmers who are HIV-positive. This will ensure that such farmers live longer and healthier lives and be better able to produce food for themselves and the growing population. This will also help to ameliorate a situation in which the more active middle-aged farmers are dying as a result of AIDS, leaving the younger and older people to farm and produce food.

10.5.4 Agricultural innovation and global warming

Shifting cultivators are not static and un-responsive to the need to improve soil fertility and develop more sustainable systems of agriculture in the face of a rapid increase in the human population, nor do they mine soils and degrade them over time. Richards (1985) observed that small-scale farmers in West Africa and, by extension in the tropics, are not 'unscientific' or 'backward', and that the farmers' environmental management is dynamic and innovative and not a mere adaptation to the ecological and socio-economic conditions. He

further stressed that although native agriculture can benefit from scientific discoveries, scientists should partner with small-scale farmers to improve existing farming methods and systems. Therefore, agricultural scientists should not seek to replace farming systems and techniques developed by native farmers with methods of farming developed in the temperate regions, where ecological conditions are strikingly different. Shifting cultivators experiment and are willing to adopt new farming technologies that meet their needs. The widespread adoption of soybean cultivation in Zimbabwe to improve soil fertility for a subsequent maize crop grown during the second growing season was referred to in Section 9.5.5. Shifting cultivators have, in fact, developed new farming techniques to suit their needs. Most of the intensified forms of shifting cultivation and alternatives were developed by the shifting cultivators themselves. An outstanding success story in this regard is the development of the *Quesungual* agroforestry system by native farmers in Honduras. This agroforestry system conserves soil moisture, lengthens the period of cultivation and improves soil nutrients and organic matter. It also has the unanticipated potential of helping to counter deforestation and the threat of global warming.

In fact, Erni (2009) argued that the land management strategies of the indigenous people of South-east Asia, through shifting cultivation, contribute more to combating global warming than do tree plantations, which are widely advocated for carbon sequestration. Shifting cultivators and other small-scale farmers in the tropics are primarily concerned with conserving or improving soil fertility, usually without using chemical fertilizers, in order to ensure that yields are sustained or improved over time to meet the needs of an increasing population. In striving to achieve this objective, they have developed ecologically sound farming systems such as the slash-and-mulch system and its variants such as the *Quesungual*, which not only substantially reduce carbon dioxide emission from farmers' fields during cultivation but convert the cultivated soil to a carbon sink by enhancing organic matter accretion over time. Further-more, in order to integrate elements of fallow vegetation into cultivated fields and ensure that organic matter and nutrient cycles do not grind to a halt during cropping, innovative shifting cultivators have embraced various forms of agroforestry.

One of the most striking and successful forms of agroforestry developed by shifting cultivators, with considerable potential in mitigating global warming through carbon sequestration in live trees, is growing field crops under tree canopies. This type of farming, which has been referred to as evergreen agriculture by Garrity (2010), is very well developed in Africa. The integration of leguminous trees, especially *Faidherbia* (*Acacia*) *albida* into farmlands in the savanna lands of West, East, Central and Southern Africa, was alluded to in Section 9.5.1. The farmers usually cultivate field crops such as maize, millet and sorghum, for several decades under and between the canopies of trees growing on farmland. The trees add organic matter and nutrients to the soil under their canopies to benefit the crops, and sequester carbon in their standing biomass and in the soil for several decades while they remain standing in the fields. They also yield useful products such as fodder for livestock during the dry season.

While evergreen agriculture is well documented for the savanna regions of Africa, it is less so for the forest areas. Most forest trees usually cast a dense shade on the soil underneath them, making the cultivation of field crops under their canopies impossible or uneconomic as a result of reduced yields. A notable exception is the oil palm, whose light canopy allows enough sunlight to reach the soil under its canopy to allow cultivation of field crops, if the density of palms is not too high. The cultivation of field crops, especially maize and cassava, beneath and between the canopies of oil palm trees by the Adja people of Mono Province of Benin Republic was discussed in Section 9.1. The Adja people also use the palm as a kind of fallow crop when the cultivation of field crops is discontinued due to the closing up of the palm canopy. In southern Nigeria, farmers usually grow field crops such as maize and cassava under the canopies of the oil palm for a few years before leaving the land fallow. The palms remain on the field throughout the fallow period, which lasts 3–7 years before the land is recultivated. In essence, the oil palms remain on the farmers' fields during both the fallow and cultivation periods. They are not felled as they provide useful products such as palm oil, kernels, brooms and

thatch to supplement the farmers' incomes. The palms sequester considerable amounts of carbon in their biomass and in the soil and help to check soil erosion and improve soil fertility while growing for several decades on farmers' fields. In south-western Nigeria, farmers plant maize and cassava beneath and between canopies of *Newbouldia laevis*. The canopies of the trees have very narrow profiles and so cast minimal shade on field crops cultivated underneath them. It is necessary for farmers in the temperate regions, especially in Europe, North America and China (regions that are the main contributors to global emissions of carbon dioxide, and hence global warming), to embrace elements of agroforestry or even evergreen agriculture, where possible, if the quest to combat global warming is to advance beyond mere rhetoric. They should also plant trees in their farms, at least in the hedgerows, so that such trees can help with the long-term sequestration of carbon dioxide.

10.5.5 Deforestation

Finally, shifting cultivation has been linked to the much-debated issues of deforestation and biomass burning in the tropics, which are considered to be major causes of greenhouse gas emission and global climate change (Nilsson, 1992; Moran, 1993). In Brazil, for instance, Lal (2005) considered deforestation to be a major source of greenhouse gases emission. It is important to observe that shifting cultivators are not the primary cause of disappearance of tropical forests, where they existed in the tropics, at least not in the last couple of decades. Owing to the enormous time and energy requirements of clearing a primary or mature rainforest or other types of tropical forest, shifting cultivators prefer to cultivate land under regenerating secondary forests of about 10–20 years of age. It is gratifying to note that the findings of Skole *et al.* (1994) indicated that most of the land clearing for shifting cultivation in the Amazon basin now takes place in secondary rather than primary forest. Neither are shifting cultivators the principal nor the only agents of biomass burning.

Let us examine the situation in Brazil, the country with the largest area of land under tropical forest and which has also undergone a very fast rate of deforestation within the last few decades, in order to ascertain the veracity of the above assertions. In the state of Acre, farmers have deforested a total of 1.4 million ha of land over 25 years, with about 80% of the deforested land being converted into pastures (Embrapa, 1999). This clearly shows that large-scale farmers who clear forest for pastures, rather than shifting cultivators, are the primary agents of deforestation. This is not an isolated case, as ranchers are active agents of deforestation in the Brazilian Amazon. Fearnside (2000) quantitatively characterized land-use change in the Brazilian Amazon in 1990, and estimated that a total of 4.0 million ha were deforested through clearing forest and secondary forest, while an astonishing 16.1 million ha were converted to pastures through biomass burning. Assuming that the 4 million ha of cleared forest land were used exclusively for arable farming by shifting cultivators, the amount of forest land converted to shifting agriculture was a mere 25% of the land converted to pastures. Also, Houghton *et al.* (1991) observed that 370 million ha of land were deforested in Latin America between 1850 and 1985, and that shifting cultivation accounted for 10% of this figure, compared to pastures, which accounted for 44% of the total land area deforested.

In Indonesia, where large expanses of land are under forest vegetation, shifting cultivators are not the primary agents of deforestation. Partohardjono *et al.* (2005) observed that timber exploitation is the main cause of deforestation of primary forests that ultimately transforms them into secondary forest, and further pointed out that shifting cultivation is of secondary importance in terms of conversion of primary forests into other types of land use.

It would seem from the foregoing that in areas where extensive natural forests still exist, shifting cultivation is not the primary cause of deforestation and conversion of primary or mature tropical forests into other types of land use. The argument that large-scale deforestation in the tropics is currently due mainly to shifting cultivation, therefore, appears to rest on a thin empirical base. This is not to deny the fact that shifting cultivation may be a primary factor of deforestation in some areas, especially in forest frontiers. Livestock farming involving the establishment of pastures, as in the Amazon of South America or logging of forests, as in Indonesia, and other factors such as infrastructural development, cash-crop farming, plantation agriculture and forestry are of pivotal importance to tropical deforestation. Finally, it is important to observe that sub-Saharan Africa has the largest area under shifting cultivation of all the tropical land masses. Yet, sub-Saharan Africa accounts for a mere 1.6% of the total global greenhouse gas emission compared to the USA and the EU, which together account for more than 50% of greenhouse emissions (*Spore*, 2008). The statistics presented above clearly show that industries with mechanized agriculture in the West and China, rather than shifting cultivation in the tropics, are the main culprits of greenhouse gas emission and of the attendant problem of global warming.

10.6 The Future of Shifting Cultivation

It is apparent from what has been discussed in Chapter 9 and the first part of this chapter that shifting cultivation has undergone considerable modification and that some alternatives have been proposed to replace it. The intensification and modification of shifting cultivation have largely been in response to a rapid increase in population and an attendant pressure on land. These changes introduced into shifting cultivation offer some clue to the future of the system. Before attempting to chart the likely course of its future development, it is important to observe that shifting cultivation is not a single system of farming. Rather, it is an amalgam of several systems, each of which is adapted to specific ecological zones, and the variants share some basic characteristics. In trying to ascertain the likely development of the system, we should note that modifications to the system will assume local or regional coloration, depending on the nature of the environment and the culture and values of the people in any particular area or region. The implication is that a blanket approach cannot be adopted to the intensification of shifting cultivation.

Intensification is inevitable and is already taking place and will continue for several decades. For a number of reasons, large-scale mechanized agriculture, based on monocultures, does not provide a rational basis for the intensification of shifting cultivation.

The Green Revolution in the West is based mainly on mechanized farm operations, including deep ploughing and the application of agrochemicals, especially inorganic fertilizers and pesticides. The intensification of agriculture and the Green Revolution in the temperate region have been achieved at great environmental cost, including the pollution of rivers and underground water and the decimation of wildlife, mainly from the toxic effects of inorganic fertilizer and pesticide residues.

In contrast, shifting cultivation is organic agriculture that requires little or no external inputs. It depends on the natural process of nutrient cycling and the input of organic matter in the form of crop residue, compost, mulch, or animal manure to maintain soil fertility. It does not result in the introduction of toxic chemicals into the environment, as with the case of the use of pesticides in conventional mechanized agriculture in temperate regions. It is pertinent to note that due to the harmful effects of pesticides on the environment, many countries including the USA have banned the use of several pesticides. Indonesia, which began the use of pesticides as part of the package of rice intensification, has also banned the use of several pesticides and has resorted to integrated pest management, which involves biological control of pests, as the national pest control strategy (Van de Fliert and Winarto, 2006). Organic agriculture is also gaining ground in the temperate regions because of the health hazards associated with pesticide and fertilizer residues in food and the environment. Agroforestry and the integration of crop and livestock farming, coupled with sound agronomic practices such as mulching, use of cover crops and composting – rather than deep ploughing and the application of mineral fertilizers and agrochemicals – hold the key to the ultimate replacement of shifting cultivation by continuous or semi-continuous cultivation. Farms that are characterized by high biodiversity, such as the small-scale farms of shifting cultivators, are more sustainable and resilient to vagaries of climate change than are large-scale farms, such as characterize farming in the temperate regions. They are also characterized by efficient energy flow and utilization and nutrient cycling, and they are largely self-sustaining as they do not depend on external, industrial inputs, especially inorganic fertilizers and pesticides.

10.6.1 Stages and processes of intensifying shifting cultivation

Shifting cultivation has undergone considerable changes from its primordial form that involved the relocation of settlements and cultivated fields after a couple of years when the fertility of the available farmland, usually within walking distance of the settlement, is depleted. This can be referred to as migratory shifting cultivation. The system of agriculture has evolved beyond the migratory stage and the likely pathway of intensification of shifting cultivation is illustrated diagrammatically in Fig. 10.2. Today, shifting cultivators usually live in permanent settlements and cultivate fields around the settlement, while leaving several fields fallow in order to restore soil fertility naturally.

This second stage in the evolution of shifting cultivation, during which farmers live in permanent settlements, can be referred to as sedentary shifting cultivation. This stage is initially characterized by long fallow periods, but as population increases over time, the length of fallow period will be considerably reduced. As pointed out in Chapter 9, intensification of shifting cultivation has resulted in considerable reduction in the long fallow periods which were characteristic of the past two to three decades. Fallow periods of up to 15 years or longer are now rare, except in remote areas, especially sparsely populated hilly or mountainous areas that lack ready access to urban markets. Today, fallow periods of 3–5 years or less are more characteristic of moderately and densely populated areas where slash-and-burn agriculture is practised. Contemporaneous with the reduction of the fallow period, farmers have adopted various management strategies to ensure that the agricultural resource base, especially the soil, is maintained in a reasonably productive state or at least is not degraded beyond the limits that would ensure its continued capacity to sustain arable food and livestock production.

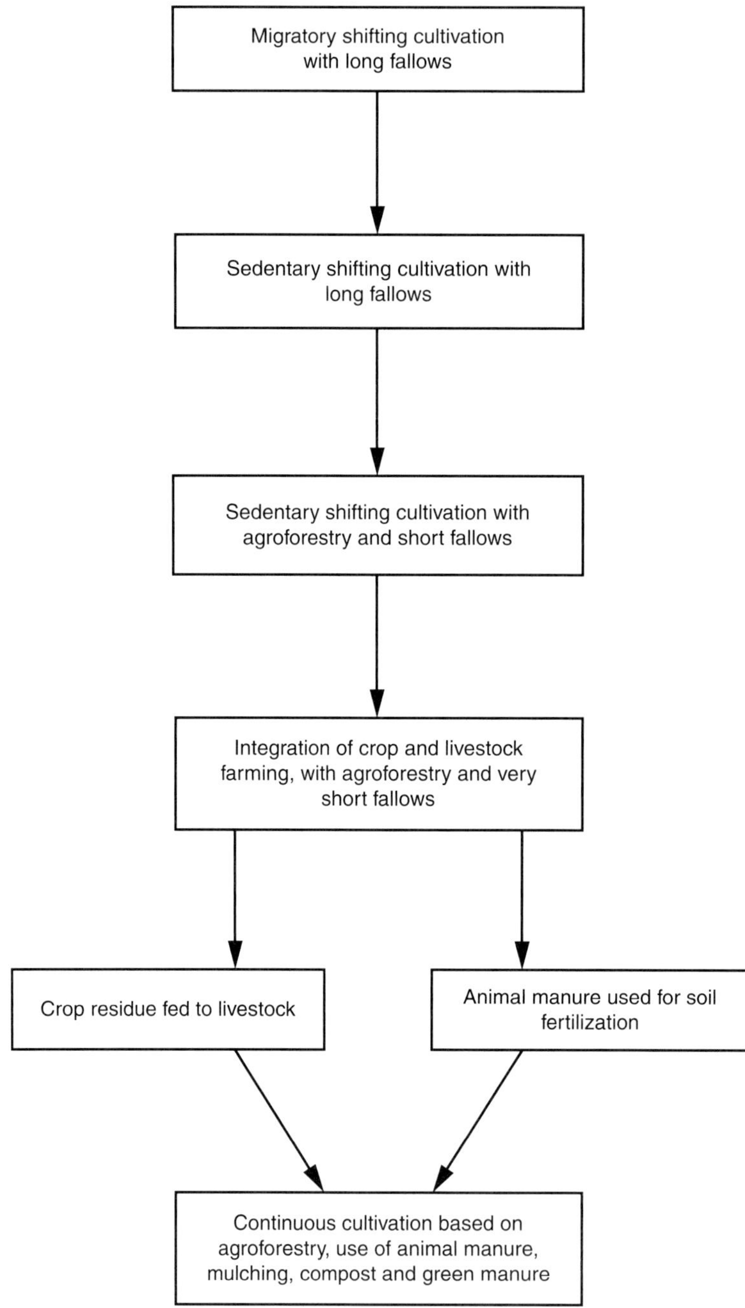

Fig. 10.2. Evolution of shifting cultivation.

One major development in this regard is the widespread adoption of agroforestry, which involves the direct planting or retention of trees, many of which have been proven to improve soil fertility or provide valuable fodder for livestock in farmers' fields. Some of the trees planted in farms or fallow vegetation to enhance its soil restorative ability include palms and legumes. The palm fallows of the Mono province of southern Benin Republic clearly demonstrate the ingenuity of shifting cultivators, not only in increasing the soil restorative capacity of fallow vegetation, but also in transforming fallow vegetation into an important source of food and income for the farmer. The Baduy people of Java, Indonesia, whose culture forbids the application of mineral fertilizers and other agrochemical inputs on their farms, have also intensified shifting cultivation through the integration of a tree legume that restores soil fertility into their farms and fallow land.

Shifting cultivation was not unique to the tropics as it was also practised in the temperate regions in the past. Farmers in Europe practised shifting cultivation during the Middle Ages (Parain, 1942) and the early European settlers in the southern part of the USA depended on the system until about the middle of the 19th century. In Japan, shifting cultivation was the main technique of land management for arable farming, and the system was the predominant land use in mountainous areas until the Second World War (Mizoguchi, 1988). In all these areas where shifting cultivation was the dominant system of agriculture in the past, permanent cultivation, usually based on fertilizer application, has emerged and replaced shifting agriculture. One can therefore hypothesize that the intensification of shifting agriculture in the tropics will ultimately terminate in continuous cultivation, as in the temperate regions. This theoretical proposition does not rest on baseless foundation but rather has empirical support. In fact, in parts of the savanna land of the tropics, where animal manure is available, continuous cultivation has become an important feature of agriculture around large settlements. One can, therefore, justifiably talk about the progression from shifting cultivation with long fallow periods to the emergence of permanent cultivation, although this trend is not evident everywhere in the humid and sub-humid tropics. Given the

dominance of low-activity clays in soils of the humid tropics and the much faster rate of organic matter decomposition following land clearing in the tropics, ultimately permanent cultivation is unlikely to be as pervasive in the tropics as in the temperate regions. Manure can be produced from wastes in urban areas. Similarly, poultries and piggeries near urban centres can become important sources of organic manure for soil fertilization in peri-urban farms. However, shifting cultivation based on agroforestry, with short fallows, is likely to persist in remote rural areas that do not have ready access to urban markets or rural areas in forest regions in Africa, where animal manure is not readily available for soil fertilization. Agroforestry is used in a restricted sense here, to imply the integration of trees with proven ability to restore soil fertility into farmlands or fallow land, rather than multi-purpose trees that may be retained by farmers for income generation, fruits, medicine, or other purposes.

The final stage of the intensification of shifting cultivation is demonstrated by farmers in the savanna regions of the tropics who have integrated agroforestry, crop and livestock farming. They use crop residue from their farms to feed livestock and use manure from their livestock to fertilize their fields. This has enabled them to reduce the fallow period to 3 years or less, and in some cases they are able to cultivate the land continuously, if there is a sufficient supply of animal manure.

10.6.2 Effects of land grabbing

It is important to point out that there are ominous signs regarding the future of shifting cultivation and cultivators in parts of Africa and in Central and South America. The main threat is the large-scale acquisition of land by multi-national corporations for plantation agriculture or forestry, which has deprived shifting cultivators of access to land for farming. In most developing countries, land in both rural and urban areas has not been surveyed, and most people do not have titles to the land they occupy and use for various purposes, including small-scale agriculture. In particular, shifting cultivators depend on customary land rights and do not have documented titles, with legal backing, to their ancestral land on which they practise slash-and-burn agriculture. Consequently, many national governments ignore shifting cultivators when leasing their land to foreign investors, although the farmers depend on such lands for their livelihood. In South and Central America, large-scale plantations of teak, gmelina and pines have been established by foreign investors, most of which are multi-national corporations, on land acquired by government from shifting cultivators without their consent and leased for the long term to the foreign investors. As a result of this development, shifting cultivators are denied access to their ancestral land and have to work as labourers on the plantations for wages. Many landless shifting cultivators, whose land was leased out to foreign investors, are not engaged as workers in the plantations established. Consequently, they and their children are sentenced to a lifetime of poverty, frustration and hopelessness.

A similar scenario is now emerging in Africa. There, as in most parts of the tropics, shifting cultivators are poorly organized and voiceless in terms of articulating their views and making inputs into government policies that affect the use of land on which they depend for their livelihood. In Senegal, government officials usually impose government land reform policies on farmers, and do not consult the farmers as key stakeholders in rural land use (Faye, 2008). Similarly, all untitled land in Benin Republic, including land in rural areas used by shifting cultivators, is deemed to be owned by the government (Le Meur, 2008), and hence the government can acquire all such land without consulting the people who farm it. The situation is very much the same in Madagascar, where an attempt by the government of President Ravalomanana to lease half of the country's total arable land to a South Korean company to produce maize and palm oil for the Korean market was stoutly resisted by his political opponents, leading to the collapse of his regime (Teyssier, 2009). Land has been acquired by foreign investors in several African countries including Ghana, Mali, Sudan and the Democratic Republic of Congo, for plantation agriculture to benefit the home countries of the investors. In the Democratic Republic of Congo, for instance, China has reportedly acquired 2.8 million ha of land for the development of oil palm plantations and another 10 million ha was leased to the Republic of South Africa (Spore, 2009b). Admittedly, the large-scale plantation projects of the foreign investors will generate employment and result in the provision of infrastructure such as roads. However, one must not lose sight of their adverse environmental effects such as deforestation, loss of biodiversity including decimation or total destruction of wildlife and medicinal plants, disruptive effects on the social and community life of indigenes, and on the system of shifting cultivation itself.

It is important to point out that there are no schools where the art and practice of shifting cultivation are taught. Rather, farmers transmit the art and science of shifting cultivation to the next generation by 'teaching' their children in the field when they participate in farm work. With the leasing of land to foreign investors for up to 99 years, at least two generations of shifting cultivators will not be able to practise the traditional methods of farming handed down to them by their forebears, let alone experiment and improve on them. The knowledge shifting cultivators have acquired over several millennia is in danger of being frittered away and lost in the bid to attract foreign investors and 'development' to rural areas and to earn foreign exchange. In areas of the tropics where large areas of arable land have been leased out in the long term, the practice of shifting cultivation and the shifting cultivators themselves face a bleak future. Farmers who practise shifting cultivation in the tropics are subsistence farmers who do not have bank accounts, nor collateral to enable them obtain loans from commercial banks. All they own is their inherited ancestral land. Taking this land from them, without providing them with land elsewhere to farm, is to mortgage their future.

In future land deals involving leasing the land of shifting cultivators to potential investors, domestic or foreign, it is imperative to consult the farmers and take their views, as well as the likely impacts of the proposed projects on the farmers and the environment and on food security, into consideration. Furthermore, the leasing out of large areas of land for plantation agriculture or plantation forestry is bound to exacerbate the problems of food insecurity in the countries or regions concerned due to shortages of land for arable farming. It follows, therefore, that it may be difficult for many tropical countries, particularly those which have leased large areas of arable land to foreign investors, to achieve the Millennium Development Goals, especially the specific goal of eradicating extreme poverty and hunger and halving the proportion of people who earn less than US$1 a day, or people who suffer from hunger, by the year 2015. In fact, the policy of leasing out land to foreign investors to the detriment of small-scale farmers will increase rather than reduce the proportion of people who suffer from hunger and acute malnutrition. An increase in crop yield or production reduces the number of people who suffer from poverty and hunger. It has been estimated, for instance, that a 1% increase in crop yield reduces the number of poor people in Africa by 0.72%, and by 0.4 and 1.9% in the short term and long term, respectively, in Asia (World Bank, 2005). The growing trend of depriving small-scale farmers in the tropics of access to land for farming will definitely worsen the problem of hunger and poverty in the developing world in the short and long term. Furthermore, it will make it increasingly difficult for the developing world to achieve the Millennium Development Goal of combating HIV/AIDS, malaria and other diseases, as the health status of people is largely influenced by their nutritional status.

The worrying and increasing trend of appropriating the land of shifting cultivators and leasing such land to foreign investors does not conform with the growing trend of devolution of the authority of control and management of natural resources to local communities through institutional arrangements known as community-based natural resource management. Local communities in several parts of Africa have demonstrated that they are capable of managing natural resources in their areas efficiently, judiciously and sustainably. The control of wildlife resources by local communities in Namibia has resulted in stoppage of abuses such as poaching, and has also re-generated the main wildlife resources. Furthermore, over 200,000 people earn a living from conservation and management of wildlife resources, which generate an estimated US$2.5 million in revenue annually (Roe et al., 2009). In Kenya, community-based wildlife management has not only boosted revenue earning from tourism, but has led to more efficient conservation of wildlife, soil and vegetal resources. It is important to stress that shifting cultivators are irrevocably committed to ensuring the sustainable management of the natural resources in their domains, especially of the soil, vegetal and water resources; not only their livelihood, but their survival and that of their children, depend on the rational and sustainable management of natural resources. In contrast, foreign companies that lease the land of shifting cultivators are primarily interested in recouping the money invested and in making a profit, even at the expense of degrading the environment. Hence it is foolhardy of politicians in tropical countries – even in those characterized by very low population densities – to lease their land to foreign investors, for these are not interested in the economic development of such tropical countries, but in repatriating profit and produce accruing from their investments to their home countries.

10.6.3 Agricultural imperialism

It is not unlikely that land grabbing in the developing countries of the tropics is an attempt by certain foreign interests to further undermine the precarious food security situation in such countries, especially countries in sub-Saharan Africa, where per capita food production has lagged behind the rate of population growth in the last two or three decades. Foreign interests, masquerading as commercial consortia which intend to bring 'development' to developing countries, may actually be pursuing a secret agenda of 'agricultural imperialism'. This term is used here in two closely related contexts: (i) it refers to the action of a country, its citizens, or its economic concerns to acquire land in another country for the purpose of agriculture or forest plantation that is intended to benefit the foreign country, to the disadvantage of the country whose land was leased to the foreign interest; and (ii) it implies the pursuit of policies to destabilize agriculture in the developing world by politically or economically powerful nations or their multinational corporations. In either case, the underlying motive is the same: to undermine agriculture and food security in developing countries, presumably in order to dump agricultural products, including genetically modified food that does not find a ready market in the West, on developing countries.

10.6.4 Long-term prospects

Finally, it is important to observe that given the predominance of low-activity clays in the soils of the humid and sub-humid tropics and the problems of maintaining adequate levels of soil organic matter, coupled with the difficulties of applying mineral fertilizers or animal manure to ensure continuous cropping, shifting cultivation is not likely to disappear completely in the tropics in the next two or three decades. Even in the savanna regions, where the prospects of continuous farming, based on animal manure application, are much brighter than in forest regions, global warming may exacerbate the occurrence of droughts leading to a reduction in the livestock population, and hence the amount of land that can be farmed continuously through manure application. Hence, one does not foresee a wholesale adoption of continuous farming in the savanna regions in the near future. Shifting cultivation is also likely to persist in remote rural areas that do not have ready access to urban markets, and areas of difficult terrain such as mountainous areas with low population densities.

References

Achard, F. and Banoin, M. (2003) Fallows, forage production and nutrient transfers by livestock in Niger. *Nutrient Cycling in Agroecosystems* 65, 183–189.

Adeoye, G.O., Sridhar, M.K.C., AdeOluwa, O.O. and Akinsoji, N.A. (2005) Evaluation of naturally decomposed solid wastes from municipal dump sites for their manurial value in southwest Nigeria. *Journal of Sustainable Agriculture* 26, 143–152.

Ahn, P.M. (1970) *West African Soils.* Oxford University Press, London.

Altieri, M.A. (2008) *Small Farms as a Planetary Ecological Asset: Five Key Reasons Why We Should Support the Revitalization of Small Farms in the Global South.* Third World Network, Penang, Malaysia.

Altieri, M.A., Ponti, L. and Nicholls, C.I. (2006) Managing pests through plant diversification. *LEISA* 22, 9–11.

Aweto, A.O. (1981) Organic matter build-up in fallow soil in a part of south-western Nigeria and its effects on soil properties. *Journal of Biogeography* 8, 67–74.

Aweto, A.O. and Ayuba, H.K. (1993) Effect of continuous cultivation with animal manuring on a sub-Sahelian soil near Maiduguri, north eastern Nigeria. *Biological Agriculture and Horticulture* 9, 343–342.

Ayarza, M.A. and Welchez, L.A. (2004) Drivers affecting the development and sustainability of the Quesungual Slash and Mulch Agroforestry System (QSMAS) on the hillsides of Honduras. Comprehensive Assessment Bright Spots Final Report. Available at: www.ik.iwmi.org/brightspots/PDF/Latin_America/QSMAS_Case_Study_Honduras.pdf (accessed 14 October 2011), pp. 187–201.

Benge, M.D. (1987) *Agroforestry System.* (mimeo) Bureau for Science and Technology, Agency for International Development, Washington, DC.

Brady, N.C. and Weil, R.R. (2002) *The Nature and Properties of Soils.* Prentice-Hall, Upper Saddle River, New Jersey.

Brams, E.A. (1971) Continuous cultivation of West African soils: organic matter diminution and the effects of applied lime and phosphorus. *Plant and Soil* 35, 401–414.

Cunningham, W.P., Cunningham, M.A. and Saigo, B. (2005) *Environmental Science: A Global Concern.* McGraw-Hill, Boston, Massachusetts.

David, S. (1995) What do farmers think? Farmers' evaluation of hedgerow intercropping under semi-arid conditions. *Agroforestry Systems* 32, 15–28.

Dudgeon, G.C. (1911) *The Agricultural and Forest Products of British West Africa.* John Murray, London.

Embrapa (1999) Reducao dos impactos ambientais da pecuaria corte no Acre. Rio Branco, Acre, Brazil.

Erni, C. (2009) Shifting the blame: Southeast Asia's indigenous peoples and shifting cultivation in the age of climate change. *Indigenous Affairs* 1, 38–49.

Evensen, C.I., Dierolf, T.S. and Yost, R.S. (1995) Decreasing rice yield in alley cropping on a highly weathered Oxisol in West Sumatra, Indonesia. *Agroforestry Systems* 31, 1–19.

Faye, J. (2008) *Land and Decentralization in Senegal*. Institute for Environment and Development, Issue paper 149, London.

Fearnside, P.M. (2000) Greenhouse emissions from land use change in Brazil's Amazon region. In: Lal, R., Kimble, J.M. and Stewart, B.A. (eds) *Global Climate Change and Tropical Ecosystems*. CRC/Lewis Publishers, Boca Raton, Florida, pp. 231–249.

Fonte, S.J., Barrios, E. and Six, J. (2008) *Earthworms, Soil Fertility, and Organic Matter Dynamics in the Quesungual Agroforestry System of Western Honduras*. Poster 53-8, 93rd ESA Annual Meeting, 3–8 August 2008, Milwaukee, Wisconsin.

Francis, C.A. (1986) *Multiple Cropping Systems*. MacMillan, New York.

Garrity, D. (2010) Growing crops under canopy. *Spore* 150, p.12.

Haas, H.J., Evans, C.E. and Miles, E.F. (1957) *Nitrogen and Carbon Changes in Great Plains soils as Influenced by Cropping and Soil Treatments*. Technical Bulletin 1164, US Department of Agriculture, Washington, DC.

Hellin, J., Welchez, L.A. and Cherrett, I. (1999) The Quezungual system: an indigenous agroforestry system from western Honduras. *Agroforestry Systems* 46, 229–237.

Holt-Gimenez, E. (2001) Measuring farms' agroecological resistance to hurricane Mitch. *LEISA* 17, 18–20.

Houghton, R.A., Lefkowitz, D.S. and Skole, D.L. (1991) Changes in the landscape of Latin America between 1850 and 1985. I. Progressive loss of forests. *Forest Ecology and Management* 38, 143–172.

Kang, B.T., Wilson, G.F. and Lawson, T.L. (1984) *Alley Cropping: A Stable Alternative to Shifting Cultivation*. International Institute of Tropical Agriculture, Ibadan, Nigeria.

Kang, B.T., Reynolds, L. and Atta-Krah, A.N. (1990) Alley farming. *Advances in Agronomy* 43, 315–359.

Kotto-Same, J., Woomer, P.L., Appolinaire, M. and Zapfack, L. (1997) Carbon dynamics in slash-and-burn agriculture and land use alternatives of the humid forest zone in Cameroon. *Agriculture, Ecosystems and Environment* 65, 245–256.

Lal, R. (2005) Soil carbon sequestration for sustaining agricultural production and improving the environment with particular reference to Brazil. *Journal of Sustainable Agriculture* 26, 23–42.

Le Meur, P.Y. (2008) *Information on Land: A Common Asset and Strategic Resource. The Case of Benin*. International Institute for Environment and Development, Issue paper 147, London.

MacLean, R.H., Litsinger, J.A., Moody, K., Watson, A.K. and Libetario, E.M. (2003) Impact of *Gliricidia sepium* and *Cassia spectabilis* hedgerows on weeds and insect pests of upland rice. *Agriculture, Ecosystems and Environment* 94, 275–288.

Mendoza, T.C. (2010) Organic agriculture: The logical sequence to modern chemical agriculture in the Philippine context. *Annals of Tropical Research* 32, 112–129.

Mizoguchi, T. (1988) Slash-and-burn field cultivation in pre-modern Japan: With special reference to Shirakawa-go. *Geographical Review of Japan* 62 (Ser. B), 14–34.

Moran, E.F. (1993) Deforestation and land use in the Brazilian Amazon. *Human Ecology* 21, 1–21.

Murdiyarso, D., Tsuruta, H., Ishizuka, S., Hairah, K. and Palm, C.A. (2005) Greenhouse gas fluxes in slash and burn and alternative land use practices in Sumatra, Indonesia. In: Palm, C.A., Vosti, S.A., Sanchez, P.A. and Ericksen, P.J. (eds) *Slash-and-Burn Agriculture: The Search for Alternatives*. Columbia University Press. New York, pp. 64–82.

Nilsson, A. (1992) *Greenhouse Earth*. John Wiley & Sons, Chichester, UK.

Parain, C. (1942) *The Cambridge Economic History of Europe* Vol 1. Cambridge University Press, Cambridge, UK.

Partohardjono, S., Pasaribu, D. and Fagi, A.M. (2005) The forest margins of Sumatra, Indonesia. In: Palm, C.A., Vosti, S.A., Sanchez, P.A. and Ericksen, P.J. (eds) *Slash-and-Burn Agriculture: The Search for Alternatives*. Columbia University Press, New York, pp. 291–304.

Quinones, M.A., Borlaug, N.E. and Dowswell, C.R. (1997) Fertilizer-based green revolution for Africa. In: Buresh, R.J., Sanchez, P.A. and Calhoun, F. (eds) *Replenishing Soil Fertility in Africa*. Soil Science Society of America, Madison, Wisconsin, pp. 81–96.

Reijntjes, C., Haverkort, B. and Waters-Bayer, A. (1992) *Farming for the Future: An Introduction to Low External-Input and Sustainable Agriculture*. Macmillan and ILEIA, Leusden, The Netherlands.

Richards, P. (1985) *Indigenous Agricultural Revolution*. Hutchinson, London.

Roe, D., Nelson, F. and Sandbrook, C. (2009) *Community Management of Natural Resources in Africa*. IIED, London.

Sanchez, P.A., Villachica, J.H., and Bandy, D.E. (1983) Soil fertility dynamics after clearing a tropical rain forest in Peru. *Soil Science Society of America Journal* 47, 1171–1178.

Sillitoe, P. (1998) It's all in the mound: fertility management under stationary shifting cultivation in the Papua New Guinea highlands. *Mountain Research and Development* 18, 123–134.

Skole, D.L., Chomentowski, W.H., Salas, W.A. and Nobre, A.D. (1994) Physical and human dimensions of deforestation in Amazonia. *BioScience* 44, 314–322.

Spore (2007) Carbon market: a golden opportunity? *Spore* 130, 8–11.

Spore (2008) Global trends: Time is running out. *Spore* (special issue), August 2008, 1–2.

Spore (2009a) Soil fertility: Feeding the land. *Spore* 139, 8–10.

Spore (2009b) Land transfer: Consulting rural communities. *Spore* 142, 1–2.

Teyssier, A. (2009) Land tenure: a framework for land transfer. *Spore* 142, 16.

The Guardian (2009) Malawi proves Obama's argument for African self-reliance. *The Guardian*, 21 July 2009, p. 35.

Uphoff, N. (2006) The system of rice intensification and its implications for agriculture. *LEISA* 22, 6–8.

Vakeesan, A., Nishanthan, T. and Mikunthan, G. (2008) Green manures: nature's gift to improve soil fertility. *LEISA* 24, 16–17.

Van de Fliert, E. and Winarto, Y. (2006) From technological packages to ecological principles. *LEISA* special issue (December), 12–14.

Wardjito (2008) Rejuvenating soils with innovative farming approaches. *LEISA* 24, 29.

Wezel, A. and Haigis, J. (2002) Fallow cultivation system and farmers' resource management in Niger, West Africa. *Land Degradation and Development* 13, 221–231.

World Bank (2005) *Agricultural Growth for the Poor: An Agenda for Development.* The World Bank, Washington, DC.

Wright, J. (2008) Organic agriculture and HIV/AIDS in Sub-Saharan Africa. Report commissioned by IFOAM, IFOAM, Bonn.

Young, A. (1976) *Tropical Soils and Soil Survey.* Cambridge University Press, Cambridge, UK.

Zoebisch, M., Gan, Y., Ellis, W., Watson, A. and Sombatpanit, S. (2008) *No-Till Farming Systems.* Special Publication 3, World Association of Soil and Water Conservation, Bangkok.

Index